Smart Farming Technologies to Attain Food and Nutrition Security

Smart Farming Technologies to Attain Food and Nutrition Security

Dr P. PARVATHA REDDY

Former Director
Indian Institute of Horticultural Research
Bangalore

CRC Press
Taylor & Francis Group
Boca Raton London New York

CRC Press is an imprint of the
Taylor & Francis Group, an **informa** business

First published 2024
by CRC Press
4 Park Square, Milton Park, Abingdon, Oxon, OX14 4RN

and by CRC Press
2385 NW Executive Center Drive, Suite 320, Boca Raton FL 33431

British Library Cataloguing-in-Publication Data
A catalogue record for this book is available from the British Library

Print edition not for sale in India.

ISBN13: 9781032690865 (hbk)
ISBN13: 9781032691022 (pbk)
ISBN13: 9781032691039 (ebk)

DOI: 10.4324/9781032691039

Typeset in Adobe Caslon Pro
by Elite Publishing House, Delhi

-EPH-

Contents

Preface

Agriculture is the main pillar of our country and occupies the center stage of India's social security and overall economic welfare. Unfortunately, the allied sector has remained neglected in the past, now fully realizing the emerging complex challenges, there is a need to set a vision to attain rainbow revolution covering the entire spectrum of activities in agriculture, which will make India a developed nation free of poverty, hunger, malnutrition, and make environmentally safe. It is important to recognize that starting from mid-1960s, Indian agriculture has made significant strides in the production of cereals, milk, fisheries, poultry, fruits and vegetables and lately in cotton, where innovations in seed technologies (by public and private sector); innovations in policies and institutions, played instrumental role in driving the transformation. Rainbow revolution with focus on allied activities and new concepts has emerged out which has to be integrated in a sustainable manner for preserving a healthy environment and enhancing ecosystem.

The multiple colors of the "Rainbow Revolution" indicate multiple farm practices such as "Green Revolution (Food grains production), White Revolution (Milk), Yellow Revolution (Oilseeds), Blue Revolution (Fisheries); Golden Revolution (Horticulture); Silver Revolution (Poultry eggs and meat), Round Revolution (Potato), Red Revolution (Livestock meat), Pink Revolution (Onion)" and so on. Thus, the concept of Rainbow revolution is an integrated development of crop cultivation, horticulture, forestry, fishery, poultry, animal husbandry, and food processing industry. The concept of Rainbow revolution in agriculture is a step towards sustainability. India has already achieved resilience in agriculture (including the horticultural sector) through effective agricultural technology generation and suggests that the country is now on the threshold of a "Rainbow Revolution" that will ensure both household nutrition security and prosperity for its people.

The information on strategies for achieving Rainbow Revolution is very much scattered and there is no book at present which comprehensively and exclusively deals with the above aspects in agriculture emphasizing integrated development of crops and allied sectors. The present book on **"Smart Farming Technologies to Attain Food and Nutrition Security"** outlines a new paradigm which aims to integrate agriculture and allied sectors in a sustainable manner which demands proper coordination, implementation and monitoring of the support policies in addition to the allocation of resources. There is a need to attain rainbow revolution covering the entire spectrum of activities in agriculture, which will make India a developed nation free of poverty, hunger, malnutrition, and make the country environmentally safe.

The book is divided into seventeen Chapters. The first chapter gives overview of the 'Rainbow Revolution' using different revolutions. The role of 'Green Revolution' in the notable increase in cereal-grains production by the adoption of modern agricultural technologies, including use of dwarf high yielding varieties, irrigation and heavy doses of chemical fertilizers and pesticides, and farm mechanization was enumerated in the second chapter. The third chapter deals with 'Evergreen Revolution' emphasizing the need to improve productivity in perpetuity without associated ecological and/or social harm. To promote integrated development of horticulture, to help in coordinating, stimulating and sustaining the production and processing of fruits and vegetables and to establish a sound infrastructure in the field of production, processing and marketing with a focus on post-harvest management to reduce losses through 'Golden Revolution' are dealt in chapter four. The fifth chapter deals with 'Yellow Revolution' to create/manage conditions that would harness the best of oil seed production, processing and storage technologies to attain self-reliance in edible oils in the foreseeable future.

Formulating strategic plans to increase and improve pulse production in India involving field demonstrations of best farming practices, incentives for adoption of modern technologies, and resource conservation and management practices to achieve 'Pulse Revolution' are dealt in sixth chapter. Seventh chapter deals with achieving 'Silver (Cotton) Fiber Revolution' through research and technology generation, transfer of technology, improvement of market infrastructure, and modernization of ginning and pressing factories. Aspects like introduction of new technologies to produce standard jute products at low cost to capture the growing international market, bringing raw material supply under control, keeping labor rate in check, and framing of proper policies to maintain a sustainable growth for achieving 'Golden Fiber (Jute) Revolution' are discussed in the eighth chapter. Ninth chapter deals with achieving 'Pink (Onion) Revolution' by developing technologies to increase the onion productivity under short day onion since Indian onions are short-day type, developing improved varieties or hybrids for abiotic stresses namely drought, heat, salinity and water logging, tolerance and biotic stresses namely anthracnose, blight, purple blotch and thrips. Intensifying research efforts in potato varietal improvement including GM varieties, enhancing availability of quality planting material, adaptation for climate change, resource use efficiency, integrated nutrient and water management, eco-friendly pest management, and post-harvest management in achieving 'Round (Potato) Revolution' are discussed in tenth chapter.

Eleventh chapter emphasizes on achieving 'Gene Revolution' through developing high yielding and more nutritious crop varieties, to improve resistance to pests including weeds and adverse conditions, or to reduce the need for fertilizers and other expensive agricultural chemicals. Achieving 'White (Milk) Revolution/Operation Flood' by creating a national milk grid linking producers throughout India (through village milk producers' co-operatives, which procure milk and provide inputs and services) to consumers in over 700 towns and cities and reducing seasonal and regional price variations while ensuring that producers get a major share of the profit by eliminating the middlemen is dealt in twelfth chapter. Thirteenth chapter deals with achieving 'Blue (Fisheries) Revolution' by tapping the total fish potential of India on both iNlands as well as in the marine sector, utilization of modern technologies and processes, and increasing productivity, encouraging deep-sea fishing through joint ventures, and improving the post-harvest marketing infrastructure including e-commerce. Fourteenth chapter deals with achieving 'Silver (Poultry egg and meat production) Revolution' through breeding for low-input technology variety of chickens, developing high yielding breeding stocks, proper nutrition, poultry health management, poultry products processing and value addition, development of infrastructure and poultry equipment. Achievement of 'Red (livestock meat production) Revolution' by breeding improved livestock, providing balanced nutrition, management of diseases, use of nanotechnology, development of artificial meat, and adaptation and mitigation of climate change effects are emphasized in fifteenth chapter.

Sixteenth chapter discusses achievement of 'Green (Renewable) Energy Revolution' by generation of renewable energy using resources such as solar, wind, biomass, hydroelectric and geothermal in order to mitigate climate change and reduce GHG emissions. Aspects such as increasing crop/

animal production, post-harvest management and value addition, transfer of technology, marketing and price realization, enhancing non-farm income, and enabling policies to achieve Rainbow Revolution are discussed in the final chapter 'The Way Ahead'.

The book is illustrated with excellent quality photographs enhancing the quality of publication. The book is written in lucid style, easy to understand language along with adoptable recommendations involving eco-friendly and sustainable practices.

This book will be of immense value to scientific community involved in teaching, research and extension activities related to strategies for achieving 'Rainbow Revolution' in enhancing farmers' income. The material can be used for teaching post-graduate courses. The book can also serve as a very useful reference to policy makers and practicing farmers. Suggestions to improve the contents of the book are most welcome (E-mail: reddypp42@gmail.com). The publisher, Elite Publishing House, New Delhi, deserves commendation for their professional contribution.

P. Parvatha Reddy

About the Author

Dr P. Parvatha Reddy obtained his Ph. D. degree jointly from the University of Florida, USA, and the University of Agricultural Sciences, Bangalore.

Dr Reddy served as the Director of the prestigious Indian Institute of Horticultural Research (IIHR) at Bangalore from 1999 to 2002 during which period the Institute was honored with "ICAR Best Institution Award". He also served as the Head, Division of Entomology and Nematology at IIHR and gave tremendous impetus and direction to research, extension and education in developing bio-intensive integrated pest management strategies in horticultural crops. These technologies are being practiced widely by the farmers across the country since they are effective, economical, eco-friendly and residue-free. Dr Reddy has about 34 years of experience working with horticultural crops and involved in developing an F1 tomato hybrid "Arka Varadan" resistant to root-knot nematodes.

Dr Reddy has over 250 scientific publications to his credit, which also include 40 books. He has guided two Ph.D. students at the University of Agricultural Sciences, Bangalore.

Dr Reddy is serving as Senior Scientific Advisor, Dr Prem Nath Agricultural Science Foundation, Bangalore. He had also served as Chairman, Research Advisory Committee (RAC), Indian Institute of Vegetable Research, Varanasi; Member, RAC of National Centre for Integrated Pest Management, New Delhi; Member of the Expert Panel for monitoring the research program of National Initiative on Climate Resilient Agriculture (NICRA) in the theme of Horticulture including Pest Dynamics and Pollinators; Member of the RAC of the National Research Centre for Citrus, Nagpur; and the Project Directorate of Biological Control, Bangalore. He served as a Member, QRT to review the progress of the Central Tuber Crops Research Institute, Trivandrum; AICRP on Tuber Crops; AICRP on Nematodes; and AINRP on Betel vine. He is the Honorary Fellow of the Society for Plant Protection Sciences, New Delhi; Fellow of the Indian Phytopathological Society, New Delhi; and Founder President of the Association for Advancement of Pest Management in Horticultural Ecosystems (AAPMHE), Bangalore.

Dr Reddy has been awarded with the prestigious "Association for Advancement Pest Management in Horticultural Ecosystems Award", "Dr G.I. D'Souza Memorial Lecture Award", "Prof. H.M. Shah Memorial Award" and "Hexamar Agricultural Research and Development Foundation Award" for his unstinted efforts in developing sustainable, bio-intensive and eco-friendly integrated pest management strategies in horticultural crops.

Dr Reddy has organized "Fourth International Workshop on Biological Control and Management of *Chromolaena odorata*", "National Seminar on Hi-Tech Horticulture", "First National Symposium on Pest Management in Horticultural Crops: Environmental Implications and Thrusts", and "Second National Symposium on Pest Management in Horticultural Crops: New Molecules and Biopesticides".

Chapter - 1

Rainbow Revolution: An Overview

1.1. INTRODUCTION

Agriculture is the main pillar of our country and occupies the center stage of India's social security and overall economic welfare. Unfortunately, the allied sector has remained neglected in the past, now fully realizing the emerging complex challenges, there is a need to set a vision to attain rainbow revolution covering the entire spectrum of activities in agriculture, which will make India a developed nation free of poverty, hunger, malnutrition, and make environmentally safe. It is important to recognize that starting from mid-1960s, Indian agriculture has made significant strides in the production of cereals, milk, fisheries, poultry, fruits and vegetables and lately in cotton, where innovations in seed technologies (by public and private sector); innovations in policies and institutions, played instrumental role in driving the transformation. Rainbow revolution with focus on allied activities and new concepts has emerged out which has to be integrated in a sustainable manner for preserving a healthy environment and enhancing ecosystem.

The multiple colors of the "Rainbow Revolution" indicate multiple farm practices such as "Green Revolution (Food grains), White Revolution (Milk), Yellow Revolution (Oilseeds), Blue Revolution (Fisheries); Golden Revolution (Horticulture); Silver Revolution (Eggs), Round Revolution (Potato), Red Revolution (Meat), Pink Revolution (Onion)" and so on. Thus, the concept of Rainbow revolution is an integrated development of crop cultivation, horticulture, forestry, fishery, poultry, animal husbandry, and food processing industry. The concept of Rainbow revolution in agriculture is a step towards sustainability. India has already achieved resilience in agriculture (including the horticultural sector) through effective agricultural technology generation and suggests that the country is now on the threshold of a "rainbow revolution" that will ensure both household nutrition security and prosperity for its people.

1.2. RAINBOW REVOLUTION

1.2.1. History

The agricultural policy of 2000 envisaged holistic development of Indian agriculture and aimed to achieve it through Rainbow revolution. Economic survey 2015-16 observed that "Indian agriculture is in a way, a victim of its own past success, especially the Green revolution". It suggested an Integral Development Program to make the agriculture sustainable and eventually developed the concept of Rainbow revolution (Fig. 1.1).

1.2.2. Objectives

» Sustainability: Agricultural practices have to be reoriented to maintain environmental sustainability as well as resource sustainability e.g.,

- Promotion of zero budget natural farming/organic farming to reduce use of chemicals in agriculture.
- Crop diversification in water stress areas like Punjab and Haryana.
- Promoting soil health through schemes like soil health cards, practices like rain water harvesting made compulsory etc.
- Promoting lab to land exhibition, investing in research and development of agricultural technologies.

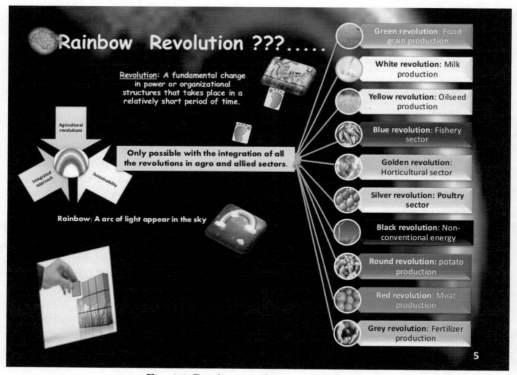

Fig. 1.1. Rainbow revolution in agriculture.

» Farm income: To increase the annual growth rate in agriculture over 4%.
» Scale of agriculture production: Objective is to increase the collective cropped area which would help in increased capital investment and use of latest agricultural technologies and machines to increase agricultural productivity. E.g. greater private sector participation through

contract farming, promote co-operative farming, leasing of farm machines, subsidizing the purchase of new machines (submission on agricultural mechanization) etc.

» Price protection to farmers: Agricultural contracts, promotion of practices like warehouse receipts, promotion of agricultural exports through improved quality of agricultural produce (latest Agriculture export policy) etc.

» Market availability to farmers: To dismantle the restrictions on movement of agricultural commodities throughout the country. Ensuring sufficient number of godowns, warehousing and cold storage facilities etc.

» Insurance protection to farmers: inclusion of all farmers in agricultural insurance scheme (PM Fasal Bhima Yojana) and eventually to cover all crops (horticulture produce is introduced lately on a pilot basis).

» Elimination of Regional disparity in agriculture: Rainfed areas are mostly reeled with low farmer income and productivity. Rainbow revolution through climate specific and farmer specific selection of crops and inclusion of allied sectors to increase farmer income is an objective.

» Incentivise the crops like pulses etc. through inclusion to MSP to attain agricultural as well as nutritional security.

1.3. GREEN REVOLUTION (FOOD GRAIN PRODUCTION)

The Green Revolution is referred to as the process of increasing agricultural production by incorporating modern tools and techniques. Green Revolution is associated with agricultural production. It is the period when agriculture of the country was converted into an industrial system due to the adoption of modern methods and techniques like the use of high yielding variety seeds, tractors, irrigation facilities, pesticides, and fertilizers. Until 1967, the government mainly concentrated on expanding the farming areas. But the rapidly increasing population than the food production called for a drastic and immediate action to increase yield which came in the form of the Green Revolution.

In the year 1965, the government of India launched the Green Revolution with the help of a geneticist, Prof. M.S. Swaminathan, now known as the father of the Green Revolution in India. The movement of the Green Revolution was a great success and changed the country's status from a food-deficient economy to one of the world's leading agricultural nations. It started in 1967 and lasted till 1978.

The Green Revolution within India led to an increase in agricultural production, especially in Haryana, Punjab, and Uttar Pradesh. Major milestones in this undertaking were the development of a high-yielding variety of seeds of wheat and rust-resistant strains of wheat.

The green revolution focused on some basic elements which include:

» Introduced high yielding variety (HYV) seeds in Indian agriculture. The HYV seeds were highly effective in regions that had rich irrigation facilities and were more successful with the wheat crop. Therefore, the Green Revolution at first focused on states with better infrastructure such as Tamil Nadu and Punjab.

» During the second phase, the high yielding variety seeds were given to other states, and crops other than wheat were also included in the plan.

» The most important requirement for the high yielding variety seeds is proper irrigation. Crops grown from HYV seeds need good amounts of water supply and farmers could not depend on monsoon. Hence, the Green Revolution has improved the irrigation systems around farms in India.

» Commercial crops and cash crops such as cotton, jute, oilseeds, etc were not a part of the plan. Green revolution in India mainly emphasized food grains such as wheat and rice.

» To enhance farm productivity, green revolution increased the availability and use of fertilizers, weedicides, and pesticides to reduce any damage or loss to the crops.

» It also helped in promoting commercial farming in the country with the introduction of machinery and technology like harvesters, drills, tractors, etc.

1.3.1. Impact

» Green Revolution has remarkably increased agricultural production. Food grains in India saw a great rise in output. The biggest beneficiary of the revolution was the wheat grain. The production increased to 55 million tons in the early stage of the plan itself.

» Green Revolution was not just limited to agricultural output, the revolution also increased per hectare yield in the case of wheat from 850 kg per hectare to an incredible 2281 kg/hectare in its early stage.

» With the introduction of the green revolution, India reached its way to self-sufficiency and was less dependent on imports. The production in the country was sufficient to meet the demand of the rising population and to stock it for emergencies. Rather than depending on the import of food grains from other countries, India started exporting its agricultural produce.

» The introduction of the revolution inhibited a fear among the masses that commercial farming would lead to unemployment and leave a lot of the labor force jobless. But the result seen was totally different, there was a rise in rural employment. The tertiary industries such as transportation, irrigation, food processing, marketing, etc. created employment opportunities for the workforce.

» The Green Revolution in India majorly benefited the farmers of the country. Farmers not only survived but also prospered during the revolution their income saw a significant rise which enabled them to shift from sustenance farming to commercial farming.

Besides the positive impact, the revolution had a gloomy side too. Some of the negative effects of the Green Revolution are stated below:

» Retardation of agricultural growth due to inadequate irrigation cover, shrinking farm size, failure to evolve new technologies, inadequate use of technology, declining plan outlay, unbalanced use of inputs, and weaknesses in credit delivery system.

» Regional dispersal of the evolution created regional inequalities. The benefits of the green revolution remained concentrated in the areas where the new technology was used. Moreover, since the revolution for the number of years remained limited to wheat production, its benefits were mostly accrued only to wheat-growing areas.

» Interpersonal inequalities between large- and small-scale farmers. The new technologies introduced during the revolution called for substantial investments which were beyond the means of a majority of small farmers. Farmers having large farmlands continued to make greater absolute gains in income by reinvesting the earnings in farm and non-farm assets, purchasing land from the smaller cultivators, etc.

1.4. EVERGREEN REVOLUTION (SUSTAINABLE FOOD GRAIN PRODUCTION)

The Evergreen Revolution should be analysed and understood from the perspective of food security, environmental sustainability, doubling farmers' income, addressing agriculture distress, etc.

1.4.1. Need for evergreen revolution

Need for evergreen revolution arose due to failures of green revolution. Important demerits of green revolution include:

» More than five decades after India launched the Green Revolution, it has not only failed to eliminate hunger but also malnutrition is at its high.

» Wheat and rice have largely displaced more nutritious pulses and other cereals such as millets in consumption.

» Soil has lost its fertility due to unscientific application of fertilizers.

» Due to mechanisation of agriculture, the likeliness for sons instead of daughters led to skewed sex ratio in Punjab and Haryana.

» Indian agriculture became cereal- centric and regionally biased.

» Water logging in fields and salinity increased due to excess irrigation.

» Farmers got burdened with debts from moneylenders, banks, etc.

Prof. M.S. Swaminathan, father of Green Revolution in India, had forewarned as early as in 1968 that "Intensive cultivation of land without conservation of soil fertility and soil structure would lead ultimately to the springing of deserts".

1.4.2. Evergreen revolution

The Green Revolution transformed the image of India from being a 'begging bowl' to 'bread basket'. However, to rectify flaws and loopholes of the Green Revolution we need to make it evergreen. Though India is now self-sufficient in many aspects of food production, it still relies on imports for crops such as pulses and oilseeds, where production has not kept pace with demand.

'Evergreen Revolution' implies productivity improvement in perpetuity without ecological and social harm. The evergreen revolution involves the integration of ecological principles in technology development and dissemination. In order to achieve the goal of food security by ending hunger and to promote sustainable agriculture, it is important that in the field of social protection as well as the scientific measures needed for achieving food and nutrition security, we should move from the green to an 'evergreen revolution' approach.

Prof. M.S. Swaminathan coined the term "Evergreen Revolution" to highlight the pathway of increasing production and productivity in a manner such that short- and long- term goals of food production are not mutually antagonistic. The logic is to produce more from less, less land, less pesticide, less water and it must be an evergreen revolution to get sustainable agriculture.

The new initiatives include use of cutting-edge technologies to increase farm productivity, promotion of climate-resilient indigenous crop varieties, launch of a nationwide program to harvest the advantages of space technology in agriculture, setting up of seed production and processing units at 'panchayat' level, increase of cropping intensity by 1 million hectares per year through the utilization of rice fallow areas for pulses and oil-seeds, and consolidation of online trading and inter-market transactions, among others.

Evergreen revolution eliminates chemical use, reduces irrigation and produces long-term, sustainable growth in yields to feed the world's growing population. There is a need to promote climate-resilient farming.

1.5. GOLDEN REVOLUTION (HORTICULTURAL CROP PRODUCTION)

The period between 1991 to 2003 is known as the period of Golden Revolution in India because, during this period, the investment planned in the horticulture segment became highly productive. The Golden revolution is related to the production of horticulture. It is a part of the important agricultural revolutions of India. Nirpakh Tutej is considered to be the Father of The Golden Revolution in India. The Government of India launched the National Horticulture Mission in the year 2005-2006 with a mandate to increase the production in the horticulture sector.

India became the world leader in the production of a variety of horticultural crops like fruits, vegetables and tubers, plantation and spice crops, ornamental, medicinal and aromatic crops. The sector emerged as a sustainable livelihood option and became the second-largest producer of vegetables and fruits in the world, after China. Economic conditions of many farmers who were engaged in horticulture improved thus improving the livelihood for many underprivileged classes.

The factors that affected the growth of horticulture sector during Golden Revolution in India are as mentioned below:

» The shift in cropping pattern in favor of crops with higher returns
» Increase in the area of harvesting
» Improvement in the techniques of cultivation.

The Horticulture exports of India marked an increase from ₹ 6308.53 crores in 2004- 2005 to ₹ 28,62861 crores in 2014-2015. This significant growth in the sector is undoubtedly attributed to the organized and planned policies of the horticulture sector under Golden Revolution.

The total area under vegetables and fruits was 11.72 million hectares and the total production stood at 150.73 million tons in the year 2005. As a result of huge spurt in the sector due to the National Horticulture Mission, the production of horticulture tremendously increased to 281 million tons from an area of 23.2 million hectares in 2015-2016.

1.6. ROUND REVOLUTION (POTATO PRODUCTION)

This revolution deals with the production of potatoes. It is one of the major parts of the Green Revolution. Round Revolution is the increase in production of potato yield. The production of potatoes was aimed to be increased at least double or triple the single annual increase. Due to the advancement in this sector, the whole agricultural sector has found a new boost. This revolution has paved its way to the increase in economic condition of the nation.

Potato (Solanum tuberosum) is one of the most important food crops after wheat, maize and rice, contributing to food and nutritional security in the world. Keeping in view the shrinking cultivable land and burgeoning population in India, potato is a better alternative to deal with the situation. The success story of over five decades of potato research in India is phenomenal. Compared to the area, production and productivity in 1949-50, the increase over this period is 550 per cent, 1745 per cent and 178 per cent, respectively. India now ranks fourth in potato area (1.48 million ha) and third in production (28.47 million tons) in the world with an average yield of 18.33 t/ha.

It was only because of indigenously developed technologies that potato in India has shown spectacular growth in area, production and productivity during the last five decades. The major achievements of potato research in India include varietal improvement, seed plot technique, tissue culture, agro-techniques, plant protection, storage, processing and value addition, computer applications, and transfer of technology.

1.7. YELLOW REVOLUTION (EDIBLE OIL SEED PRODUCTION)

The Yellow Revolution was launched to increase the production of edible oilseeds in the country to meet the domestic demand. The revolution that was launched in 1986-1987 to increase the production of edible oil, especially mustard and sesame seeds to achieve self-reliance is known as the Yellow Revolution. Sam Pitroda is known as the father of the Yellow Revolution in India. Yellow Revolution targets nine oilseeds that are groundnut, mustard, soybean, safflower, sesame, sunflower, niger, linseed, and castor.

The Yellow revolution had the implantation of hybrid mustard and sesame seeds which significantly increased the production of edible oil which was also due to the use of improved technology for oil production. The Revolution marked the beginning of an entirely new era with floating sunflowers in fields of Punjab, created many opportunities, and also helped in covering socio-economic differences in the country. The oil production in India was about 12 million tons when the revolution started which doubled in 10 years to about 24 million tons. Along with the use of the hybrid seed, various other measures were taken such as the increase in agricultural land to about 26 million hectares, and use of modern technological inputs.

1.7.1. Features of the yellow revolution

>> The yellow revolution included incentives to farmers who were also provided processing facilities that included irrigation, fertilizers, pesticides, transportation facility, minimum support price, warehousing, etc.

>> Under the revolution, many boards such as the National Dairy Board were entrusted with responsibilities to promote oilseed production. The NDB has the responsibility to increase groundnut oil production in Gujarat. Similarly, the National Oilseeds and Vegetable Oils Development Board were responsible for the production enhancement of oilseeds in non-traditional areas.

>> Oilseeds Production Trust was established to popularize the four major oilseeds that are mustard, groundnut, soybean, and sunflower. Also, there were about 3000 oilseed societies established with 13 lakh farmers and 25 hectares of cultivable land in different states of the country.

>> Though India achieved self-sufficiency in oil production in the next ten years, sadly the output of India does not meet its consumption. To meet demand, India started to import oilseeds from other countries. India imported about 5 million tons in 2007 from many countries like Malaysia, Argentina, Brazil, etc.

>> The hour calls for a second yellow revolution. Oilseeds are major crops and meet the country's needs for edible oils. Also, technical advancement in dryland farming is needed to maximize productivity and farm income. If the country achieves self-sufficiency in oilseeds it would have a significant impact on agriculture and the economy and would also reduce dependence on foreign markets.

1.8. GENE REVOLUTION

In case of fiber, release of Bt cotton technology in 2002, the only genetically modified (GM) crop released so far in the country, paved the way for the Gene Revolution. It has made the country, the largest producer with an estimated 37.2 million bales production in 2017-2018 [United States Department of Agriculture (USDA), Foreign Agricultural Service, 2017; Agriwatch, 2018] and second largest exporter of cotton in the world with 93.14% cotton area is under Bt cotton (Directorate of Cotton Development, 2017). In an ongoing impact evaluation study by Indian Council for Research on International Economic Relations (ICRIER), it is estimated that the cumulative gain from import savings, extra raw cotton export and extra yarn export—compared to the business-as-usual scenario—between 2003-2004 to 2016-2017 is USD 67.4 billion at the all-India level (Gulati and Juneja, 2018).

1.9. GOLDEN FIBER REVOLUTION (JUTE PRODUCTION)

The Golden Fiber revolution in India is related to jute production. Jute is a natural fiber with a golden, soft, long, and silky shine. It is the cheapest fiber procured from the skin of the plant's stem. Because of its color and high cash value, Jute is known as a golden fiber. During the industrial revolution, jute started being used as a raw material in the fabric industry and until today, the processed jute is used for making strong threads and jute products. The British East India Company started jute cultivation commercially and used jute-woven bags to distribute food grains all over the world for the same reason.

India is the largest Jute producing country with annual production estimated to be around 1.986 million tons. The improvement in crop cultivation and the use of technology in jute farming has made India prominent in global production of Jute. West Bengal accounts for almost 50% of the country's total jute production. Other major jute producing regions in the country include Uttar Pradesh, Bihar, Assam, Meghalaya, and Orissa.

India, along with the major producers, is also the largest consumer of jute and jute products in the world. So much so that it had to import around 3,37,000 tons of jute and jute products in the year 2011 to meet the domestic demands.

1.9.1. Need for golden fiber revolution

India's first jute factory was established at Kolkata in 1854. India had a monopoly both in the production of raw jute as well as the jute products before independence. Post-independence, the jute industries in India are damaged due to several problems. The problems faced by Jute industries are mentioned below:

>> During the partition of India, most of Jute mills remained in India while major Jute producing areas went to Bangladesh (then East Pakistan). Thus, created the problem of a shortage of raw material. Despite the efforts of the Government to increase the area under Jute, India is not self-sufficient in raw material and has to import the same from Bangladesh and other countries.

>> The mills and machinery in the Jute industry in India are obsolete and need technological advancements. The Government of India in 2006 launched a Jute Technology Mission with four mini Missions that included jute research, development of raw jute agriculture and its extension; processing and marketing of raw jute. But this mission was unable to achieve targets and used the allocated funds.

>> The jute industry in India has become stagnant, inefficient and unproductive due to non-diversification and over-dependence on jute sacks. The industries face serious competition in

the global market with countries such as Bangladesh, South Korea, Philippines, Japan, etc. The industry also loses competitiveness due to sickness in the Jute mills, obsolete technology, higher prices, etc.

» Jute industries in India are also facing decreasing demands for their produce. Jute products are fast losing market to synthetic fibres, plastic, and similar substitute products. To protect the Jute industry, Parliament enacted the Jute Packaging Mandatory Act, 1987. The jute industry is reeling under the crisis caused by the shutdown of Jute mills.

1.9.2. Significance of the golden fiber revolution

» Jute increases the organic fertility of the soil for other crop plantations.

» Organized processing and cultivation of jute help farmers to earn and save substantial money from carbon credits

» the valuation of the carbon credit of jute cultivation is pinned at farmers' savings made in purchasing inorganic fertilizer.

» Renewable resource application of Jute has made it a key raw material in the paper industry.

» Burning one tonne of plastic bags emits 63 gigajoule of heat and 1,340 tons of carbon dioxide whereas burning one tonne of jute bags emits only 2-gigajoule heat and 150 kg carbon dioxide.

1.10. WHITE REVOLUTION (MILK PRODUCTION)

The Government of India initiated Operation Flood known as the White Revolution during 1970 with a motive of increasing milk production to make the country one of the largest producers of milk in the world. It created a national milk grid linking producers throughout India to consumers in over 700 towns and cities and reducing seasonal and regional price variations while ensuring that producers get a major share of the profit by eliminating the middlemen. At the bedrock of Operation Flood stands the village milk producers' co-operatives, which procure milk and provide inputs and services, making modern management and technology available to all the members.

White Revolution had the objectives as stated below:

» Creating a flood of milk by increased production

» Increase the incomes of the rural population

» Provide milk to consumers at fair prices

The revolution associated with a sharp increase in milk production in the country is called the White Revolution in India also known as Operation Flood. White revolution period intended to make India a self-dependent nation in milk production. Today, India is the world's largest producer of milk and Dr Verghese Kurien is known as the father of the White Revolution in India.

1.10.1. Significance of operation flood

» The White Revolution in India helped in reducing malpractice by traders and merchants. It also helped in eradicating poverty and made India the largest producer of milk and milk products.

» Operation Flood empowered the dairy farmers with control of the resource created by them. It helped them in directing their own development.

» To connect milk producers with the consumers of more than 700 cities and towns and throughout the country, a 'National Milk Grid' was formed.

» The revolution also reduced regional and seasonal price variations ensuring customer satisfaction and at the same time. Also, it ensured that the producers get a major share of the price that customers pay.

» Improved the living standards of the rural people and led to the progress of the rural economy.

Towards the end of White Revolution or Operation Flood, 73,930 dairy cooperatives had set up that links more than 3.5 crore dairy farmer members. At present, due to the White Revolution, there are several hundred Cooperatives in India that are working very efficiently. Hence, the revolution is the cause of the prosperity of many Indian villages. India today is the largest producer of milk in the world with 176.4 million tons in 2017-18, up from 17 million tons in 1950-1951, leaving USA (97 million tons) and China (45 million tons) behind (Department of Animal Husbandry, Dairying and Fisheries, 2017).

1.11. BLUE REVOLUTION (FISHERIES PRODUCTION)

Blue Revolution also called as Neel or Nili Kranti Mission in India was launched in 1985-1990 during the 7th Five-Year Plan (1985-1990) for integrated development and management of fisheries. The main objective was to develop, manage, and promote fisheries to double the farmers' income. Blue Revolution in India was started in India with the vision to achieve the economic prosperity of India keeping in view the sustainability, bio-security, and environmental concerns. Later, during the 8th Five Year Plan (1992-97), the Intensive Marine Fisheries Program was launched, and eventually, the fishing harbors in Vishakhapatnam, Kochi, Tuticorin, Porbandar, and Port Blair were also established over the time.

1.11.1. Objectives

The objectives of the Blue Revolution are mentioned below:

» Completely tapping the total fish potential of India on both islands as well as in the marine sector and to triple the production by the year 2020.

» Transforming the fisheries sector into a modern industry through the utilization of new technologies and processes.

» Doubling the income of the fishers through increased productivity and improving the post-harvest marketing infrastructure including e-commerce, technologies, and global best innovators.

» To ensure the active participation of the fishers and the fish farmers in income enhancement.

» Tripling the export earnings by the year 2020 with a major focus on the benefits covering the institutional mechanisms.

» Developing the nutritional and food security of the nation.

1.11.2. Outcomes

The Blue Revolution in India along with the Fish Farmers Development Agency (FFDA) brought an improvement in the aquaculture and fisheries sector with the introduction of new techniques of rearing, marketing, exporting, and fish breeding.

Some of the major outcomes of the Blue Revolution in India are mentioned below:

1. Currently, the Indian Fisheries Sector reached a production of 11 million tons of fish during

2016-2017 from a limit of 0.75 million tons during 1950-1951 [Press Information Bureau (PIB). 2017].

2. At present, India is the second largest producer of fisheries in the world and sea food exports constitute the second largest share in the total agri-exports from India after rice, accounting for more than USD 7 billion in 2017-2018 [Press Information Bureau (PIB). 2017], which gave a real boost to fishery sector's growth.

3. India is recorded to achieve an average annual growth of 14.8% as compared to the global average percentage of 7.5 in the production of fish and fish products.

4. The fishery has become India's largest agricultural export over the last five years with a growth rate of 6% – 10%.

5. India has become the world's second-largest producer of fish with exports worth more than 47,000 crore rupees.

6. The fisheries and aquaculture production contributes 1% and 5% to India's GDP and Agricultural GDP, respectively.

1.12. SILVER REVOLUTION (POULTRY EGG AND MEAT PRODUCTION)

Silver Revolution deals with the increase in poultry farming and egg production. It has been able to meet the needs of the citizens successfully. The application of veterinary science and other technologies has allowed the development of this sector of agriculture.

Since 2000-01, India's poultry industry experienced significant shift in the structure and scale of operation leading to Silver Revolution. Policy innovations such as liberalization of imports of grandparent poultry stock [Mehta, 2003], vertical integration of poultry operations and contract farming model between large integrators and small farmers, transformed the poultry sector from a mere backyard activity into a major organized commercial one (Mehta and Nambiar, 2007; Manjula and Saravanan, 2015). Globally, India is today the number three in layers (eggs) production at 88 billion eggs (accounting for 5% of the world production) and number five in broilers (poultry meat) production at 3.4 million tons (accounting for 3% of the world production), with almost 80% production coming from organized commercial farms (Department of Animal Husbandry, Dairying and Fisheries, 2017).

In addition, the poultry meat consuming population in the country has risen dramatically from a meagre 8% in 1993-1994 to 38% in 2011-2012 (Gulati and Verma, 2016).

1.13. RED REVOLUTION (MEAT PRODUCTION)

The term Red Revolution denotes the revolution in the technologies used in the meat processing sector in the country. Durgesh Patel is known as the Father of the Red Revolution. The modernization of the meat processing sector in India is the Red Revolution.

The challenges faced by the Red revolution are as mentioned below.

» Despite the huge livestock population in India, it accounts for only around 2% of the global market.

» Standardizing the quality and safety aspects of meat and poultry

» Creating standard policies for meat production and export

» Providing meat testing facilities

» Providing cold storages for the growth of the meat and poultry processing sector

>> Infrastructure facilities for modern slaughter houses

>> Increased investment in the sector and more hygienic method for meat and poultry processing

Regardless of the number of challenges mentioned above, the meat and poultry processing sector in India has great potential for growth. The present per capita consumption of meat is around 6 grams a day which will improve to 50 grams a day by the next decade or so. Such a phenomenal increase in meat consumption ensures that the sector will witness tremendous growth.

Red revolution which is focusing on modernisation of meat processing units has attained the number 1 position in the world in exports of buffalo meat in 2012, exporting approximately 1.5 million metric tons of beef, according to the United States Department of Agriculture (USDA) Foreign Agricultural Service. The major importers are from Middle East and South East Asian countries.

In a report titled the 'Indian Meat Industry Perspective', the Food and Agriculture Organization of the UN (FAO) has recommended an outlined four steps framework that India must adopt and adhere to it for the success of Red revolution are as follows:

>> Setting up state of the art meat processing plants.

>> Developing technologies to raise male buffalo calves for meat production.

>> Increasing the number of farmers rearing buffalo under contractual farming.

>> Establishing disease-free zones for rearing animals.

1.14. GREEN (RENEWABLE) ENERGY REVOLUTION

Energy has a key role in economic and social development but there is a general lack of rural energy development policies that focus on agriculture. Agriculture has a dual role as an energy user and as an energy supplier in the form of bioenergy. This energy function of agriculture offers important rural development opportunities as well as one means of climate change mitigation by substituting bioenergy for fossil fuels.

India is witnessing exponential growth in the renewable energy sector, helping it effectively overcome the pressing challenges of growing energy demand and alarming growth in air pollution levels. As a committed partner to the Government of India, Hitachi is harnessing its green energy solutions, to augment India's clean energy capabilities. Through its intelligent solar inverters and collaborative partnership in delivering mega hydropower projects, Hitachi is ensuring that millions of people get access to clean, affordable electricity that improves their Quality of Life.

1.15. INNOVATIVE AGRICULTURAL EXTENSION

The first color I visualize that our focus must be on the farmer. Not just as a passive recipient of knowledge, as in the traditional paradigm of 'Last Mile in the Technology Transfer", but as an active Co-creator of knowledge in a multi-way knowledge exchange paradigm. This would ensure access to personalised solutions to the farmer, besides setting a reality-oriented agenda for agricultural research itself.

The second color depicted synchronised delivery of information, knowledge and inputs. This takes us beyond just transfer of knowledge to actually applying it on the ground. Inputs here mean credit, seeds, nutrients, crop protection chemicals, access to farm equipment, risk management solutions, etc.

The third color called for an alignment with the Government's flagship agriculture program RKVY, to be able to create a force multiplier. This in turn means building the capacity for village and district level planning. At the same time, there is also a need for redesigning Government programs to ensure that they do not unduly distort markets; because at the end of the day, we are relying on a market economy.

The fourth color sought integration of agricultural extension into the full value chain of that produce. This approach will figure a way to create value by serving the changing needs of an evolving consumer. This also will consciously extend the role of agricultural extension into the vital post-harvest arena.

The fifth color I demanded accountability of the system to all its stakeholders. An outcome-oriented system instead of input or output as the only metrics. Today, by making the extension service free, we may have taken away the farmers' right to demand quality. Even a token payment will work wonders in ensuring accountability!

The sixth color appealed that we be sensitive to the non-renewable natural resources. Agricultural extension must promote Conservation Agriculture as much as possible. All of us know the three dimensions of agriculture vis-à-vis climate change:

» Agriculture is a part of the problem, causing climate change through Greenhouse Gas emissions (methane from flooded paddy fields and ruminants like cows, nitrous oxide from the soils, CO_2 from fossil fuels used in farm equipment, etc.).

» Agriculture is also one of the most vulnerable sectors impacted by climate change (fall in productivity due to changing weather patterns).

» Agriculture can be an important part of the solution to climate change (through emission reductions, carbon sequestration, increasing soil organic matter, etc.)

Finally, the seventh color stressed the need for a proactive agenda on gender issues in agricultural extension. Analyze gender roles in farming systems, enable gender sensitive agricultural extension methodologies.

1.16. BENEFITS OF RAINBOW REVOLUTION

» It keeps a check on the needs of the citizens and provide accordingly.

» It has been an aid for the production of the best products.

» It make better supply to the consumers.

» It is a boon for the Indian agricultural and husbandry sector.

» It aims at increasing environmental sustainability as well as development of resources.

» It promotes organic farming to decrease the use of chemicals and fertilizers.

» It has made practices like rain water harvesting compulsory.

» It aims at improving soil testing techniques and other agricultural technologies. It promotes soil health schemes.

» It takes care of the income of the farmers. It keeps a check on annual growth in agricultural sector.

» It keeps the farmers informed about miscellaneous plans and programs of the government for their growth and development.

» It ensures proper market facilities for the farmers without any restrictions on the movement of products.

» It supplies the farmers with adequate number of go-downs and warehouses. It promotes the agricultural exports with the help of good quality agri-products. It keeps on maintaining nutritional values of the products.

» Rainbow Revolution aims at interlinking all sectors of agriculture and husbandry for better efficiency of generating and producing crops and other products.

1.17. OUT-OF-BOX THINKING

Agriculture sector in India requires an out-of-box thinking keeping in mind the criticality of this sector.

» The scientific/modem genetic engineered farming today is crucial for India's future.

» The markets for the farmers are distorted, not enabling them to get the best prices and there is a need to connect the farmer directly with the markets what is referred as 3F's F(Farmer) • F(Firm) • F(Fork). Today, internet technology provides not only for domestic access but also for global access.

» Contract farming which also allows for direct contact of the farmer with the market. Under this, the land is with the farmer except that production of a crop is under a 'contract' with a buyer directly who also has the responsibility of providing necessary inputs and also picking up the produce whenever ready.

» A serious thought would have to be afforded to corporate farming which allows private sector players to enter into agricultural activities. It is not true that this step would lead to greater marginalization and exploitation of the small and marginal farmers. There is a larger take away in the form of increased productivity, commercialization, diversification, greater value-addition, greater and efficient use of land, building an efficient supply chain, increased investment and readily absorbed modern technologies. Almost 40 per cent of food products are wasted and destroyed in the absence of supply chain which can easily be plugged by the large corporates resulting in increased supply of food products and this would lead to lower prices in the markets.

» What is required today is complete mapping of soils across the length and breadth of the country, superimposed with historical data of the climate, rainfall, crop suitability and then decide on the cropping pattern. Today, technology information is available to allow for soil, climate-based cropping pattern and not on traditional and historical-based cropping pattern.

» Land under non-agriculture use (for setting up the special economic zones, setting up power plants, building roads, etc.) has increased from 3 per cent in 1950—51 to over 11 per cent presently. This makes increasing productivity not only important but an absolute 'Must'.

» There is an increasing trend amongst farmers in the belief that agriculture as non-viable and unprofitable enterprise. The increasing cost of production has led farmers to sell their land to the industrial activities. In recent times, the government is also declaring large land area as non-agriculture to support industrial growth. Herein, lies the challenges of balancing both, but larger challenge would be to re-establish agricultural activities as not only viable but also as a profitable commercial proposition.

» The land reforms which have been an avowed objective since Independence but little has been done and still lesser achieved. There is a need for this to be prioritized by the state governments. Further efforts should be made to computerize land records such as the 'Bhoomi Project in Karnataka' and web-based land records under the 'Dharitree Project in Assam'. India also has large waste land area which could be given to rural and less privileged people on ownership basis at free of cost for integrated farming -cum -forestry operations. This would serve the objective of utilization of waste land besides giving the landless farmers a source of livelihood.

» Current agriculture sector is starved of investment and it receives as little as 0.3 per cent of GDP. There is an urgent need to step up public investment in irrigation, roads, power and public health.

» Today strategy for agriculture sector would have to be broken down to the last unit which is the village or at best district level. Issues at each district level would need to be prioritized and then efforts should be made for their resolution.

What the agriculture sector needs is not another green or rainbow or evergreen revolution but a renaissance which is rebuilding the agriculture sector.

1.18. FUTURE THRUSTS

» Identification of area to promote rainbow revolution.

» Establishment of large number of agro-industries, processing mills, value added services, etc.

» Non-productive or waste land should be utilized for fuel generation purpose through PPP mode.

» Knowledge and awareness need to be transferred not only production side, but also consumer side.

» Human resource development should be given thrust for capacity building of farmers, horticulture entrepreneurs/ supervisors and field functionaries.

» Public and private extension should give higher priority to process innovations, product innovation, natural resource management (NRM) practices.

» To make these institutional changes, public extension systems must become more decentralized, farmer-led and market-driven.

1.19. CONCLUSION

Prof. M.S. Swaminathan gave the call for evergreen revolution and this is possible only through the integrated and holistic approach to agricultural development. It involves leading innovation in agriculture, enabling farmers big and small, driving a sustainable intensification of agriculture, enhancing human health, etc. With uncertainty of climate, low farm income and growth, nutritional issues – rainbow revolution is a way of addressing these issues in the most comprehensive way making agriculture not only profitable but also sustainable.

Despite of the limitations regarding present scenario, Rainbow revolution is a proper solution, because it integrates all agriculture and allied sectors in a sustainable manner. This demands proper coordination, implementation and monitoring of the support policies in addition to the allocation of resources. The farmers had low to medium nature of favorable attitude towards IPM and green manuring i.e. organic farming and the level of sensitivity to minimize global warming problem was medium among the agricultural research scholars. Motivating factors like education, economic status, cosmopolitanises, extension media exposure, risk orientation and level of aspiration had positive and highly significant association with perception towards using agrochemicals. Appropriate and effective institutional vehicles with PPP are essential to develop and integrate all on farm and non-farm sector policy and interventions in a sustainable manner. Agricultural extension and rural advisory services are necessary to bring positive attitude, knowledge, technologies, and services towards sustainability among human resources involved in agricultural activities.

Chapter - 2

Green Revolution
(Food Grain Production)

2.1. INTRODUCTION

The Green Revolution was the notable increase in cereal-grains production in Mexico, India, Pakistan, the Philippines, and other developing countries in the 1960s and 1970s. This trend resulted from the introduction of hybrid strains of wheat, rice, and corn (maize) and the adoption of modern agricultural technologies, including irrigation and heavy doses of chemical fertilizer. The Green Revolution was launched by research establishments in Mexico and the Philippines that were funded by the governments of those nations, international donor organizations, and the U.S. government.

2.1.1. Back ground

In 1943, India suffered from the world's worst recorded food crisis; the Bengal Famine, which led to the death of approximately 4 million people in eastern India due to hunger. Even after independence in 1947, until 1967, the government largely concentrated on expanding the farming areas. But the population was growing at a much faster rate than food production. This called for an immediate and drastic action to increase yield. The action came in the form of the Green Revolution. The green revolution in India refers to a period when Indian Agriculture was converted into an industrial system due to the adoption of modern methods and technology such as the use of HYV seeds, tractors, irrigation facilities, pesticides and fertilizers.

One of the chronic problems India faced following independence was insufficiency of food. With the separation from Burma (now Myanmar) in 1937, India became deficient in food. Food problem became even more acute after the partition of the sub-continent into India and Pakistan in 1947, presenting a series of challenges to India's agricultural sector. Although there was a sharp rise in grain production after independence, it was not sufficient enough to meet the food requirements of a growing population. The shortage of grain production in the face of an increasing population resulted in food imports and a rise in the prices of grains. This necessitated the "Green Revolution", which occurred primarily as a result of technological breakthroughs, improved water supplies and better agricultural practices. In addition, increased mechanization of agricultural operations and the use of plant protection measures also contributed to the emergence of the "Green Revolution" in India.

In the year 1965, the government of India launched the Green Revolution with the help of a geneticist, Prof. **M.S. Swaminathan** now known as **the father of the Green revolution (India)**. The movement of the green revolution was a great success and changed the country's status from a food-deficient economy to one of the world's leading agricultural nations. It started in 1967 and lasted till 1978. The Green Revolution within India led to an increase in agricultural production, especially in Haryana, Punjab, and Uttar Pradesh. Major milestones in this undertaking were the development of a high-yielding variety of seeds of wheat and rust-resistant strains of wheat.

2.2. REASONS FOR ADOPTION OF GREEN REVOLUTION IN INDIA

India was among the first developing countries to adopt farming strategies under the "Green Revolution" in the mid-1960s. This has been sustained and expanded throughout the country. Indeed, India became self-sufficient in food production within a relatively short span of time after launching the "Green Revolution." The decision to adopt the "Green Revolution" was precipitated by the unprecedented drought condition in 1966. As a result, the country had to import large quantities of food grains from foreign countries at an enormous cost or seek food aid from friendly countries. Cereal imports, which averaged about 5.9 million tons per year during the early 1960s, reached a record high of 10.4 million tons in 1966. Limited foreign exchange meant that India had no alternative but to seek food aid from friendly countries. For example, the United States supplied 8.4 million of the 10.4 million tons India imports in 1966. The remainder was received in the form of wheat and wheat products from Canada, then USSR, and Australia.

This situation prompted the government to step in and devise a new policy – a new agricultural strategy for increasing the agricultural production within the shortest possible time and for minimizing fluctuations in agricultural production on account of unfavorable weather conditions. Thus, India's Ministry of Agriculture announced the New Strategy of Agricultural Development in August 1965. This new strategy came to be known as the "Green Revolution."

The Indian agriculture, which was stagnant and backward for centuries, underwent a vast change due to adoption of the New Agricultural Development Strategy (the Green Revolution). The wider application of systematic and modern technological innovations brought about revolutionary changes in the methods of farming in India. This combined with improved supplies of water and selection of breeds resulted in tremendous increases in yield per acre for many crops. Indeed, the term Green Revolution was appropriate to describe the results in India.

Due to the success of the "Green Revolution," cereal imports in India were generally negligible in the 1970s, except massive crop failures in 1979-80 led to renewed imports of 2.3 million tons of food grains in 1981-82 (De Janvry and Subbarao, 1986). By the end of the 20th century, India had achieved self-sufficiency by producing enough food, for example, 212 million tons of food grains in the year 2001-2002.

2.3. GREEN REVOLUTION

2.3.1. Objectives

The objectives of the "Green Revolution" were to:

» Make available the required inputs in sufficient quantities.
» Encourage investment in fertilizer factories and the manufacturing of agricultural equipment.
» Identify and co-ordinate agricultural research activities to raise productivity.
» Intensify agricultural extension service in selected areas.

- » Provide adequate credit to the farmers who are willing to grow varieties of cereals and adopt the appropriate farm practices.
- » Implement a production-oriented cereal price policy.
- » The short-term objective was to address India's hunger crisis during the second Five Year Plan.
- » The long-term objectives included overall agriculture modernization based on rural development, industrial development, infrastructure, raw material, *etc.*
- » To provide employment to both agricultural and industrial workers.
- » Developing crop varieties which could withstand extreme climates and diseases.
- » Globalization of the Agricultural World by spreading technology to non-industrialized nations and setting up many corporations in major agricultural areas.

2.3.2. Features

- » Introduced high yielding variety (HYV) seeds of wheat and rice in Indian agriculture.
- » The HYV seeds were highly effective in regions that had rich irrigation facilities and were more successful with the wheat crop. Therefore, the Green Revolution at first focused on states with better infrastructure such as Tamil Nadu and Punjab.
- » During the second phase, the high yielding variety seeds were given to other states, and crops other than wheat were also included in the plan.
- » The most important requirement for the high yielding variety seeds is proper irrigation. Crops grown from HYV seeds need good amounts of water supply and farmers could not depend on monsoon. Hence, the Green Revolution has improved the irrigation systems around farms in India.
- » Green revolution in India mainly emphasized food grains such as wheat and rice. Commercial crops and cash crops such as cotton, jute, oilseeds, *etc.* were not a part of the plan.
- » To enhance farm productivity, green revolution increased the availability and use of fertilizers, weedicides, and pesticides to reduce any damage or loss to the crops.
- » It also helped in promoting commercial farming in the country with the introduction of machinery and technology like harvesters, drills, tractors, etc.
- » Expansion of Farming Areas: Although the area of land under cultivation was being increased from 1947 itself, this was not enough to meet the rising demand. The Green Revolution aided in this quantitative expansion of farmlands.
- » Double-cropping System: Double cropping was a primary feature of the Green Revolution. The decision was made to have two crop seasons per year instead of just one. The one-season-per-year practice was based on the fact that there is only one rainy season annually. Water for the second phase now came from huge irrigation projects. Dams were built and other simple irrigation techniques were also adopted.
- » Using seeds with improved genetics: Using seeds with superior genetics was the scientific aspect of the Green Revolution. The Indian Council for Agricultural Research developed new strains of high yield variety seeds, mainly wheat and rice, millet and corn.
- » Main crops included in the Revolution were wheat, rice, jowar, bajra and maize. Non-food grains were excluded from the ambit of the new strategy. Wheat remained the mainstay of the Green Revolution for years.

Mexican research team, led by U.S. Plant Pathologist Norman Borlaug developed varieties of wheat that grew well in various climatic conditions and benefited from heavy doses of chemical fertilizer, more so than the traditional plant varieties. The Mexican dwarf wheat was first released to farmers in 1961 and resulted in a doubling of the average yield. Borlaug described the twenty years from 1944 to 1964 as the "silent revolution" that set the stage for the more dramatic "Green Revolution" to follow.

In the 1960s, many observers felt that widespread famine was inevitable in the developing world and that the population would surpass the means of food production, with disastrous results in countries such as India. Borlaug with great persuasion introduced hundreds of tons of seeds of Mexican dwarf wheat varieties to jump-start production in India during 1965. By the 1969–1970 crop season, 35 percent of India's 35 million acres of wheat were sown with the Mexican dwarf varieties or varieties derived from them. New production technologies were also introduced, such as a greater reliance on chemical fertilizers and pesticides and the drilling of thousands of wells for controlled irrigation. Government policies that encouraged these new styles of production provided loans that helped farmers adopt it. Wheat production in India went from 12.3 million tons in 1965 to 20 million tons in 1970 to become self-sufficient in cereal production by 1974.

As important as the wheat program was, however, rice remains the world's most important food crop, providing 35–80 percent of the calories consumed by people in Asia. The International Rice Research Institute in the Philippines was to do for rice what the Mexican program had done for wheat. The new rice varieties (for the same reasons as the wheat) were dwarf (for the same reasons as the wheat), or more disease-resistant, or more suited to tropical climates. Scientists crossed thirty-eight different breeds of rice to create IR8, which doubled yields and became known as "miracle rice." IR8 served as the catalyst for what became known as the "Green Revolution". By the end of the twentieth century, more than 60 percent of the world's rice fields were planted with varieties developed by research institutes and related developers. A pest-resistant variety known as IR36 was planted on nearly 28 million acres, a record amount for a single food-plant variety.

In addition to Mexico, Pakistan, India, and the Philippines, countries benefiting from the Green Revolution included Afghanistan, Sri Lanka, China, Indonesia, Iran, Kenya, Malaya, Morocco, Thailand, Tunisia, and Turkey. The Green Revolution contributed to the overall economic growth of these nations by increasing the incomes of farmers (who were then able to afford tractors and other modern equipment), the use of electrical energy, and consumer goods, thus increasing the pace and volume of trade and commerce.

The success of the Green Revolution also depended on the fact that many of the host countries— such as Mexico, India, Pakistan, the Philippines, and China—had relatively stable governments and fairly well-developed infrastructures. These factors permitted these countries to diffuse both the new seeds and technology and to bring the products to market in an effective manner.

The Green Revolution could not have been launched without the scientific work done at the research institutes in Mexico and the Philippines. The two original institutes have given rise to an international network of research establishments dedicated to agricultural improvement, technology transfer, and the development of agricultural resources, including trained personnel, in the developing countries.

The leader of a Mexican research team, U.S. Plant Pathologist Norman Borlaug, was instrumental in introducing the new Mexican dwarf wheat to India and Pakistan and was awarded the Nobel Peace Prize in 1970.

However, the rates at which production increased in the early years of the program could not continue indefinitely, which caused some to question the "sustainability" of the new style. For example, rice yields per acre in South Korea grew nearly 60 percent from 1961 to 1977, but only 1 percent from 1977 to 2000 (Brown et al., *State of the World 2001*, p. 51). Rice production in Asia as a whole grew an average of 3.2 percent per year from 1967 to 1984 but only 1.5 percent per year from 1984 to 1996 (Dawe, p. 948). Some of the levelling-off of yields stemmed from natural limits on plant growth, but economics also played a role. For example, as rice harvests increased, prices fell, thus discouraging more aggressive production. Also, population growth in Asia slowed, thus reducing the rate of growth of the demand for rice. In addition, incomes rose, which prompted people to eat less rice and more of other types of food.

The Green revolution followed 3 major strands in its attempts to transform agriculture - social, biochemical and mechanical (Fig. 2.1).

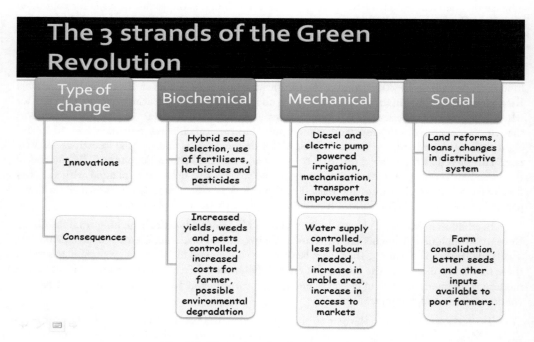

Fig. 2.1. Three strands of green revolution

2.4. GOVERNMENT SCHEMES UNDER GREEN REVOLUTION

Prime Minister Narendra Modi approved the Umbrella Scheme Green Revolution – 'Krishonnati Yojana' in the agriculture sector for the period of three years from 2017 to 2020 with the Central Share of Rs. 33,269.976 crores. The Umbrella scheme "Green revolution- Krishonnati Yojana" comprises 11 Schemes under it and all these schemes look to develop the agriculture and allied sector in a scientific and holistic manner so as to increase the income of farmers by increasing productivity, production, and better returns on produce, strengthening production infrastructure, reducing the cost of production and marketing of agriculture and allied produce. The 11 schemes that are part of the Umbrella Schemes under the Green revolution include (Fig. 2.2):

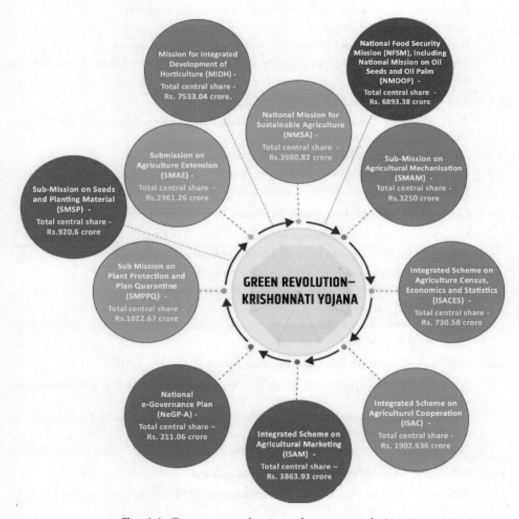

Fig. 2.2. Government schemes under green revolution.

2.4.1. MIDH – Mission for Integrated Development of Horticulture

» It aims to promote the comprehensive growth of the horticulture sector.
» Enhance the production of the sector.
» Improve nutritional security.
» Increase income support to household farms.

2.4.2. NFSM – National Food Security Mission

This includes NMOOP – National Mission on Oil Seeds and Oil Palm.

» Aims to increase the production of wheat, pulses, rice, coarse cereals and commercial crops.

- » Productivity enhancement, and area expansion in a suitable manner.
- » Enhancing farm level economy.
- » Restoring soil fertility and productivity at the individual farm level.
- » Aims to reduce imports and increase the availability of vegetable oils and edible oils in the country.

2.4.3. NMSA – National Mission for Sustainable Agriculture

- » Aims to promote sustainable agriculture practices that are best suitable to the specific agro-ecology.
- » Focusing on integrated farming.
- » Appropriate soil health management.
- » Synergizing resource conservation technology.

2.4.4. SMAE – Submission on Agriculture Extension

Aims to strengthen the ongoing extension mechanism of State Governments, local bodies, etc.

- » Achieving food security and socio-economic empowerment of farmers.
- » To forge effective linkages and synergy amongst various stakeholders.
- » To institutionalize program planning and implementation mechanism.
- » Support HRD interventions.
- » Promote pervasive and innovative use of electronic and print media, interpersonal communication, and ICT tools, etc.

2.4.5. SMSP – Sub-Mission on Seeds and Planting Material

- » Aims to increase the production of quality seed, upgrade the quality of farm-saved seeds and increase SRR.
- » Strengthen the seed multiplication chain, and promote new methods and technologies in seed production, processing, testing, etc.
- » Strengthen and modernize infrastructure for seed production, storage, quality, certification, etc.

2.4.6. SMAM – Sub-Mission on Agricultural Mechanisation

- » Aims to increase the reach of farm mechanization to small and marginal farmers and to the regions where availability of farm power is low.
- » To promote 'Custom Hiring Centres' to offset the adverse economies of scale arising due to small landholding and high cost of individual ownership.
- » to create hubs for hi-tech and high-value farm equipment.
- » To create awareness among stakeholders through demonstration and capacity building activities.
- » To ensure performance testing and certification at designated testing centres located all over the country.

2.4.7. SMPPQ – Sub Mission on Plant Protection and Plan Quarantine

- » Aims to minimize loss to quality and yield of agricultural crops from insects, pests, weeds, etc.
- » To shield our agricultural bio-security from the incursions and spread of alien species.

» To facilitate exports of Indian agricultural commodities to global markets.

» To promote good agricultural practices, particularly with respect to plant protection strategies.

2.4.8. ISACES – Integrated Scheme on Agriculture Census, Economics, and Statistics

» Aims to undertake the agriculture census.

» Undertake research studies on agro-economic problems of the country.

» Study the cost of cultivation of principal crops.

» Fund conferences, workshops, and seminars involving eminent agricultural scientists, economists, experts so as to bring out papers.

» Conduct short term studies, improve agricultural statistics methodology.

» Create a hierarchical information system on crop condition and crop production from sowing to harvest.

2.4.9. ISAC – Integrated Scheme on Agricultural Cooperation

» Aims to provide financial assistance for improving the economic conditions of cooperatives.

» Remove regional imbalances.

» Speed up cooperative development in agricultural processing, storage, marketing, computerization, and weaker section programs.

» Ensuring the supply of quality yarn at reasonable rates to the decentralized weavers and help cotton growers fetch a remunerative price for their produce through value addition.

2.4.10. ISAM – Integrated Scheme on Agricultural Marketing

» Aims to develop agricultural marketing infrastructure.

» To promote innovative technologies and competitive alternatives in agriculture marketing infrastructure.

» To provide infrastructure facilities for grading, standardization, and quality certification of agricultural produce.

» To establish a nationwide marketing information network.

» To integrate markets through a common online market platform to facilitate pan-India trade in agricultural commodities, etc.

2.4.11. NeGP-A – National e-Governance Plan

» Aims to bring farmer-centric and service-oriented programs.

» To improve access of farmers to information and services throughout the crop-cycle and enhance the reach and impact of extension services.

» To build upon, enhance and integrate the existing ICT initiatives of the Centre and States.

» To enhance efficiency and effectiveness of programs through providing timely and relevant information to the farmers for increasing their agriculture productivity.

2.5. FACTORS RESPONSIBLE FOR THE GREEN REVOLUTION/ INTERVENTIONS

The Green Revolution was driven by a technology revolution, comprising a package of modern inputs – irrigation, improved seeds, fertilizers, and pesticides – that together dramatically increased crop production. But its implementation also depended on strong public support for developing the technologies, building up the required infrastructure, ensuring that markets, finance, and input systems worked, and that ensured farmers had adequate knowledge and economic incentive to adopt the technology package.

There were several factors that were responsible for the success of the "Green Revolution" in India. The main ones are briefly described below:

2.5.1. High yielding variety of seeds/ improved seeds

This was the main scientific aspect of the "Green Revolution." The Indian Council for Agriculture Research, which was established by the British in 1929 yet was not known to have done any significant research, was reorganized first in 1965 and then in 1973. It developed new strains of high yield value (HYV) seeds, mainly wheat and rice, but also millets and corn. The most noteworthy HYV seed was the K68 variety for wheat. In addition, other high-yielding varieties of seeds were produced by agricultural universities and research centers in India. For example, in the case of wheat, S-308, Kalyan and Sona 227, etc.; in the case of rice, IR-7, IR-8, Massuri, Padma, Jaya, etc., were of high-yield variety of seeds. These seeds ensured a higher yield per acre for the farmer. The discovery and use of standard high-yielding variety of seeds, which considerably raised agricultural productivity, is primarily responsible for the success of the "Green Revolution."

Irrigation and fertilizer helped raise cereal yields, but their full impact was only realized after the development of high-yielding varieties. Scientists sought to develop cereal varieties that were more responsive to plant nutrients, and that had shorter and stiffer straw that would not fall over under the weight of heavier heads of grains. They also wanted tropical rice varieties that could mature more quickly and grow at any time of the year, thereby permitting more crops to be grown each year on the same land. Varieties also needed to be resistant to major pests and diseases that flourish under intensive farming conditions and to retain desirable cooking and consumption traits.

International Rice Research Institute (IIRI) in the Philippines developed semi-dwarf varieties that met most of these requirements and could be grown under a wide range of conditions. Similar achievements were made for wheat after Norman Borlaug (later awarded a Nobel Prize for his work) crossed Japanese semi-dwarf varieties with Mexican wheat varieties at what is now known as the International Center for Maize and Wheat Improvement (CIMMYT) in Mexico.

The adoption of high-yielding varieties occurred quickly and by 1980 about 40 percent of the total cereal area in India was planted to modern varieties (World Bank, 2007). This had increased to about 80 percent of the cropped area by 2000 (Gollin et al., 2005).

2.5.2. Double-cropping in existing farmland

Double cropping was a primary feature of the "Green Revolution." Due to the early maturity of new seeds, it became possible to grow two or three crops in a year from a plot of land, instead of just one crop. The one-season per-year practice was based on the fact that there is only one natural monsoon per year that brings in rainfall. So, there had to be two "monsoons" per year. One would be the natural monsoon and the other an artificial monsoon. The artificial monsoon came in the form of huge irrigation facilities. Dams were built to arrest large volumes of natural monsoon water, which until then were being wasted, and simple irrigation techniques were used to water the fields. This practice contributed to increased agricultural production in India.

2.5.3. Extensive irrigation facilities

The provision of irrigation facilities constituted yet another important component of the "Green Revolution." An extensive irrigation facility made it possible to provide water to farmers, and ensures better use of land and multiple cropping. The additional land brought under irrigation increased from 1.37 million hectares in 1968-69 to 81.0 million hectares in 1991-92.

India had already investing heavily in irrigation prior to the Green Revolution and by 1970 around 18.4 percent of the agricultural land was already irrigated (Table 2.1). In India, there were 10.4 million hectares of canal irrigated land in 1961, and 4.6 million hectares of tank irrigated land (Evenson *et al.*, 1999). Significant additional investments were made across India during the Green Revolution era, and the irrigated area grew from 18.4 percent to 31.8 percent of the agricultural area between 1970 and 1995 (Table 2.1).

Table 2.1. Indicators of input use during the Green Revolution in India

Indicators of input use	Increase	
	1970	**1995**
Irrigated area (% of agricultural area)	18.4	31.8
Fertilizer (kg/ha)	13.7	81.9
Annual growth rate in agricultural work force, 1967–82	1.59	
Annual growth rate in agricultural land area, 1967–82	0.19	

Source: Rosegrant and Hazell (2000).

2.5.4. Use of fertilizers

Increased use of fertilizers also contributed to significant increases in agricultural output. Use of chemical fertilizers in India increased from 1.8 million tons in 1968-69 to about 12.7 million tons in 1991-92.

Like irrigation, fertilizer use across India was also growing prior to the Green Revolution. In 1970, 13.7 kg of plant nutrients were applied per hectare of agricultural land and average use grew rapidly to reach 81.9 kg/ha by 1995 (Table 2.1).

2.5.5. Use of modern machinery

India's farming practice was dominated by traditional tools and methods. The increasing use of machinery and other modern equipment such as tractors, pump sets, power tillers, tube wells, harvesters etc., during the "Green Revolution" enabled multiple cropping and the growing of high-yielding varieties of crops in the country.

2.5.6. Plant protection scheme

Protecting plants by using pesticides and other such devices was another important aspect of the "Green Revolution." The area covered under India's plant protection scheme increased from 17 million hectares in 1965-66 to about 66 million hectares in 1991-92.

2.5.7. Expansion of farming areas

The expansion of farming areas was also an important factor for the success of the "Green Revolution" in India. Expansion of areas of land under cultivation had actually started right after the attainment of independence in 1947. The "Green Revolution" continued the trend at an accelerated rate. In 1960, the total area covered under the high-yielding-varieties program was a negligible 1.9 million hectares. By 1970, it reached 15.4 million hectares, by 1980, 43.1 million hectares and by 1990, it reached nearly 65.0 million hectares. Such spectacular increases in the areas of land under cultivation contributed to the success of the "Green revolution."

2.5.8. Miscellaneous factors

In addition to the above described factors, improvements in storage, food processing and marketing facilities, as well as government support price policies, also contributed to the success of the "Green Revolution" in India.

2.5.8.1. Public investment and policy support: The Green Revolution required a supporting economic and policy environment. The need to educate farmers about the new technology, rapidly expand input delivery and credit systems so they could adopt the new inputs, and increase processing, storage, trade and marketing capacities to handle the surge in production, was considered too large a challenge for the private sector on its own at the time, especially if small farmers were to participate.

Many Indian farmers lacked the financial resources needed to acquire the seeds, equipment and fertilizers. Thus, more attention had to be paid to making available adequate credit facilities to ordinary farmers to alleviate the lack of financial resources.

Technology package was profitable for farmers. To achieve these ends, governments across Asia actively intervened in launching and implementing the Green Revolution. Some but not all public interventions were market mediated, and all were backed by substantial public investments in agricultural development (Djurfeldt and Jirström, 2005).

Governments also shored up farm credit systems, subsidized key inputs—especially fertilizer, power, and water—and intervened in markets to ensure farmers received adequate prices each year to ensure the technologies were profitable. Many governments used their interventions to ensure that small farmers participated in the Green Revolution, and did not get left behind. Agricultural growth led by small farms proved not only to be more efficient but also more pro-poor; a win–win proposition for growth and poverty reduction.

2.6. RESULTS OF THE GREEN REVOLUTION

The results of the "Green Revolution" of India can be categorized into economic, social and political aspects.

2.6.1. Economic aspect

From the economic aspect, the "Green Revolution" resulted in a record grain output of 131 million tons in 1978-79. This achievement established India as one of the world's biggest agricultural producers. No other country in the world that attempted the "Green Revolution" recorded such a level of success. By the end of the 1970s India transformed itself from a net importer to a net exporter of food. Yield per unit of farmland increased by more than 30 percent between 1947 and 1979, when the "Green Revolution" was considered to have delivered its goods. The crop area under the high-yield varieties (HYV) grew from mere seven percent to 22 percent of the total cultivated area during the 10 years of the "Green Revolution." More than 70 percent of the wheat crop area, 35 percent of the rice crop area and 20 percent of the millet and corn crop areas, used the HYV seeds. Crop areas under HYV needed more water, more fertilizer, more pesticides, and certain other chemicals. This spurred the growth of the local manufacturing sector. Such industrial growth created new jobs and contributed to the country's gross domestic product (GDP). The increased emphasis on irrigation created the need for new dams to harness monsoon water. The water stored was used to create hydroelectric power. This in turn boosted industrial growth, created jobs and improved the quality of life of the people in villages.

2.6.2. Political aspect

India paid back all loans it had taken from the World Bank and its affiliates for the purpose of the "Green Revolution." Politically, this improved India's credit worthiness in the eyes of the lending agencies. Some developed countries, especially Canada, which were facing a shortage in agricultural labor, were so impressed by the results of India's "Green Revolution" that they asked the Indian government to supply them with farmers experienced in the methods of the Green Revolution. Many farmers from the states of Punjab and Haryana, in northern India, were thus sent to Canada where they settled.1 These people remitted part of their incomes to their relatives in India. This not only helped the relatives but also added, albeit modestly, to India's foreign exchange earnings.

2.6.3. Social aspect

Socially, the "Green Revolution" created plenty of jobs not only for agricultural workers but also industrial workers by creating lateral facilities such as factories and hydro-electric power stations, as previously stated. In short, because of the "Green Revolution," India transformed itself from a starving nation to an exporter of food. This earned admiration for India, especially in the Third World.

With respect to the success of the "Green Revolution" in India, it is usually the increased production in rice and wheat, particularly in the northern and north-western parts of the country, which is often cited. However, it should also be mentioned here that it has been argued that during the period of the "Green Revolution," sorghum yields outperformed rice and wheat in India. In fact, all the three key African cereals - maize, millet, and sorghum - did very well in some Indian states during the same period. This clearly demonstrates that impressive increases in yields is possible due to the use of fertilizers and HYVs even under conditions of low irrigation (Lipton, 1985).

2.7. IMPACTS

2.7.1. Positive impacts

» Tremendous increase in crop produce: It resulted in a grain output of 131 million tons in the year 1978-79 and established India as one of the world's biggest agricultural producers. The crop area under high yielding varieties of wheat and rice grew considerably during the Green Revolution.

» Green revolution has remarkably increased Agricultural Production. Food grains in India saw a great rise in output. The biggest beneficiary of the revolution was the wheat grain. The production increased to 55 million tons in the early stage of the plan itself.

» Green revolution was not just limited to agricultural output, the revolution also increased per hectare yield in the case of wheat from 850 kg per hectare to an incredible 2281 kg/hectare in its early stage (Table 2.2).

Table 2.2. Average yield per hectare of crops during 1950-51 to 1999-2000

Year	Yield/ha (kgs)		
	Rice	Wheat	Pulses
1950-51	668	663	441
1960-61	1013	851	539
1970-71	1123	1307	524
1980-81	1336	1630	433
1990-91	1740	2281	578
1992-93	1744	2327	573
1995-96	1855	2493	552
1999-2000	1986	2778	635

» With the introduction of the green revolution, India reached its way to self-sufficiency and was less dependent on imports. The production in the country was sufficient to meet the demand of the rising population and to stock it for emergencies. Rather than depending on the import of food grains from other countries, India started exporting its agricultural produce.

» The introduction of the revolution inhibited a fear among the masses that commercial farming would lead to unemployment and leave a lot of the labor force jobless. But the result seen was totally different, there was a rise in rural employment. The tertiary industries such as transportation, irrigation, food processing, marketing, etc. created employment opportunities for the workforce.

» The green revolution in India majorly benefited the farmers of the country. Farmers not only survived but also prospered during the revolution, their income saw a significant rise which enabled them to shift from sustenance farming to commercial farming.

» Reduced import of food-grains: India became self-sufficient in food-grains and had sufficient stock in the central pool, even, at times, India was in a position to export food-grains. The per capita net availability of food-grains has also increased.

» Benefits to the farmers: The introduction of the green revolution helped the farmers in raising their level of income. Farmers plowed back their surplus income for improving agricultural productivity. The big farmers with more than 10 hectares of land were particularly benefited by this revolution by investing large amounts of money in various inputs like HYV seeds, fertilizers, machines, etc. It also promoted capitalist farming.

» Industrial growth: The revolution brought about large-scale farm mechanization which created demand for different types of machines like tractors, harvesters, threshers, combines, diesel engines, electric motors, pumping sets, etc. Besides, demand for chemical fertilizers, pesticides, weedicides, etc. also increased considerably. Several agricultural products were also used as raw materials in various industries known as agro-based industries.

» Rural employment: There was an appreciable increase in the demand for labor force due to multiple cropping and use of fertilizers. The green revolution created plenty of jobs not only for agricultural workers but also industrial workers by creating related facilities such as factories and hydroelectric power stations.

2.7.2. Negative impacts

» Retardation of agricultural growth due to inadequate irrigation cover, shrinking farm size, failure to evolve new technologies, inadequate use of technology, declining plan outlay, unbalanced use of inputs, and weaknesses in credit delivery system.

» Regional dispersal of the evolution created regional inequalities. The benefits of the green revolution remained concentrated in the areas where the new technology was used. Moreover, since the revolution for the number of years remained limited to wheat production, its benefits were mostly accrued only to wheat-growing areas.

» Interpersonal inequalities between large- and small-scale farmers. The new technologies introduced during the revolution called for substantial investments which were beyond the means of a majority of small farmers. Farmers having large farmlands continued to make greater absolute gains in income by reinvesting the earnings in farm and non-farm assets, purchasing land from the smaller cultivators, etc.

» Non-food grains left out : Although all food-grains including wheat, rice, jowar, bajra and maize have gained from the revolution, other crops such as coarse cereals, pulses and oilseeds were left out of the ambit of the revolution. Major commercial crops like cotton, jute, tea and sugarcane were also left almost untouched by the green revolution.

» Limited coverage of HYVP: High yielding variety program (HYVP) was restricted to only five crops: Wheat, rice, jowar, bajra and maize. Therefore, non-food grains were excluded from the ambit of the new strategy. The HYV seeds in the non-food crops were either not developed so far or they were not good enough for farmers to risk their adoption.

» Regional disparities: Gren revolution technology has given birth to growing disparities in economic development at inter- and intra-regional levels. It has so far affected only 40 per cent of the total cropped area and 60 per cent is still untouched by it. The most affected areas are Punjab, Haryana and western Uttar Pradesh in the north and Andhra Pradesh and Tamil Nadu in the south. It has hardly touched the Eastern region, including Assam, Bihar, West Bengal and Orissa and arid and semi-arid areas of Western and Southern India. The green revolution affected only those areas which were already better placed from an agricultural point of view. Thus the problem of regional disparities has further aggravated as a result of the green revolution.

» Excessive usage of chemicals: The green revolution resulted in a large-scale use of pesticides and synthetic nitrogen fertilizers for improved irrigation projects and crop varieties. However, little or no efforts were made to educate farmers about the high risk associated with the intensive use of pesticides. Pesticides were sprayed on crops usually by untrained farm laborers without following instructions or precautions. This causes more harm than good to crops and also becomes a cause for environment and soil pollution.

» Water consumption: The crops introduced during the green revolution were water-intensive crops. Most of these crops being cereals, required almost 50% of dietary water footprint. Canal systems were introduced, and irrigation pumps also sucked out the groundwater to supply the water-intensive crops, such as sugarcane and rice, thus depleting the groundwater levels. Punjab is a major wheat- and rice-cultivating area, and hence it is one of the highest water depleted regions in India.

» Impacts on soil and crop production: Repeated crop cycle in order to ensure increased crop production depleted the soil's nutrients. To meet the needs of new kinds of seeds, farmers increased fertilizer usage. The pH level of the soil increased due to the usage of these alkaline chemicals. Toxic chemicals in the soil destroyed beneficial organisms, which further led to the decline in the yield.

»   Unemployment: Except in Punjab, and to some extent in Haryana, farm mechanization under the green revolution created widespread unemployment among agricultural laborers in the rural areas. The worst affected were the poor and the landless laborers.

»   Health hazards: The large-scale use of chemical fertilizers and pesticides such as Phosphamidon, Methomyl, Phorate, Triazophos and Monocrotophos resulted in in a number of critical health illnesses including cancer, renal failure, still born babies and birth defects.

2.7.3. Ecological and societal impacts

The major ecological and societal impacts of the Green Revolution can be summarized as follows:

»   Loss of landraces that were indigenous to our country.

»   The loss of soil nutrients making it unproductive.

»   Excessive use of pesticides increases the presence of its residues in foods and environment (Bowonder, 1979; Yadav *et al.*, 2015).

»   The farmers shift to unsustainable practices to obtain more yield.

»   Increased rates of suicide among farmers.

»   Unable to withstand the increasing expenses for farming and debts, small farmers sold their lands to large commercial farmers.

»   Unable to withstand the food inflation and economic crisis, the farmers left farming resorting to other occupation.

2.8. COMPETING VIEWS OF DEVELOPMENT

Supporters noted that the green revolution increased crop yields. India, for example, produced more wheat and rice, which helped avoid famines and save foreign exchange currency. Critics, however, charged that the green revolution increased inequalities: rich farmers became richer and poor farmers became poorer. Critics also complained that the green revolution encouraged increased environmental problems through the use of fertilizers, pesticides, and irrigation.

There were problems with both perspectives on the green revolution policies. Critics avoided providing realistic alternatives for solving national food deficits, and supporters avoided noting that poor individuals continued to be hungry, despite the increased supplies.

As successful as the green revolution was, the wholesale transfer of technology to the developing world had its critics. Some objected to the use of chemical fertilizer, which augmented or replaced animal manure or mineral fertilizer. Others objected to the use of pesticides, some of which are believed to be persistent in the environment. The use of irrigation was also criticized, as it often required drilling wells and tapping underground water sources, as was the encouragement of farming in areas formerly considered marginal, such as flood-prone regions in Bangladesh. The very fact that the new crop varieties were developed with foreign support caused some critics to label the entire program imperialistic. Critics also argued that the green revolution primarily benefited large farm operations that could more easily obtain fertilizers, pesticides, and modern equipment, and that it helped to displace poorer farmers from the land, driving them into urban slums. Critics also pointed out that the heavy use of fertilizer and irrigation causes long-term degradation of the soil.

Proponents of the Green Revolution argued that it contributed to environmental preservation because it improved the productivity of land already in agricultural production and thus saved millions of acres that would otherwise have been put into agricultural use. It is estimated that if cropland productivity had not tripled in the second half of the twentieth century, it would have been necessary to clear half of the world's remaining forest-land for conversion to agriculture (Brown, *Eco-Economy*).

The green revolution was a change in agricultural practices with secondary social and political effects. Both industrialized and less-industrialized countries adopted the practices. Almost all wheat and rice grown today originated in the green revolution.

2.9. EXPECTED AND UNEXPECTED CONSEQUENCES

Rockefeller adopted the survey team's recommendations. Borlaug's group employed traditional and novel scientific methods to produce high-yielding semi-dwarf wheat varieties that exceeded all expectations. Semi-dwarf varieties are stalky plants that can hold a heavy head of grain. These varieties, used with plentiful water, fertilizers, and pesticides, produced dramatically high crop yields. Interest in semi-dwarf varieties spread quickly, especially where food security was a concern. The Indian government asked Borlaug to help it develop wheat varieties for India; these were ultimately credited with preventing a major famine (Perkins, 1997). Governments lauded the social good of the technology that allowed them to import less food despite growing populations and green revolution science was soon extended to other staple grains, especially rice. Rice-producing countries around the world adopted these new rice varieties as readily as had wheat producers. Those who adopted Green Revolution technologies often experienced increases in their standards of living, although in some places, government-mandated food prices sometimes undercut the economic benefits of higher yields (Leaf, 1984).

The fears of critics were also realized, especially in the early years. Medium-sized and large farms could adopt the new technologies easily, and their high yields led to declining food prices. While urban populations benefited, small farmers watched the profits from their own harvests decrease. Some smaller farmers were able to adopt the technologies and improve their standards of living, but others were forced into rural labor or to move to the cities. Because people went hungry despite growing food supplies, critics argued that the Green Revolution could create food, but not relieve hunger (Sen, 1981). They pointed to regional inequities, as areas suited to Green Revolution grains and favored by government attention flourished, while poorer regions fell behind. For critics, the Green Revolution failed the test of social justice (Shiva, 1991).

Later, unanticipated environmental effects fed ongoing debates about social justice. The issue of monocropping highlights the environmental angle. Monocropping (producing a single crop in a field) helps produce uniform, high-yielding crops. However, it also produces microenvironments in which crops are more vulnerable to pests. Scientists responded by recommending heavy use of pesticides, with serious systemic consequences: sometimes toxic levels of pesticide exposure for farm laborers (who were often those disenfranchised by the Green Revolution), and rapid adaptation by pests requiring constant innovation and resulting in higher prices. Extensive monocropping sometimes led to less diversity in local food supplies, which critics have argued disproportionately affected the nutrition of the poor. In Green Revolution areas, the poor have come to depend almost exclusively on grains, decreasing the nutritional value of their diet (Shiva, 1993). In each critique, the question of justice, whether for the poor or for future generations, is the central concern.

2.10. GAINS AND LIMITATIONS

The gains and limitations of green revolution are presented in Table 2.3.

Table 2.3. Gains and limitations of green revolution

Gains	Limitations
Yield per unit of farmland improved by more than 30 per cent between 1947 (when India gained political independence) and 1979. Crop areas under high-yielding varieties needed more water, more fertilizer, more pesticides, and certain other chemicals. This spurred the growth of the local manufacturing sector. The increase in irrigation created need for new dams to harness monsoon water. The water stored was used to create hydro-electric power. This in turn boosted industrial growth, created jobs and improved the quality of life of the people in villages. India transformed itself from a starving nation to an exporter of food.	Even today, India's agricultural output sometimes falls short of demand. India has failed to extend the concept of high-yield value seeds to all crops or all regions. In terms of crops, it remain largely confined to food grains only, not to all kinds of agricultural produce. There are places like Kalahandi (Orissa) where famine-like conditions have been existing for many years and where some starvation deaths have also been reported. Of course, this is due to reasons other than availability of food in India, but the very fact that some people are still starving in India (whatever the reason may be), brings into question whether the Green Revolution has failed in its overall social objectives though it has been a resounding success in terms of agricultural production.

2.11. RECONSIDERATIONS

The attention that critics have paid to social justice, while sometimes questioned by supporters of the Green Revolution, have not fallen on deaf ears. The agency responsible for the scientific development of Green Revolution crops, the Consultative Group on International Agricultural Research (CGIAR), has responded vigorously. Scientists have decreased the amounts of pesticide needed, reducing risk to farm workers and lowering the cost of inputs. They increased the number of food crops for which they have developed high-yielding varieties, including some crops traditionally cultivated by the poor. Scientists have given attention to developing high yielding crops using less water, an important consideration in arid regions. In the 1990s, scientists began to research ways to introduce Green Revolution technologies to the poor regions of Africa that had been previously bypassed.

Advocates have also argued that making Green Revolution technologies socially just is not only the responsibility of scientists, but also of regional and national governments (Hazell, 2003). In places where agricultural credit is accessible, more small farmers have been able to retain or expand their land and benefit from the technologies. Such efforts are not lost on critics, but neither have they quieted the criticism that Green Revolution technologies promote injustice. Supporters are equally steadfast that Green Revolution technologies produce social goods that outweigh shortcomings. A widely agreed-upon ethical judgment of the Green Revolution remains unlikely, because the complex social and environmental consequences of this technology continue to unfold.

2.12. CONCLUSION

Overall, the Green Revolution was a major achievement for many developing countries, specially India and gave them an unprecedented level of national food security. It represented the successful adaptation and transfer of the same scientific revolution in agriculture that the industrial countries had already appropriated for themselves. However, lesser heed was paid to factors other than ensuring food security such as environment, the poor farmers and their education about the know-how of such chemicals. As a way forward, the policymakers must target the poor more precisely to ensure that they receive greater direct benefits from new technologies and those technologies will also need to be more environmentally sustainable. Also, taking lessons from the past, it must be ensured that such initiatives include all of the beneficiaries covering all the regions rather than sticking to a limited field.

Chapter - 3

Evergreen (Sustainable Food Grain Production) Revolution

3.1. INTRODUCTION

Agriculture continues to be a source of livelihood for majority of Indian population and contributed about 14 per cent to the gross domestic product (GDP) of the country in 2014-15. About 60% population of India is dependent on agriculture for their livelihood. Interest in farming community in agriculture is reported to be declining and consequently, agricultural workers, including cultivators and farm laborers, are moving away from agriculture. Studies have also reported that agrarian distress is increasing owing to low farm income and inequality in income between agriculture and non-agriculture sectors, which is a matter of concern. The farmer's average per capita income is 30 to 40% of their neighboring urban counter parts. The number of farmers' suicides increased from 10,700 in 1995 to 18,200 in 2005; an increase of 70 per cent in 11 years.

A number of initiatives have been taken up by the present Government to improve the performance of Indian agriculture. For the first time, Honourable Prime Minister of India has set a target of "Doubling Farmers' Income by 2022" (coinciding with 75th year of independence). This goal has enthused and fuelled motivation among the stakeholders and channelized the efforts in a holistic manner.

Doubling of farmers' income (DFI) goal was also coupled with many new and well thought out schemes on insurance for mitigating losses (*Pradhan Mantri Fasal Bhima Yojana*), ensuring effective marketing through unified national agricultural marketing platform (*e-National Agricultural Market*), and improving soil health *via* promoting organic farming through *Paramparagat Krishi Vikas Yojana* for maximizing the gains from farming (*Soil Health Card*), and *Pradhan Mantri Krishi Sinchayi yojana* - A dedicated micro- irrigation fund to be set up in National Bank for Agriculture and Rural Development (NABARD) to achieve the goal, '*per drop more crop*'.

The pathway for doubling of farmers' income requires consideration of different dimensions related to enhancement in agricultural production along with providing efficient markets and improved marketing facilities. As area expansion is limited, production enhancement can be done through bridging yield gaps in crops through adoption of efficient and effective cultivation practices, crop diversification with focus on high value crops, further improvements in the total factor productivity, proper irrigation management along with other factors leading to productivity improvements (GOI, 2007; Evenson *et al.*, 1998; Chand *et al.*, 2012; Birthal *et al.*, 2007). The losses in India's agricultural produce is estimated to be Rs. 926,510 million, approximately US$13 billion (Moloney, 2016), indicating that there is need for post-harvest management, better infrastructure and proper management to prevent these losses. The efficient marketing network would be the key factor for monetization of the output and realization of better gains.

3.2. EVERGREEN REVOLUTION

The agricultural policy of 2000 envisaged holistic development of Indian agriculture and aimed to achieve through evergreen revolution. Economic survey 2015-16 observed, "Indian agriculture is in a way, a victim of its own past success, especially the Green revolution". It suggested an integral development program to make the agricultural sustainability and evergreen revolution as a concept was developed eventually.

According to Prof. Swaminathan, an eminent agricultural scientist of international repute, India needs an "evergreen revolution" through consolidation of local resources and skills to enhance food security. "The ever-green revolution will be triggered by farming systems that can help produce more from the available land, water and labor resources without either ecological or social harm" (Swaminathan, 2000). Achievement of food security should involve a healthy mix of improving technical competence coupled with stable food and agricultural policies as well as good governance to ensure stronger implementation. Reducing regional disparities, cross learning and knowledge sharing can also go a long way in improving food security.

3.2.1. Objectives

» Sustainability: Agricultural practices have to be reoriented to maintain environmental sustainability as well as resource sustainability (e.g. promotion of zero budget natural farming/ organic farming to reduce use of chemicals in agriculture).

» Crop diversification in water stress areas like Punjab and Haryana.

» Promoting soil health through schemes like soil health cards, practices like rain water harvesting made compulsory, *etc.*

» Promoting lab to land exhibition, investing in research and development of agricultural technologies.

» Farm income: To increase the annual growth rate in agriculture over 4%.

» Scale of agriculture production: Objective is to increase the collective cropped area which would help in increased capital investment and use of latest agricultural technologies and machines to increase agricultural productivity [e.g. greater private sector participation through contract farming, promote cooperative farming, leasing of farm machines, subsidizing the purchase of new machines (sub mission on agricultural mechanization) etc.]

» Price protection to farmers: Agricultural contracts, promotion of practices like warehouse receipts, promotion of agricultural exports through improved quality of agricultural produce (latest Agriculture export policy), *etc.*

» Market availability to farmers: To dismantle the restrictions on movement of agricultural commodities throughout the country. Ensuring sufficient number of godowns, warehousing and cold storage facilities, *etc.*

» Insurance protection to farmers: Inclusion of all farmers in agricultural insurance scheme

(*Pradhan Mantri Fasal Bhima Yojana*) and eventually to cover all crops (horticulture produce is introduced lately on a pilot basis).

» Elimination of Regional disparity in agriculture: Rainfed areas are mostly reeled with low farmer income and productivity. Evergreen revolution through climate specific and farmer specific selection of crops and inclusion of allied sectors to increase farmer income is an objective.

» Harness the potential of Indian agriculture which hitherto unexplored (e.g. Blue revolution is a part of Indian foreign policy with cooperation with other countries. Incentivize the crops like pulses etc. through inclusion to MSP to attain agricultural as well as nutritional security).

3.3. ADDRESSING YIELD GAPS

3.3.1. Field crops

There exist huge yield gaps in agricultural sector which vary between 6 to 300 per cent in cereals, 5 to 185 per cent in oilseeds and 16 to 167 per cent in sugarcane in different states (GOI, 2007). Such gaps exist at two levels—one, between the best scientific practices and the best farm practices and second, between the best farm practices to the average farmer practices and are caused by a number of environmental factors.

The estimates portray that yield gap vary from one-fourth to one-third within the paddy farms. Sorghum farms in Maharashtra and Karnataka, and pearl millet farms in Rajasthan still exhibit yield gap as high as 50 per cent. The estimates of yield gap for chickpea in Madhya Pradesh stands more than 30 per cent, and by 45 per cent in Rajasthan and Maharashtra. Cash crops, which are input intensive, also exhibit yield gap of around 30-50 per cent.

The yield gaps between potential and actual yields of wheat could have several reasons, as shown in Fig. 3.1

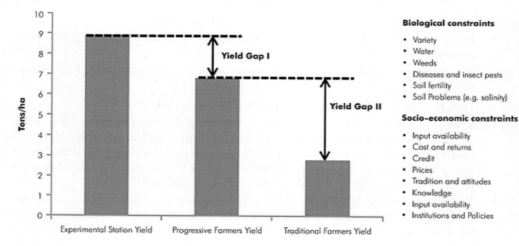

Fig. 3.1. Yield gap analysis of wheat in Tunisia

These scenarios offer us to increase output, thereby income, by using the existing improved technologies. Technology adoption helps in reducing yield gap at farm level. The estimates of yield gap for major crops across states are given in Table 3.1.

Table 3.1. Yield gap estimates, adoption of improved and hybrid seeds along with area under irrigation for selected crops and states

Crop	State	Crop yield gap (%)	Adoption of improved and hybrid seeds (%)	Area under irrigation (% of cropped area)
Rice	West Bengal	33	98	48.2
	Uttar Pradesh	29	100	83.1
	Andhra Pradesh	29	95	96.8
	Punjab	28	100	99.6
Wheat	Uttar Pradesh	27	98	98.4
	Punjab	19	100	98.9
	Madhya Pradesh	33	100	90.8
	Haryana	23	96	99.5
Sorghum	Maharashtra	53	59	9.5
	Karnataka	56	66	11.5
Pearl millet	Rajasthan	50	78	3.3
	Uttar Pradesh	35	83	8.9
Chickpea	Madhya Pradesh	32	100	57.9
	Rajasthan	46	50	49.2
	Maharashtra	45	84	24.2
Pigeon pea	Maharashtra	61	70	1.5
	Madhya Pradesh	36	52	1.6
	Karnataka	59	23	5.1
Maize	Andhra Pradesh	33	99	49.5
	Karnataka	45	98	36.0
	Bihar	58	67	65.2
Cotton	Gujarat	47	-	58.7
	Maharashtra	45	-	2.7
	Andhra Pradesh	38	-	13.9
Sugarcane	Uttar Pradesh	25	-	95.1
	Maharashtra	41	-	100
	Karnataka	35	-	100

Note: Estimates of yield gap and seed use are obtained for 2011-12 to 2013-14. Yield at 90th percentile is used as bench mark in computing the estimates. Irrigation figures correspond to the year 2012-13.

Source: Yield gaps and seeds use are estimates based on Ministry of Agriculture data (various years); irrigation coverage is based on Agricultural Statistics at a Glance (2015).

In fact, the wide gaps between and within developing and developed countries indicate that global food production still can be increased.

3.3.2. Horticultural crops

It has been observed that in most of the horticultural crops the productivity gap between national level and leading states is more than 100% as compared to demonstration plots or experimental station yields. Even though India is leading in productivity of some horticultural crops like grapes, cassava, ginger and turmeric, still there is scope to increase productivity in other horticultural crops as compared to other countries (Table 3.2).

Table 3.2. Productivity of major horticultural crops in India *vis-à-vis* other countries

Horticultural crop	National (tons/ha)	World highest (tons/ha)	Country	Increase over India (%)
Banana	34.00	52.54	Costa Rica	54.50
Mango	6.75	12.50	Brazil	85.00
Citrus	8.40	25.00	Brazil	297.60
Grape	25.69	25.69	India	---
Apple	7.68	84.01	Austria	993.88
Pear	5.59	283.19	Austria	4966.01
Potato	17.05	43.87	Netherlands	157.00
Tomato	17.08	26.69	USA	70.45
Eggplant	16.08	17.48	Japan	8.60
Chilli	9.18	14.40	Spain	44.50
Onion	13.41	51.19	USA	73.80
Garlic	4.93	17.71	USA	72.20
Okra	9.59	6.47	Jordan	17.78
Peas	9.14	8.35	Lithuania	20.00
Coconut (copra)	0.51	0.61	Indonesia	20.00
Black pepper	0.315	2.95	Malaysia	836.00
Cardamom	0.175	0.25	Guatemala	42.80
Ginger	3.47	3.47	India	---
Turmeric	3.92	3.92	India	---
Cassava	31.44	31.44	India	---
Sweet potato	8.94	24.24	Japan	171.00

In India, prospects of increasing production of horticultural crops by increasing land under cultivation are limited. Hence, it is essential to increase productivity of horticultural crops in order to meet the future demand and fulfillment of country's commitment to the sustainable horticultural development with the objective of nutritional security at individual level, poverty alleviation and employment generation through horticulture sector. Major focus on research should be devoted to productivity enhancement technologies like development of high yielding varieties/F1 hybrids, use of micro-irrigation, integrated nutrient management, fertigation, protected cultivation of ornamental and vegetable crops, use of biotechnological approaches, use of bioregulators, post-harvest management, integrated weed, pest, disease and nematode management. This calls for efficient transfer of technology and training of the stakeholders both private and public coupled with providing appropriate inputs, technical knowledge and infrastructure facilities.

3.3.2.1. Vegetable crops: There is significant gap in the national average productivity of different vegetable crops as compared to the world average productivity as well as the potential productivity for the respective vegetable crops (Table 3.3). The per capita availability of vegetables of 210 g/caput/day is still behind the recommended quantity of 285 g/caput/day. To meet the per capita requirement of vegetables at the recommended level, the vegetable production has to be increased substantially to match the population growth rate of 1.8% per year. To provide the balanced diet to Indian population, the production level has to be increased to 200 million tons of vegetables by 2020 as against approximately 125 million tons of present vegetable production. Hybrid vegetable technology is one of the better options particularly at this juncture because of the fact that growing of vegetables is four times more remunerative than cereals. Hence in coming decade, hybrid vegetable technology has to go a long way in this country to meet the challenges.

Table 3.3. Vegetable productivity scenario (tons/ha)

Vegetable crop	India (Average)	World (Average)	Potential productivity	Maximum productivity
Tomato	14.07	26.69	60-80	70.45 (USA)
Eggplant	16.08	17.48	40-50	34.7 (Japan)
Chilli	9.18	14.4	30-40	44.5 (Spain)
Okra	9.59	6.47	15-20	17.78 (Jordan)
Peas	9.14	8.35	18-20	20 (Lithuania)
Melons	20.48	20.95	30-40	45.83 (Cyprus)
Cucurbits	9.72	12.97	25-30	41.33 (Israel)
Cucumber	6.67	16.98	40-50	67.67 (Korea)
Watermelon	12.75	27.13	30-40	40.96 (Spain)
Cabbage	21.43	21.1	30-40	42.59 (Japan)
Cauliflower	17.14	18.36	35-40	45.25 (New Zealand)
Onion	10.38	17.53	40-50	60.33 (Korea)
Garlic	4.17	12.37	15-20	23.23 (Egypt)

Thus there is a vast scope for improving the productivity of these crops at national level. The critical gap between the experimental yield and national average can be minimized by a strong extension service, training and demonstration. There is an urgent need to recast the vegetable production strategy so that vegetable production may get impetus. Further for enhancing the vegetable production in the country, our target should be to achieve 25 tons/ha productivity by 2020, which is presently around 16 tons/ha.

3.3.2.2. Spice crops: Considerable increase in productivity has been reported from research stations especially in black pepper, turmeric and coriander (Table 3.4).

Table 3.4. Potential for productivity increase in spice crops (tons/ha)

Spice crop	National	Progressive farmer	Research station	Abroad	Scope for increase
Black pepper	0.315	2.000	2.445	2925 (Malaysia)	1.500
Cardamom	0.154	1.625	0.450	250 (Guatemala)	1.000
Ginger	3.477	5.500	8.250	---	2.000
Turmeric	3.912	6.200	10.700	---	2.000
Coriander	0.591	---	1.900	515 (Morocco)	1.000
Cumin	0.578	---	2.000	---	0.500
Clove	0.400	---	1.100	---	0.700
Nutmeg	0.600	---	0.885	---	0.250
Cinnamon	0.200	---	0.400	---	0.200
Fennel	1.800	---	2.500-3.000	---	0.700
Fenugreek	600-800	---	1.500-2.000	---	0.700

The effective transfer of technology attempts should be made to identify the main production constraints and to workout appropriate and economic strategies to overcome the same (Peter, 1999).

Key strategies to reduce yield gaps, therefore, include:

- » Increasing the efficiency of technology transfer.
- » Use of recommended practices: sowing date, seed rate, fertilizer amount, rotation, use of proper farm machinery, and disease and pest management practices.
- » Proper targeting of varieties to production zones.
- » Timely availability of inputs: quality seed, water, and fertilizers.
- » Government intervention and policies to strengthen input availability and crop marketing.

3.4. GOVERNMENT'S SEVEN-POINT STRATEGY FOR DOUBLING FARMERS' INCOME

To improve the economic condition of Indian farmers, Prime Minister Narendra Modi has set the target of doubling their income by 2022. The agriculture ministry is working on a seven-point strategy towards this end.

» Stepping up irrigation
» Effective use of inputs
» Reducing post-harvest losses
» Value addition
» Creation of a national farm market (e-NAM)
» Introduction of crop insurance scheme
» Promotion of ancillary activities

3.4.1. Stepping up irrigation

The first step is to increase productivity. It means focusing on irrigation, and accordingly the Union Government has increased the irrigation budget. India has 142 million hectares agriculture land, out of which only 48% is under institutional irrigation. With the objective of providing water to every field, *Pradhan Mantri Krishi Sinchai Yojana* was launched on July 1, 2015, and, to provide an end-to-end solution in irrigation supply chains, water resources, network distribution as well as farm level application. The comprehensive approach that combines irrigation with water preservation has been adopted. The objective is "*More Crop per Drop*". In addition, the aim is to complete pending medium and large irrigation projects on a priority basis in the next four years. Water harvesting, management, and watershed development projects have been put on the fast-track.

3.4.2. Effective use of inputs

The second factor is effective use of inputs, which means increasing production through improved seeds/planting material, organic farming, soil health care and other schemes. For the first time, a scheme *Paramparagat Krishi Vikas Yojana* has been launched for organic farming. Similarly, the government has curbed illegal use of urea and ensured its adequate supply through Neem-Coated Urea scheme. In addition, the Soil Health Card Scheme has helped reduce cultivation cost and increase production by curbing misuse of fertilizers. Farmers are also getting timely information and advisory services through new technologies such as space technology and online and telecom facilities via the Kisan Call Centre and Kisan Siddha App.

3.4.3. Reducing post-harvest losses

The next critical factor is reducing post-harvest losses. One of the biggest problems of farmers is storage after harvesting; as a result, they are forced to sell their products at a lower cost. Therefore, the government is encouraging farmers to use warehouses and avoid distress sales. Loans against negotiable warehouse receipts are being provided with interest subvention benefits. The focus is on storage facilities and integrated cold chains in rural areas.

3.4.4. Value addition

Value addition is being encouraged as a critical factor for augmenting income. The government has launched the *Pradhan Mantri Kisan Samara Yojana*. Under this, food-processing capabilities will be developed by working on forward and backward linkages of agro-processing, benefitting 2 million farmers and creating employment opportunities for about 0.5 million.

3.4.5. Creation of a national agricultural market (e-NAM)

In agriculture marketing, The Electronic-National Agriculture Market has been launched with three reforms and so far, 455 markets have been linked to it. Online trading has begun on various markets.

In addition, a model APMC Act has been circulated, which includes private market yards and direct marketing. Farmers are also being organized as Farmer Producer Organization. This helps them achieve economy of scale and increase bargaining power.

3.4.6. Introduction of crop insurance scheme

The *Pradhan Mantri Fasal Bhima Yojana* (PMFBY) helps to reduce the possible risks. The scheme is a shield for farmers' income. The lowest rate has been fixed for *kharif* and *rabi* crops. Maximum rate is 2% and 1.5%, respectively. The scheme covers standing crops as well as pre-sowing to post-harvest losses, and 25% of the claim is settled immediately online. Under PMFBY, many states are using remote sensing technology and drones to estimate losses and settle claims. To reduce climate change impact, various tolerant species and animal species have been developed. Contingency plans for affected districts have also been prepared.

3.4.7. Promotion of ancillary activities

Focus on agro-allied activities is critical. Focus is being given to horticulture, dairy (white revolution), poultry, bee keeping, fisheries (blue revolution), agroforestry, integrated farming and rural backyard poultry development to increase the income of the farmers. The farmers' income is increased through the above ancillary activities. Partially, it will be done through poultry, bee keeping, animal husbandry, dairy development, and fishery. The farmers are being encouraged to utilize uncultivated areas on periphery and boundary to grow trees for wood and to produce solar cells. More emphasis is being given to horticulture, agro-forestry, and integrated farming.

3.5. STRATEGIES FOR EVERGREEN REVOLUTION

3.5.1. Varietal improvement/Role of genotypes in productivity enhancement

Role of Indian Council of Agricultural Research (ICAR) Institutes/ State Agricultural Universities (SAUs) is extremely crucial in developing and spreading the use of better yielding varieties suitable for different technologies which can contribute to increasing farmers' income. Besides, the development of improved varieties/hybrids of food crops and their cultivation are central to increased farm production and consequently national food and nutritional security. Several high-yielding varieties of cereals, oilseeds, pulses, horticultural crops, forage crops and commercial crops were released from ICAR/ SAU institutions for cultivation in different production ecologies of the country. Bio fortified rice variety 'CR Dan 310' was commercialized successfully in the Indo-Gangetic Plains belt and 'Saran Sheryl', a new rice variety for drought–prone conditions was released. To ensure a faster spread to farmers' fields, several tons of breeder, foundation, certified, truthfully labelled seed and planting material were produced.

The semi dwarfing gene in rice (*sd-1*) is one of the most important genes deployed in modern rice breeding. Its recessive character results in a shortened culm with improved lodging resistance and a greater harvest index, allowing for the increased use of nitrogen fertilizers. The semi dwarf cultivar 'IR 8', which produced record yields throughout Asia and formed the basis for the development of new high-yielding, semi dwarf plant types. Rice cv. 'Pusan Basmati 1121' has the productivity of 4.0-4.5 t/ha and matures in 140-145 days, a fortnight earlier than Tarboro Basmati. It requires low input and provides high yield with better quality rice for export.

Similarly, in wheat "Morin 10," a cultivar from Japan, provided two very important genes, *Rht1* and *Rht2*, that resulted in the reduced height (or dwarf) in wheat with rust resistance.

The transfer of thick stem and higher sugar content and other desirable characters from the noble cane to Indian cane, which is commonly referred as mobilization, that increased cane yields and sugar content of sugarcane.

Development and release of high yielding varieties and hybrids in cereals (rice. maize, sorghum, pearl millet and cotton) and pulses (chickpea, red gram, green gram, black gram, lentil, field pea) was responsible for tremendous increase in yield and brought about green revolution.

Genetically modified (GM) crops can play a major role in increasing the productivity of global agriculture, helping farmers to meet the food, feed, and other demands of a rapidly rising world population, while saving water and forest lands. The use of GM crops for nearly two decades has consistently increased harvests, saved farmers money, and reduced the use of pesticides. First generation GM crops featured input-conserving traits, like resistance and tolerance. New GM crop varieties in the pipeline will offer traits such as drought and heat tolerance and improved nutritional quality, healthier food, and nutrient use efficiency. More than 20 countries grow GM crops for their high yields. Such crops are resistant to herbicides and some are said to be resistant to even the strongest herbicides.

3.5.2. Total factor productivity

A significant contributor to output growth would be the total factor productivity (TFP). Murali *et al.* (2012) revealed that more technological progress and hence more improvement in productivity was recorded after introduction of sugarcane variety 'CO-86032' than pre-introduction of variety 'CO-86032' period. 'CO-86032' variety is an early season variety which performs well in all soil types and extremely well in garden land conditions, yielding good quality cane with higher yield having multi-rationing capacity and can be grown throughout the year. The annual TFP growth over the whole period is 7.6 per cent. Government schemes like *Rashtriya Krishi Vikas Yojana*, (RKVY), National Food Security Mission (NFSM) and *Pradhan Mantri Fasal Bhima Yojana* (PMFBY) will facilitate attainment of desired growth in output as these schemes aim at holistic development of agriculture and allied sector, aim at accelerating production of crops mainly responsible for ensuring the food security along with soil fertility, and compensating farmers for crop losses/damages along with ensuring credit flow to farmers. Chand (2016) opined that TFP growth, which is mainly contributed by agricultural R&D, extension services, new knowledge, efficient practices like precision farming, is required to follow annual increase of 3.0 per cent.

3.5.3. Increase in crop intensity

Cropping intensity refers to raising a number of crops from the same field during one agricultural year; it can be expressed through a formula.

Cropping intensity = Gross cropped area / Net sown area x 100

Thus, higher cropping intensity means that a higher proportion of the net sown area is being cropped more than once during one agricultural year. This also implies higher productivity per unit of arable land during one agricultural year. Various measures to raise cropping intensity are discussed below:

3.5.3.1. Irrigation: Irrigation has played an important role in raising the cropping intensity in northern states where it has risen considerably. Irrigation helps to raise the cropping intensity by enabling rising of crops during the dry season also.

3.5.3.2. Fertilizers: The need to leave the land fallow for some period to regain the lost nutrients can be dispensed with by using fertilizers and following some other suitable cropping practices.

3.5.3.3. Crop rotation: It is the suitable arrangement of successive crops in such a way that the different crops draw nutrients in different proportions or from different strata. For instance, if legumes (pulses, gram, etc.) or certain oilseeds (groundnut) are sown just before the cereals, they fix the atmospheric nitrogen in soil, which can be absorbed by the cereals.

3.5.3.4. Mixed cropping: This works on similar principles. In this case, wheat and barley or wheat and gram or barley and gram are grown together to maintain a balance of consumption between different nutrients.

3.5.3.5. Relay cropping: This means simultaneous sowing of different crops with different nurturing periods in the same field and harvesting them one after the other. For instance, highly fertilizer-intensive crops like sugarcane and tobacco can be followed by cereals, in order to utilize the residual nutrients.

3.5.3.6. Selective mechanization: Use of tractors, tillers, threshers, etc. can save critical time between raising two crops, thus enabling the sowing of more than one crop.

3.5.3.7. Use of fast maturing varieties: These varieties can enable growing of more than one -crop within one growing season.

3.5.3.8. Appropriate plant protection: These measures include the use of pesticides, seed treatment, weed control, rodent control measures, etc. These measures are effective when all the farmers in an area take these up collectively.

Increase in crop intensity at the same rate as observed in the recent past has the potential to raise farmers' income by 3.4 per cent in 7 years and 4.9 per cent in ten years; this can turn out to be much higher as the possibilities for taking second crop are brightening (Chand, 2016).

3.5.4. Gains from irrigation/ Impact of irrigation on crop yield and income

The irrigated area in the country increased by 11 per cent between Triennium (TE) 2006-07 and TE 2013-14. The irrigation intensity expressed as the ratio of gross irrigated area (GIA) to gross cropped area (GCA), increased by 8 per cent during the above mentioned period. The states like Madhya Pradesh, Chhattisgarh, Karnataka, Bihar, Gujarat and Rajasthan have shown appreciable increase in GIA and thereby increase in irrigation intensity. The growth performance in Gross State Domestic Product (GSDP) of these states has also been much ahead of other states.

Considering the potential of micro-irrigation in saving of water and nutrients along with productivity enhancement, a lot of emphasis is being given to micro-irrigation in the country. Micro-irrigation can bring substantial increase in productivity and also result in water saving (GOI, 2009; 2014). According to the report of Task Force on Irrigation, an increase in productivity on adoption of micro-irrigation ranged from 3 per cent each in cowpea and cabbage to 27 per cent in gram. At the same time, micro-irrigation resulted in water saving of 16 per cent in lucerne to 56 per cent each in pearl millet and barley. According to another study, only 9.2 per cent of potential area of 42.23 million hectares is currently covered under micro-irrigation (Palanisami and Raman, 2012). Therefore, strategies for enhancing micro-irrigation coverage would be beneficial in enhancing the productivity and income of the farm household and need special attention. Other schemes recently started like soil health card will also bring change in output through qualitative and quantitative growth in output.

3.5.5. Protected cultivation

Protected cultivation is a cropping technique for growing horticultural crops under protective structures to shield them from pests and weather for assured, climate-resilient and enhanced production of quality products (Fig. 3.2). Naturally ventilated polyhouse technology is suitable for peri-urban areas where high value vegetables like tomato, bell pepper, parthenocarpic cucumber, and flowers like rose, chrysanthemum and gerbera can be grown easily.

Fig. 3.2. Protected cultivation

The yield under polyhouse cultivation can be achieved to the level of 5-8 times as compared to the open crop cultivation. Various trials conducted at Agricultural Research Centers in India indicated that bell pepper (planted in mid-September), cucumber (planting in mid-October) and tomato (November planting) under polyhouse produced 1060 kg, 1460 kg and 1530 kg per 100 square meters. The duration of these crops were 4- 9 months and more than 90% of total yield were obtained during off-season (during winter before the start of summer) which fetches significantly higher market price (2-4 times than normal season). Further, the crop duration can be extended up to July–August with the application of micro-irrigation and fertigation and yield can be achieved to the level of 20-25 kg/m^2. Therefore, it is possible to harvest a single crop round the year with minimum additional inputs and higher income can be generated.

3.5.6. High value agriculture

The importance of high value agriculture is increasing day by day. In the past two decades there have been substantial changes in the patterns of production in India. There is a shift in production from food grains to high value agriculture such as fruits and vegetables, milk and milk products. The study reveals the fact that due to shift in demand pattern toward high value crops, the farmers have also responded to market signals and gradually shifting production mix to meet the growing demand for high value commodities. This has reflected trough the changing share of high value crops in total value of output from agriculture.

Agricultural goods with high economic value are generally covered under high value agriculture (HVA). Transformation in favor of high value commodities (HVCs) is driven mainly due to the changing food consumption pattern and income elasticity. Demand-supply balance is crucial for retaining high value status of HVCs. Shifting agricultural resources to higher value options has been the new strategy for the agricultural development in the last decade or so, favoring crop diversification. Demands for fruits and vegetables will continue to grow.

Considering the type of growth that the livestock and poultry sector has maintained, it is the right time to attempt high value animal production to encash both the domestic and export market potential. Livestock is to be seen as the main source of livelihood in semi-arid and arid zones of the country. Recognizing the relatively high growth in fish production, post-harvest technology for fish is another important aspect of high value aquaculture. Globalization may further create opportunities for export of high value commodities.

3.5.7. Integrated pest management (IPM)

The major contributions of IPM relate to validation and dissemination of IPM in the targeted crops (rice, cotton, pulses, oilseeds, vegetables and fruits). During 2008 to 2014, area covered under IPM programs in different target crops increased from 658 to 1587 ha. Adoption of IPM resulted in reduction in the use of pesticides for pest management without compromising the productivity of crops. The IPM module developed in basmati rice was found to be very effective. IPM practices helped in increasing cotton productivity by 20-25 per cent and also gave significant reduction in mealy bug infestation. A major impact of IPM was observed in improving productivity of pigeon pea in Karnataka.

3.5.8. Resource use efficiency (RUE)

Resource use efficiency may also contribute significantly to the savings on cost front and thus enhancing the revenues to the farmers.

3.5.8.1. Zero–tillage technology: In zero tillage (ZT) technology, soil is not plowed, but sowing of crop is done by using a specially designed zero till seed-cum-fertilizer drill/planter, which disturbs soil to the least possible extent. At the time of seeding, fertilizers are simultaneously placed beneath the seeds. Several modern seeding machines, such as happy seeder, turbo seeder, multi-crop planter, roto-double disc planter are necessary for sowing in residue-laden conditions. ZT proves better for direct seeded rice, maize, soybean, cotton, pigeon pea, green gram, cluster bean, pearl millet during *kharif* season and wheat, barley, chickpea, mustard and lentil during *rabi* season. Wheat sowing after rice can be advanced by 10-12 days by adopting this technique compared to conventionally tilled wheat, and wheat yield reduction caused by late sowing can be avoided. ZT provides opportunity to escape wheat crop from terminal heat stress. ZT reduces cost of cultivation by nearly Rs 2,500-3,000/ha through reduction in cost of land preparation, and reduces diesel consumption by 50-60 liters per hectare. ZT reduces water requirement of crop and the loss of organic carbon by oxidation. ZT reduces *Phalaris minor* weed problem in wheat. The carbon status of soil is significantly enhanced in surface soil (0-5 cm), particularly under crop residue retention with zero tillage.

3.5.8.2. Raised bed planting technology for enhancing crop productivity: Raised bed planting is a promising technique of crop establishment during *kharif* season. It increases the productivity of crops like cotton, maize, pigeon pea, green gram, soybean, cowpea, vegetables, etc., which are grown in *kharif* and prone to water logging. Raised bed planting increases grain yield and economic returns, improves resource use efficiency and reduces weed problem. Bed planting system helps in efficient use of water under rain-fed as well as irrigated conditions because of optimum water storage and safe disposal of excess water. Furrow irrigated raised-bed system (FIRBS) of wheat usually saves seed by around 25 per cent, water by 25-30 per cent and nutrients by 25 per cent without affecting wheat grain yield. It reduces weed populations on the top of beds and lodging of wheat crop. The productivity of cotton-wheat, pigeon pea-wheat and maize-wheat systems is higher under ZT bed planting with crop residue. Cotton-wheat cropping system under ZT broad bed with residues of both crops gave higher system productivity and net returns than that in the transplanted rice-conventional till wheat cropping system. Therefore, it can be an alternative option for rice-wheat system under irrigated conditions.

3.5.8.3. Direct-seeded rice: Direct-seeded rice (DSR) avoids water required for puddling and reduces overall water demand compared to conventional puddled transplanted rice (TPR). DSR is a labor, fuel, time and water saving technology, which gives comparable yield as that of TPR. Soil health is maintained or improved, and fertilizer and water-use efficiencies are higher in DSR (saving of 30-40% irrigation water). Therefore, DSR is a technically and economically feasible alternative to TPR. In North Indian conditions, summer green gram can be adopted before DSR. It gives grain yield of 0.8-1.0 t/ha and usually adds 40-60 kg N/ha in soil, reducing N requirement for the subsequent crop.

Efficient resource use is the basis for achieving universal food security and poverty reduction strategies particularly in the rural areas. To attain optimality in the use of farm resource inputs, farmers should be advised to reduce their use of labor, seed/planting material, water, fertilizers, pesticides and capital and increase their use of land resources. Two pivotal components required to enhance crop productivity include: (i) the development of integrated soil–crop systems management (ISSM), which will address key constraints in existing crop varieties (Figs. 3.3 and 3.4), and (ii) the production of new crop varieties that offer higher yields but use less water, fertilizer or other inputs and are more resistant to drought, heat, submersion, and pests and diseases.

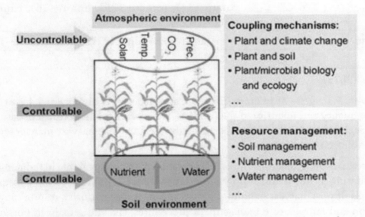

Fig. 3.3. Integrated soil–crop systems management approach

Fig. 3.4. Performance of an integrated soil–crop systems management (ISSM) (Yield=14.6 t ha^{-1}, Radiation use efficiency (RUE)=56 kg yield/kg of N) compared to farmer's practices (FP) (Yield=6 t ha^{-1}, RUE=20 kg yield/kg of N).

3.5.9. Reducing post-harvest losses

A recent estimate has been given that the country lost the output worth approximately Rs. 926,510 million during 2012-13 at 2014 prices (Jha *et al.*, 2015). Such losses basically happen on account of ineffective post-harvest management at critical stages of value-chains. Out of the total production of fruits and vegetables, wastage and losses account for 20 to 22 per cent. Fruits, vegetables and livestock account for about 54 per cent of these losses (Table 3.5).

Table 3.5. Extent of post-harvest losses across commodity groups

Crop/ commodity	Monetary losses (Rs. million)	Major crops/segments in term of monetary losses (%)
Cereals	206,980	Paddy (50), wheat (38), others (12)
Pulses	38,770	Chickpea (63), pigeon pea (25), others (12)
Oilseeds	82,780	Soybean (65), mustard (18), others (16)
Fruits	166,440	Mango (43), banana (23), citrus (9), apple (8), others (16)
Vegetables	148,420	Potato (34), tomato (25), onion (16), cauliflower (8), others (18)
Plantation crops	93,250	Sugarcane (60), coconut (22), others (18)
Total	736,340	

Source: Jha *et al.* (2015).

Post-harvest management in crops, livestock and fisheries is highly crucial as it is responsible for maintaining quality and quantity of the produce. It becomes further critical in case of perishable commodities. Such losses need to be saved through effective post-harvest management, improved marketing and value chain networks.

Post-harvest handling of fresh fruits and vegetables plays a critical role in reducing post-harvest losses and facilitating a continuous supply of high quality fresh produce to the consumers. Though the fundamentals of post-harvest handling: harvesting at proper maturity, minimizing injury, using proper sanitation and temperature management procedures, etc. still remain important, many new technologies developed and refined in recent years continue to make possible an even expanding supply of fresh products. The demand from consumers for quality and safe products and global concerns for reducing deterioration of the environment are inducing changes in the way technology is being used to minimize post-harvest losses and maintain product quality. These concerns have laid emphasis on development of new technologies that are more environmentally sustainable and economically competitive. The use of more natural chemicals or physical treatments to replace synthetic chemicals and increasing the efficiency of older, more traditional methods in combination with newer bio-control treatments have attracted the attention of researchers during the recent past.

Mission for Integrated Development of Horticulture (MIDH) is a Centrally Sponsored Scheme for the holistic growth of the horticulture sector and it aims at regionally differentiated strategies in accordance with comparative advantage of various agro-climatic regions, encourages aggregation of farmers, enhances horticulture production, augments farmers' income and strengthens nutritional security along with skill development and employment generation in horticulture and post-harvest management (Government of India, 2014). Along with this, the schemes like Rural Godown Scheme aim to facilitate storage of cereal crops by building storage capacity at farmers' doors.

Ensuring public-private partnership can provide further boost to effective post-harvest management especially in rural areas.

3.5.10. Encouraging value addition and processing

The performance of food processing has not been commensurate with production growth performance. In India, only 2 per cent of vegetable production and 4 per cent of fruit production is being processed, whereas the extent of fruit processing is very high in some countries such as Brazil (70 per cent), Malaysia (83 per cent), Philippines (78 per cent) and Thailand (30 per cent) (FAO, 2014).

Food processing takes place in the corporate [7-8 per cent share in the gross value added (GVA) manufacturing] as well as household (20-25 per cent) sector. Food processing usually takes place in the form of grain processing, manufacturing of animal feed, dairy processing, meat, fish, fruits, vegetables and oil processing (Fig. 3.5). Grain milling and processing along with other food products account for the major chunk of processing in both household and corporate sector. Extent of horticulture and livestock processing remains quite low in both corporate as well as household sector. Stringent efforts need to be made to promote the value-addition of livestock (including dairy) and horticultural products. Many self-help groups (SHGs) are emerging in dairy processing in different areas which may be scaled-up with proper policy and infrastructure and institutional support.

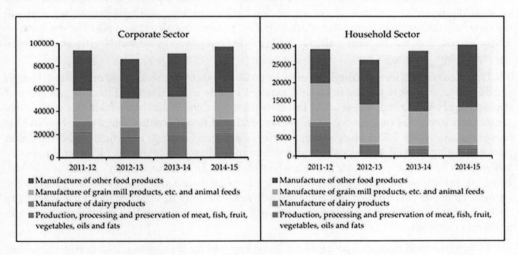

Fig. 3.5. Food processing in the corporate as well as household sector

3.5.11. Improving price realization to the farmers

The issue of high price volatility in agricultural commodities in domestic as well as international market has assumed critical importance. Fruits and vegetables seem to exhibit highest price volatility among all agricultural commodities. Some commodities in this category, like onion, have created crisis situation in the economy many a times due to the extreme volatility in their prices.

It has been established that prices will be one of the major sources of growth even if the status quo in the production is maintained. It does not imply here that prices need to be increased essentially; it implies that we need to improve farmers' share in consumer price and need to minimize during the chain costs, margins and inefficiencies. In general, medium and large farmers received higher price as compared to relatively marginal farm size categories. It was interesting to note that marginal and small farmers received comparatively higher price from cooperative and government agency in both season crops.

The studies have reported that farmers' share in consumer price remains very low, this share has been reported to be especially very low in case of horticultural commodities. Producers stand to gain when improved marketing efficiency increases demand and prices for their products (Landes, 2010). As has been established, output growth only will not drive sufficient growth for farmers; rather it will be through efficient marketing arrangements that the income of farmers would be enhanced. Many innovative marketing arrangements have shown that farmers' share can be magnified and the marketing costs and margins of the chain can be handled efficiently.

The idea behind Minimum Support Price (MSP) was to give guaranteed prices and assured market to the farmers and save them from the price fluctuations. This calls for the assurance of remunerative and stable price environment for farmers and price policy in the form of minimum support price announced by the government is one of the initiatives in this direction.

3.5.11.1. Contract farming: There is a need to adopt contract farming as part of multi-pronged strategy to enhance farmers' income. In contract farming, private agro-processing and exporting companies enter into an agreement with farmers to purchase a specified quantity of an agricultural commodity on mutually agreed terms. This type of integration between growers and agro-processing units will not only help farmers in getting a better price for their produce but also reduce post-harvest losses to a large extent. Contract farming is already being carried out privately, but state legitimacy is needed to help farmers in a bigger way.

3.5.11.2. Reducing price volatility: The price instability needs to be checked and minimized through suitable interventions to ensure stable and required returns to the farmers. Effective procurement strategies by NAFED and proper use of Price Stabilization Fund would be useful in this regard. This is extremely important that concerted efforts are continued to achieve this objective. The e-NAM is latest initiative which will provide national unified agricultural markets to farmers and bring better price realization through connectivity, transparency and enhanced integration. However, an ex-ante analysis may be conducted in those markets which have already been connected through the e-NAM. Ensuring private sector participation can bring competition and will provide added gains to the farmers. Effective post-harvest management will yield not only in terms of increased availability rather it may help farmers fetch remunerative prices for their produce. The gains from trade can further be much larger.

Objective conditions and policies need to be imposed for the commodities having significant trade potential. For this, networking is required among academic, research institutions and practicing organizations for proper technical supervision and guidance.

3.5.12. Transfer of technology

Extension services are important in spreading the technologies and the associated benefits. Kisan mobile advisory (KMA), an initiative by the ICAR, sent 93,949 short text messages, 14,788 voice messages and 1,180 both SMS and voice messages to benefit 22.394 million farmers on various aspects of agriculture based on input provided by 557 Krishi Vigyan Kendras (KVKs). The processes of technology assessment and refinement are as important as the technology generation prior to transfer at the field level. During 2015-16, 2,652 technology interventions were assessed across 4,003 locations by laying out 27,008 trials on the farmers' fields. In all, 22.875 million quality planting material of elite species of different crops were produced and provided to 1.838 million farmers.

For enhancing farmers' income, it is important that existing pool of agricultural knowledge regarding technologies and processes is channelized and implemented for the benefit of its stakeholders. The strategies covering various policies and strategies related to technology development, refinement and spread/adoption; effective implementation of marketing reforms; expansion of value addition and processing capacity; inculcating entrepreneurial skills among the educated farm youth and effective trade policies need to be prepared.

3.6. CONCLUSION

Evergreen Revolution requires not only the interventions and development in the agricultural sector but also requires the strong linkages with manufacturing and service sector to transform the *agricultural units to agricultural enterprises*. Thus, it is not going to be an isolated game that would transform the face of Indian agriculture. Rather, it will be putting all forces together for the holistic development of this sector to provide it more modern and professional orientation.

Prof. M.S. Swaminathan gave the call for evergreen revolution and this is possible only through the integrated and holistic approach to agricultural development. It involves leading innovation in agriculture, enabling farmers big and small, driving a sustainable intensification of agriculture, enhancing human health etc. With uncertainty of climate, low farm income and growth, nutritional issues – Evergreen Revolution is a way addressing these issues in the most comprehensive way by making agriculture not only profitable but also sustainable.

Chapter -4

Golden (Horticultural Crop Production) Revolution

4.1. INTRODUCTION

THE UN General Assembly designated 2021 as the International Year of Fruits and Vegetables (IYFV). The official slogan of the campaign 'Fruits and vegetables, your dietary essentials' underlines the importance fruits and vegetables have in nutrition and overall health. The objectives of IYFV are to raise awareness of the health benefits of fruit and vegetable consumption; advocate for healthy diets through increased consumption of fruits and vegetables; promote international efforts to boost fruit and vegetable production and value chains in a sustainable and safe way; bring in focus on the need to reduce losses and waste in fruit and vegetable supply chains from production to consumption; and invite relevant stakeholders to strengthen the capacities of developing countries to adopt innovative approaches and technologies in combating loss and waste of fruits and vegetables. The IYFV 2021 is a step towards achieving UN Sustainable Development Goals.

Horticulture sector is an integral element for food and nutritional security in the country. Horticulture is the main segment, while its various sub-segments are fruits, vegetables, aromatic and herbal plants, flowers, spices and plantation crops. All these are regarded as the essential ingredients of economic security. The wide range of agro-climatic conditions of India is conducive for growing a large variety of horticultural crops, including, root and tuber crops, mushroom, ornamental crops, plantation crops like coconut, areca nut, cashew and cocoa. The Government of India has recognized horticulture crops as a means of diversification in agriculture in an eco-friendly manner through efficient use of land and optimum utilization of natural resources. Horticulture seeks to create ample opportunities for employment, particularly for unemployed youths and women folk. India has maintained leadership in the production of many commodities like mango, banana, acid lime, coconut, areca nut, cashew, ginger, turmeric and black pepper. Presently, it is the second largest producer of fruits and vegetables in the world. India is next only to China in area and production of vegetables and occupies prime position in the production of cauliflower, second in onions and third in cabbage in the world. India has also made noticeable advancement in the production of flowers. Further, it is the largest producer, consumer and exporter of spices. India is home to a wide variety of spices like black pepper, cardamom (small and large), ginger, garlic, turmeric, chilli and a large variety of tree and seed spices. Almost all the States in the country grow one or more spices.

Horticultural crops form an important part of wholesome food containing carbohydrates, proteins, minerals, dietary fibers and vitamins in adequate amount. They produce more edible energy and protein per unit area and time than many other food crops. There is an urgent need to increase horticultural productivity in order to provide nutritional security to the fast-growing population of the country. Among the horticultural crops, fruits recorded a three-fold increase in area and four-fold increase in production over the last five decades.

4.2. HORTICULTURE SCENARIO IN INDIA

The area and production trend shows that there has been a sharp rise in production of horticulture crops with production during the year 2017-18 (Final) reaching 311.7 million tons which is 3.7% higher than the previous year and 10% higher than the past 5 years' average production. Production of fruits is estimated at 97.35 million tons which is 4.8% higher than previous year. Production of vegetables is estimated at about 187.5 million tons which is about 3.5% higher than the previous year. With an increase of 3.7%, the production of onion during the year 2017-18 is estimated at 23.26 million tons as against 22.4 million tons in 2016-17. Production of potato in the year 2017-18 (Final) is estimated at 51.3 million tons as against 48.6 million tons in 2016-17 (5.6% higher than 2016-17). India is the largest producer of ginger and okra amongst vegetables and ranks second in production of potatoes, onions, cauliflowers, brinjal, cabbages, *etc.* Amongst fruits, the country ranks first in production of bananas (25.7%), papayas (43.6%) and mangoes (including mangosteens and guavas) (40.4%).

4.3. GAP IN POTENTIAL AND REALIZED YIELDS

Even though India is leading in productivity of some horticultural crops like grapes, cassava, ginger and turmeric, still there is scope to increase productivity in other horticultural crops compared to other countries (Table 4.1).

Table 4.1. Productivity of major horticultural crops in India *vis-à-vis* other countries

Horticultural crop	National (tons/ha)	World highest (tons/ha)	Country	Increase over India (%)
Banana	34.00	52.54	Costa Rica	54.50
Mango	6.75	12.50	Brazil	85.00
Citrus	8.40	25.00	Brazil	297.60
Grape	25.69	25.69	India	---
Apple	7.68	84.01	Austria	993.88
Pear	5.59	283.19	Austria	4966.01
Potato	17.05	43.87	Netherlands	157.00
Tomato	17.08	26.69	USA	70.45
Eggplant	16.08	17.48	Japan	8.60
Chilli	9.18	14.40	Spain	44.50
Onion	13.41	51.19	USA	73.80
Garlic	4.93	17.71	USA	72.20

Okra	9.59	6.47	Jordan	17.78
Peas	9.14	8.35	Lithuania	20.00
Coconut (copra)	0.51	0.61	Indonesia	20.00
Black pepper	0.315	2.95	Malaysia	836.00
Cardamom	0.175	0.25	Guatemala	42.80
Ginger	3.47	3.47	India	---
Turmeric	3.92	3.92	India	---
Cassava	31.44	31.44	India	---
Sweet potato	8.94	24.24	Japan	171.00

In India, prospects of increasing production of horticultural crops by increasing land under cultivation is limited. Hence, it is essential to increase productivity of horticultural crops in order to meet the future demand and fulfilment of country's commitment to the sustainable horticultural development with the objective of nutritional security at individual level, poverty alleviation and employment generation through horticulture sector.

4.3.1. Potential for enhancing productivity

4.3.1.1. Vegetable crops: The per capita availability of vegetables (210 g/caput/day) is still behind the recommended quantity (285 g/caput/day). To meet the per capita requirement of vegetables at the recommended level of 285 g/caput/day, the vegetable production has to be increased substantially to match the population growth rate of 1.8% per year. To provide the balanced diet to Indian population, the production level has to be increased to 200 million tons of vegetables by 2020 as against approximately 125 million tons of present vegetable production. Hybrid vegetable technology is one of the better options particularly at this juncture because of the fact that growing of vegetables is four times more remunerative than cereals. Hence in coming decade, hybrid vegetable technology has to go a long way in this country to meet challenges.

There is significant gap in the national average productivity of different vegetable crops as compared to the world average productivity as well as the potential productivity for the respective vegetable crops (Table 4.2). Thus there is a vast scope for improving the productivity of these crops at national level. The critical gap between the experimental yield and national average can be minimized by a strong extension service, training and demonstration. There is an urgent need to recast the vegetable production strategy so that vegetable production may get impetus. Further for enhancing the vegetable production in the country, our target should be to achieve 25 tons/ha productivity by 2020, which is presently around 16 tons /ha.

Table 4.2. Vegetable productivity scenario (tons/ha)

Vegetable crop	India (Average)	World (Average)	Potential productivity	Maximum productivity
Tomato	14.07	26.69	60-80	70.45 (USA)
Eggplant	16.08	17.48	40-50	34.7 (Japan)
Chilli	9.18	14.4	30-40	44.5 (Spain)
Okra	9.59	6.47	15-20	17.78 (Jordan)

Peas	9.14	8.35	18-20	20 (Lithuania)
Melons	20.48	20.95	30-40	45.83 (Cyprus)
Cucurbits	9.72	12.97	25-30	41.33 (Israel)
Cucumber	6.67	16.98	40-50	67.67 (Korea)
Watermelon	12.75	27.13	30-40	40.96 (Spain)
Cabbage	21.43	21.1	30-40	42.59 (Japan)
Cauliflower	17.14	18.36	35-40	45.25 (New Zealand)
Onion	10.38	17.53	40-50	60.33 (Korea)
Garlic	4.17	12.37	15-20	23.23 (Egypt)

4.3.1.2. Spice crops: Considerable increase in productivity has been reported from research stations especially in black pepper, turmeric and coriander (Table 4.3). The effective transfer of technology attempts should be made to identify the main production constraints and to workout appropriate and economic strategies to overcome the same (Peter, 1999).

Table 4.3. Potential for productivity increase in spice crops (kg/ha)

Spice crop	National	Progressive farmer	Research station	Abroad	Scope for increase
Black pepper	315	2000	2445	2925 (Malaysia)	1500
Cardamom	154	1625	450	250 (Guatemala)	1000
Ginger	3477	5500	8250	---	2000
Turmeric	3912	6200	10700	---	2000
Coriander	591	---	1900	515 (Morocco)	1000
Cumin	578	---	2000	---	500
Clove	400	---	1100	---	700
Nutmeg	600	---	885	---	250
Cinnamon	200	---	400	---	200
Fennel	1800	---	2500-3000	---	700
Fenugreek	600-800	---	1500-2000	---	700

4.3.2. Interventions needed for enhancing productivity

It has been observed that in most of the horticultural crops the productivity gap between national level and leading states is more than 100% as compared to demonstration plots or experimental station yields. Major focus on research should be devoted to productivity enhancement technologies like development of high yielding varieties/F1 hybrids, use of micro-irrigation, integrated nutrient management, fertigation, protected cultivation of ornamental and vegetable crops, use of biotechnological approaches, use of bioregulators, post-harvest management, integrated weed, pest, disease and nematode management. This calls for efficient transfer of technology and training of the stake holders both private and public coupled with providing appropriate inputs, technical knowledge and infrastructure facilities.

4.4. CONSTRAINTS

The Indian horticulture sector is facing severe constrains such as low crop productivity, limited irrigation facilities and underdeveloped infrastructure support like cold storages, markets, roads, transportation facilities, *etc.* There are heavy post-harvest and handling losses, resulting in low productivity per unit area and high cost of production. Common problems encountered in production and productivity of horticultural crops include:

- » Inadequate availability of disease free, high quality planting material.
- » Micro-propagation techniques are under exploited.
- » Slow dissemination and adaptability of improved high yielding cultivars/hybrids.
- » Inadequate facility for identification of nutrient deficiency and disorders.
- » Lack of diseases and pests' outbreak forecast service.
- » Unavailability of refined intensive integrated production systems.
- » Lack of quality standards.
- » Lack of technologies in value addition.
- » Lack of post-harvest management technology and infrastructure.
- » Weak database and poor market intelligence.
- » Poor marketing practices and infrastructure.
- » Instability of prices, with no support price mechanism.
- » Inadequate technical manpower/human resource in farming system.
- » Poor credit supply, high rate of interest coupled with inadequate crop insurance scheme.
- » Late implementation of government policies and schemes.
- » Absence of horticultural crop suitability map of India based on agro climatic conditions depicting most suitable areas for optimum productivity of a particular crop.

However, on the other hand, India's long growing season, diverse soil and climatic conditions comprising several agro-ecological regions provide ample opportunity to grow a variety of horticulture crops. Thus, efforts are needed in the direction to capitalize on our strengths and remove constrains to meet the goal of moving towards a formidable horticultural growth in India. The foreign trade policy in 2004-11 emphasized the need to boost agricultural exports, growth and promotion of exports of horticultural products.

4.5. GOLDEN REVOLUTION

Horticultural development had not been a priority in India until recent years. It was in the post-1993 period that a focused attention was given to horticulture development through an enhancement of plan allocation and knowledge-based technology. Despite of this decade being called a "golden revolution" in horticultural production, the productivity of horticultural crops has increased only marginally from 7.5 tons per hectare in 1991-92 to 8.4 tons per hectare in 2004-05 (NHB, 2005). The National Horticulture Mission was launched in 2005-06 by the Government of India with a mandate to promote integrated development of horticulture, to help in coordinating, stimulating and sustaining the production and processing of fruits and vegetables and to establish a sound infrastructure in the field of production, processing and marketing with a focus on post-harvest management to reduce losses. The area and production trend shows that there has been a sharp rise in area and production of horticulture crops with an area of 25.4 million ha and production reaching 311.7 million tons during the year 2017-18 (Final). As a result of this huge spurt in horticulture produce, India has become the second largest producer of fruits and vegetables in the world, next only to China.

4.5.1. Current scenario

Horticulture, over the past few years, has made remarkable progress in terms of expansion in area under different crops, increase in productivity, crop diversification, technological interventions for production and post-harvest and forward linkages through value addition and marketing. A significant increase in area has been achieved in vegetables and flowers. Unprecedented growth has been achieved in off-season vegetable production and floriculture, making these sectors as upcoming opportunities. Concept of greenhouse has become extremely popular. There has been a tremendous success in large scale of tomato in open field as well as under protected condition. Critical areas like pest and disease management have been given principal focus with intensification of efforts on containing decline problems especially in orange and large cardamom. Massive rejuvenation programs in orange and re-plantation in large cardamom has brought back some ray of hope in the direction of revival of these crops. A number of nurseries have been established for producing healthy and quality planting materials of orange and large cardamom. The tissue culture laboratories, in private sector, have been able to supplement the requirement of quality planting materials.

Infrastructure development to augment marketing and post-harvest management are being accorded top priority, Integrated Pack House has been set up at Ramapo with facilities for grading, packaging and treatment of flowers. Another such facility is coming up at Mali to cater to the needs of South and West districts. Cold rooms have been set up in various production clusters to aid production and marketing. One Integrated Processing Unit for ginger has been established at Birding, Rashtriya Krishi Vikash Yojna (RKVY) aids to counter the uncertainties of market fluctuations in West districts. The Model Floriculture Centre at Namli has been strengthened with facilities for production of gerbera, rose and lileum. Another Model Floriculture Centre has been set up at Maniram, South district. The Cymbidium Development Centre has been operationalized and is fully equipped with a tissue culture laboratory and training hall for production of large quantities of planting materials and imparting trainings. Technology Mission for Integrated Development of Horticulture, sponsored by the Government of India has played the most vital role in integrating all ongoing initiatives. In addition to this flagship programed, other initiatives under RKVY, TSP/SCSP, BADP and State Plan/Non-Plan have made noteworthy contribution in supplementing the programers spearheaded through Technology Mission. The main success factors for successful programed implementation are use of high-quality planting materials, adoption of clusters and constant monitoring.

The most significant development that happened in the last decade is that horticulture has moved from rural confines to commercial production and this changing scenario has encouraged private sector investment in production system management. The last decade has seen technological infusion like micro-irrigation, precision farming, greenhouse cultivation, and improved post-harvest management impacting the development, but during the process various issues have emerged.

4.6. TECHNOLOGICAL INTERVENTIONS TO ENHANCE PRODUCTIVITY

4.6.1. High yielding varieties and F1 hybrids

During the last seven decades (since 1950), over 1500 improved varieties/hybrids of about 130 horticultural crops have been developed. As a result, productivity of banana and potato has gone up three times each and cassava two times. In the last four years itself, India has recorded 8% increase in potato production. Regular-bearing mango hybrids, export quality grapes, multiple-disease resistant vegetable hybrids, high-value spices, and tuber crops of industrial uses have been developed. Tomato varieties resistant to leaf curl virus, bacterial and Fusarium wilts have been developed. Improved varieties/hybrids have revolutionized the horticulture sector. For instance, high-yielding varieties of sweet potato viz. Gauri, Sankar and Sree Bhadra have brought a revolution by minimizing malnutrition, improving nutritional security in the Kandhamal district of Orissa state.

4.6.1.1. High yielding varieties: Vegetable production is still dominated by the locally available genotypes or inferior land races all over the country. It is estimated that under eggplant, approximately 32.2% area, cauliflower 46.71%, chilli 60%, gourds 77.56%, melons 70%, okra 14.62% and tomato 18.49% area are under unidentified local varieties. Thus, there exists a need to replace the local cultivars with the improved high yielding and disease resistant varieties. As a result of multi-disciplinary, multi-location testing of new research materials during the last two decades, 119 improved varieties in 16 major vegetable crops have been identified and recommended for cultivation in various agro-climatic regions of the country

4.6.1.2. F1 Hybrids: F1 hybrids are well known for increasing vegetable production due to their high yield potential, earliness, quality and resistance attributes. Heterosis breeding in vegetable crops in India has received serious attention only in recent years. The first F1 hybrid of tomato (Karnataka Hybrid) and capsicum (Bharat) were released for commercial cultivation in 1973 by a private seed company M/s Indo-American Hybrid Seeds followed by 28 other hybrids in nine vegetable crops.

F1 hybrids have been developed in tomato, brinjal, cabbage, cauliflower, capsicum, chilli, cucumber, carrot, muskmelon and watermelon. Presently, Karnataka, Maharashtra, Gujarat, Andhra Pradesh, Uttar Pradesh and Madhya Pradesh are leading producers of hybrid tomato. The entire cabbage production in southern parts of Maharashtra and West Bengal is under F1 hybrids.

4.6.1.3. Resistant varieties: Research on breeding for disease/pest resistance has resulted in the release of several varieties. 'Pusa Sawani' variety of okra developed as resistant to yellow-vein-mosaic virus is the first example of successful disease resistance breeding in vegetable crops in India. Okra hybrids of Indo-American Hybrid Seeds, Bangalore, such as Varsha, Vijay and Vishal are resistant to YVMV. In case of indigenous hybrids of brinjal, gourds, melons and cucumber, more area can be covered if they carry additional attributes like disease/pest resistance or any special quality attribute. Several varieties have also been released which are resistant to abiotic stresses such as heat and moisture.

4.6.1.4. Nutrition rich varieties: More than 800 million people worldwide are thought to be malnourished. In the developing world, around 192 million children and about 2000 million others experience malnutrition, particularly the pregnant women leading to infant mortality. British, American and Australian researchers are all working on development of a range of fruit and vegetable varieties that the industry claims could counter malnutrition and illness.

4.6.2. High density planting

Accommodation of the maximum possible number of plants per unit area to get the maximum possible productivity and profit per unit of the tree volume without impairing the soil fertility status is called High Density Planting (HDP). Basically, the availability of a dwarf plant is the first and foremost pre-requisite for establishing any high-density orchard. HDP is one of the methods to enhance productivity per unit area both in short duration and perennial horticultural crops. In perennial, it is more useful since it permits efficient use of land and resources, realizing higher yields, net economic returns per unit area, easy canopy management suited for farm mechanization and cultural operations, efficient spray and weed control, improvement in fruit quality and easy and efficient harvest of high-quality produce, *etc.*

Planting of fruit trees rather at a closer spacing than the recommended one using certain special techniques with the sole objective of obtaining maximum productivity per unit area without sacrificing quality is often referred as high-density planting. This technique was first established in apple in Europe during sixties and now majority of the apple orchards in Europe, America, Australia and New Zealand are grown under this system. In this system, four planting densities are recognized for apples viz., low HDP (< 250 trees/ha), moderate HDP (250-500 trees/ha), high HDP (500 to 1250 trees/ha) and ultra-high HDP (>1250 trees/ha). Recently, super high-density planting system has been also established in apple orchards with a plant population of 20,000 trees per ha. In some orchards, still closer, planting of apple trees is followed (say 70,000 trees/ha) which is often referred as 'meadow orchards.'

4.6.2.1. Advantages of HDP

The advantages of HDP are:

- » Early cropping and higher yields for a long time; the average yield in apple is about 5.0 tons/ha under normal system of planting and it is about 140.0 tons/ha under HDP.
- » Reduced labor costs.
- » Improved fruit quality.

Characteristics of HDP are:

- » The trees of HDP should have maximum number of fruiting branches and minimum number of structural branches.
- » The trees are generally trained with a central leader surrounded by nearly horizontal fruiting branches.
- » These branches should be so arranged and pruned in such a way that each branch casts a minimum amount of shade on other branches.
- » The height should be one and half its diameter at the base. A key to successful HDP depends upon the control of tree size.
- » This is achieved by
- » Use of size controlling rootstocks. In apple, dwarfing rootstocks and intermediate stocks like MM 106, MM 109, and MM 111 are used to control the size of the plant. In pears, Quince-A, Adam, and Quince-C are commonly used as dwarfing rootstocks.
- » Use of spur type scions – In temperate fruit crops like apple, the cultivars can be classified into a spur type or non-spur type. The spur types which have restricted annual growth are alone suitable for HDP.
- » Training and pruning methods to induce dwarfness – Under Indian conditions, apple trees trained under spindle bush, dwarf pyramid, cordon systems are found to contain the growth of the trees appreciably for HDP systems.
- » Mechanical device and use of chemicals to control size – Growth regulators such as diaminozide, ethephon, chlormaquat and paclobutrazal are extensively used to reduce shoot growth by 30 to 50%. This results in increased flowering in the subsequent years and may be useful in encouraging earlier commercial fruit production in strongly vegetative fruitful young trees. Besides chemical manipulation, mechanical devices employing the use of spreaders and

tying down the branches to make them grow from near horizontal to an angle of 45° from the main stem are also some of the standard practices to control tree size.

4.6.3. Micro-irrigation

4.6.3.1. Water losses under various irrigation systems: In traditional surface irrigation methods, the losses in water conveyance and application are large (Table 4.4). These losses can be considerably reduced by adopting drip and sprinkler irrigation methods. Among all the irrigation methods, the drip irrigation is the most efficient and it can be practiced in a large variety of crops, especially in vegetables, orchard crops, flowers and plantation crops.

Table 4.4. Water losses under various irrigation systems

Method of irrigation	Water losses (%)
Surface	30-45
Gate pipe	15-20
Sprinkler irrigation	5-10
Drip irrigation	3-6

Micro-irrigation system is an irrigation system with high frequency application of water in and around the root zone of plant system, which consists of a network of pipes along with a suitable emitting device. With micro-irrigation the yields of crops are better in addition to the saving of water.

Micro irrigation can be classified as drip irrigation and sprinkler irrigation. In drip irrigation, water is applied near the plant root through emitters or drippers, on or below the soil surface, at a low rate varying from 2-20 liters per hour. The soil moisture is kept at an optimum level with frequent irrigations. Drip irrigation results in a very high water application efficiency of about 90-95 per cent. A typical drip irrigation systems is shown in Fig. 4.1.

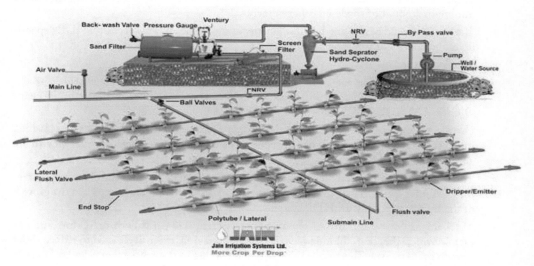

Fig. 4.1. A typical drip irrigation system

Drip irrigation system is suitable for all orchard and vegetable crops. This system has also been successfully employed for close growing crops including onion and okra. The National Committee on Plasticulture Applications in Horticulture (NCPAH), Ministry of Agriculture, Government of India (GOI), has estimated a total of 27 million hectares area in the country that has the potential for drip irrigation application.

There is a need to promote increased irrigation water use efficiency. Participatory action research program initiated by the Ministry of Water Resources, Government of India, during 2007-09 in over 2000 villages all over the country to assess the impact of water saving technologies has shown that yield and income can be increased by 50 to 100% in most crops by using water saving technologies.

4.6.3.2. Advantages of micro-irrigation: The advantages of micro-irrigation are of two types – qualitative and quantitative. The qualitative advantages are production of quality horticultural produce and increase in water and fertilizer use efficiency. Excellent soil health could be maintained due to optimum soil moisture content, accelerated microbial activities and maximization of biomass decomposition converting into humus. The quantitative advantages are related to growth of micro-irrigation in terms of area coverage and overall increase in the productivity. Quantity of water or fertilizer saved due to imposition of drip system, the energy conservation and reduced labor cost are the other advantages. Data on increase in crop yields and percent increase over traditional surface irrigation methods by using micro-irrigation system are presented in Table 4.5. It shows that micro-irrigation could save water ranging from 14% to 84% with enhancement of yield from 13% and 514% under varied agro-climatic and crop conditions. The higher yield under micro-irrigation system is attributed to the fact that during crop growth favorable soil moisture (generally at ⅓ bar) is maintained and the required volume of soil mass is wetted through irrigation water, which provides optimum environment for root growth of the plant.

Table 4.5. Water saving and yield increase with drip irrigation system in horticultural crops

Horticultural crop	Yield (tons/ha)		% water saving	% increase in yield
	Surface irrigation	Drip irrigation		
Beet root	5.7	8.9	79.1	56.1
Bitter gourd	32.0	43.0	56.6	34.4
Brinjal	91.0	148.0	61.9	62.6
Broccoli	140.0	195.0	14.3	39.3
Cauliflower	171.0	274.0	33.3	60.2
Chilli	42.3	60.9	61.7	44.0
Cucumber	155.0	225.0	55.6	45.2
Okra	100.0	113.1	84.0	13.1
Onion	284.0	342.0	50.0	20.4
Potato	172.0	291.0	54.2	69.2
Radish	10.5	11.9	76.1	13.3
Sweet potato	42.4	58.9	60.3	38.9
Tomato	61.8	88.7	78.5	43.5
Banana	575.0	875.0	45.0	52.2

Grape	264.0	325.0	47.2	23.1
Papaya	130.0	230.0	67.9	76.9
Pomegranate	34.0	67.0	23.8	97.0
Water melon	82.1	504.0	65.3	513.9

4.6.4. Fertigation

Among the various factors that are responsible for increasing the productivity of horticultural crops, a regular and optimum supply of water and fertilizers at appropriate times is very crucial as it enables the crops to attain higher growth rate and give increased yields. As both water and fertilizers are costly inputs and are becoming scarce in recent years, there is an immediate need for adoption of techniques with which these inputs can be utilized more efficiently. Fertigation is one such technique of applying fertilizers to the crops along with water through micro irrigation systems (drip/sprinkler) on a continual basis in controlled manner so as to allow steady uptake of nutrients by the plants and to effect savings in the costly inputs including water and fertilizers. Salient features of the research conducted on fertigation of horticultural crops using different combinations of water-soluble fertilizers in increasing the productivity are presented in Table 4.6.

Table 4.6. Saving in fertilizer and increase in horticultural crop yield under fertigation

Horticultural crop	Saving in fertilizer (%)	Increase in yield (%)
Banana	20	11
Onion	40	16
Potato	40	30
Tomato	40	33
Okra	40	18
Broccoli	40	10

4.6.4.1. Advantages of fertigation: The fertigation allows applying the nutrients exactly and uniformly only to the wetted root volume, where the active roots are concentrated. This remarkably increases the efficiency in the application of the fertilizer, which allows reducing the amount of applied fertilizer. This not only reduces the production costs but also lessens the potential of ground water pollution caused by the fertilizer leaching. Fertigation allows adapting the amount and concentration of the applied nutrients in order to meet the actual nutritional requirement of the crop throughout the growing season. In order to make a correct planning of the nutrients supply to the crop according to its physiological stage, we must know the optimal daily nutrient consumption rate during the growing cycle that results in maximum yield, quality and production. These functions are specific for each crop and climate, and were determined in different experiments for the main crops in Israel like tomatoes, cucumbers, melons, *etc.* The optimal curve of consumption of nutrients defines the minimal application rate of a certain nutrient that is required to maintain a constant nutrient concentration in the soil solution. These data constitute the basis for the recommendations given by the Israeli Soil Extension Service for the farmers regarding the fertigation regime for different crops.

There are three beneficial effects of fertigation:

i) Maximizing water and fertilizer use efficiency

- » Delivers water and nutrients to the root zone.
- » Increases the availability and uptake of nutrients.
- » Decrease nutrient losses from leaching.
- » Accurate and uniform application of water and nutrients.
- » Maximum saving of water and nutrients.

ii) Maximizing crop and soil productivity

- » Higher yields and high quality produce.
- » Steep-sloped and marginal lands can be made cultivable.

iii) Minimizing production cost

- » Economy in labor, water and nutrients.

iv) Other advantages

- » The saving of energy and labor.
- » The flexibility of the moment of the application (nutrients can be applied to the soil when crop or soil conditions would otherwise prohibit entry into the field with conventional equipment).
- » Convenient use of compound and ready-mix nutrient solutions containing also small concentrations of micronutrients which are otherwise very difficult to apply accurately to the soil.
- » The supply of nutrients can be more carefully regulated and monitored. When fertigation is applied through the drip irrigation system, crop foliage can be kept dry thus avoiding leaf burn and delaying the development of plant pathogens.

4.6.5. Integrated nutrient management

Integrated nutrient management refers to maintaining the soil fertility and plant nutrient supply to an optimum level for sustaining the desired crop productivity through optimization of the benefits from all possible sources of plant nutrients in an integrated manner. It is a holistic approach, to know what exactly is required by the plant for an optimum level of production, in what different forms these nutrients should be applied in soil and at what different timings in the best possible method, and how best these forms should be integrated to obtain the highest productive efficiency on the economically acceptable limits in an environment friendly manner.

The encouraging results obtained by conjunctive use of fertilizers, organic manure, green manure, biofertilizers and fertigation can provide leads to the future strategies for the rational use of fertilizers and manures for enhancing the productivity of horticultural crops without detrimental effect to the environment.

4.6.5.1. Nutrient use efficiency: The crops do not efficiently utilize fertilizers applied under traditional methods of irrigation. The efficiency of different nutrients and methods of application is presented in Table 4.7.

Table 4.7. Efficiency of different nutrients and methods of application

Element	Method of application	Efficiency (%)
N	Fertigation	90-95
	Surface Flow	50-70
	Broadcasting (Solid fertilizers)	30-50
P	Fertigation	40-45
	Surface Flow	10-20
K	Fertigation	75-80
	Surface Flow	< 50

4.6.5.2. Optimum leaf nutrient status: At optimum leaf nutrient status, the growth, yield and quality of the crop are satisfactory and there is no need to make any changes in the schedule of manures and fertilizers. Changes in nutrient concentration in the specified plant part do not increase or decrease growth or production. Leaf is the principal site of plant metabolism and changes in the nutrient supply are reflected in composition of the leaf. The concentrations of the nutrient in the leaf at a specific growth stage are related to the performance of the crop. The leaf nutrient levels required for optimum production in some important fruit crops have been presented in Table 4.8. Efforts in the nutritional management program should be directed towards the maintenance of these levels in tree leaves before flowering.

Table 4.8. Optimum leaf nutrient status of some important fruit crops

Fruit crop	Nutrients											Reference
	%						ppm					
	N	P	K	Ca	Mg	S	B	Fe	Mn	Zn	Cu	
Mango	1.23	0.66	0.54	1.71	0.91	0.12	---	171	66	25	12	Samra *et al.*, 1978
Banana	3.29	0.44	3.11	2.12	0.24	---	---	---	---	---	---	Ramaswamy, 1971
Citrus	2.50	0.14	0.90	4.20	0.43	0.25	---	90	112.5	62.5	10.5	Reuther *et al.*, 1962
Guava	1.70	0.50	1.30	2.10	0.46	---	10	---	---	77.5	75	Singh, 1981
Grape	1.75	0.56	1.66	0.94	---	0.20	---	---	125	92	---	Bhargava & Chadha, 1993
Litchi	1.47	1.03	---	---	---	---	---	---	---	---	---	Ghosh *et al.*, 1986
Sapota	1.66	0.08	0.80	0.83	0.48	0.07	---	100	39.3	15.7	6.7	Annapurna *et al.*, 1988
Papaya	1.66	0.50	5.21	1.81	0.67	0.38	---	---	---	---	---	Bhargava *et al.*, 1990
Pineapple	1.40	0.16	3.70	---	---	---	---	---	---	---	---	Chadha & Reddy, 1993
Pomegranate	2.50	0.19	1.47	---	---	---	96	199	196	55	---	Shende, 1977 Bhambal, 1987
Ber	2.50	0.25	1.82	1.21	0.45	---	---	---	---	---	---	Khanduja & Garg, 1984

4.6.6. Bioregulators

Ever since the discovery that the plant hormones have the ability to influence several physiological processes in plants, attempts have been continuously made to manipulate the growth and development process in different horticultural crops to streamline the production. As the years passed, besides the natural compounds, a number of synthetic compounds were also found to have growth regulatory roles, which are now collectively known as bio-regulators. Many of the research accomplishments have led to the commercial applications.

The use of bioregulators to increase the vitality and therewith the productivity of horticultural crops is a distinct possibility. In several experiments, the positive effect of different bioregulators used (Humate, Lactate, *Bacillus subtilis*) could be discovered if one of the growth factors was at a critical level. Humates are effective in nutrient and salt control as well as subtropical temperature. Lactates have a good effect if the pH-value is unstable. Microorganisms like *Bacillus subtilis* are effective under different stress conditions. They have an unspecific positive effect in particular if the nutritional status is suboptimal. However, horticultural crops with long vegetative or fruiting phase are often confronted with different stress factors. A combined application of bioregulators could be beneficial in such complex stress situations in order to improve the productivity of the horticultural crops.

Bioregulators have been used for many years to alter the behavior of fruit or fruit trees for the economic benefit of the fruit grower. Control of vegetative vigor, stimulation of flowering, regulation of crop load, reduction of fruit drop, and delay or stimulation of fruit maturity and ripening are important examples of processes in fruit and fruit trees that can be regulated with exogenous applications of bioregulators.

The use of plant bioregulators in ornamental crops is more prevalent than in edible crops. The most common types of regulators used are gibberellins (GAs, most often GA$_3$), ethylene and their antagonists. GAs are used to enhance elongation of many cut flowers and to promote bud break, thus producing more flowering shoots. They are commonly used to promote flowering of long day (LD) and cold requiring plants grown under marginal inductive conditions and of autonomous-flowering plants of the Araceae. GA also enhance post-harvest life of certain flowers and foliage plants. The main use of ethylene in ornamentals is to promote flower formation, especially in bromaliads and certain bulbous plants. When desired, ethylene also promotes lateral branching. The largest use of bioregulators in ornamental crops is probably that of GA antagonists - the growth retardants. They are commonly used in pot plant production to achieve more compact and attractive structure they promote flowering in certain woody ornamentals. Endogenous and external ethylene is the main factor reducing post-production life and promoting flower and foliage abscission of many ornamentals. Silver-thio-sulphate (STS) is the main ethylene antagonist used to protect plants from ethylene damage. Concern about the impact of STS on the environment has promoted a search for effective alternatives. Those presently available, however, are less effective than STS. An effective chemical may be a volatile ethylene binding-site inhibitor such as diazocyclopentadiene. Manipulation of cytosolic Ca^{2+} level may become a novel means for controlling flowering, flower senescence and geotropic bending of flowers.

4.6.6.1. Induction of flowering: Bioregulators have been used to overcome the problem of alternate bearing in mango and apple. It has been reported that paclobutrazol (cultar) has good potential both as foliar spray and soil application in overcoming biennial bearing in mango. Soil application of paclobutrazol at 10 g mixed in 5 liters of water in the ring of 15 cm depth dug 60 cm away around the tree trunk in August resulted in appreciable flowering in the following expected "off" year in Alphanso (Rao *et al.*, 1997, Srihari and Rao, 1996).

Alternation in cropping from year to year has always been a problem in apple production. Chemical thinning techniques have been used for half a century to address this problem but with limited success. Removal of young fruitlets alone is not always sufficient. The use of gibberellic acid (GA) has been evaluated as a technique for reducing flower formation in the low crop year to see if the alternating cycle can be effectively interrupted with a combination of flowering control in the low crop year and good thinning practices in the high crop year. Consistent reductions have been obtained in flower formation in the 'Fuji' cultivar when GA is applied shortly after flowering.

Evening out cropping from year to year would greatly benefit the profitability of the tree fruit industry by diminishing the impact of reduced prices in heavy crop years and improving returns to more growers in what would have been light crop years. The use of bioregulators to reduce biennial bearing appears promising at this time, but much more work is necessary before it can be recommended for commercial practice.

4.6.6.2. Promoting the growth and development of fruits: Bioregulators help in improving the fruit size of banana and grapes. Spraying of 100 ppm GA_3 twice at 25 and 50 days after the shooting resulted in an increase of 32.5, 23.9 and 16.98% of length, girth and weight of the fingers, respectively in banana cv. Rajapuri (syn. Munavalli) which is popular in Belgaum district of Karnataka (Rao, 1977).

Application of GA_3 has caused multi-fold increase in berry size and consequently productivity has been raised from 6 to 8 tons/ha to 15 to 18 tons/ha in Thompson Seedless cv. of grapes.

Bioregulators have been used in improvement of color and ripening in grapes. Rao and Shankarnarayanan (1982) could overcome the problem of uneven ripening in Gulabi cv. of grape by dipping the clusters twice at 5 days interval during the lag phase in 250 ppm ethrel containing 2% urea. Use of 50 ppm 2,4,5-TP was also found effective in this respect (Hanumashetti *et al.*, 1981).

Bioregulators help in controlling the berry drop in Anab-e-Shahi cv. of grape. Spraying of 100 ppm NAA or PCPA a week before harvest reduced the berry drop (Rao *et al.*, 1968). Mixing of 100 ppm BA along with 100 ppm NAA was much more effective. NAA helps in thickening of the pedicel attachment.

4.6.7. Canopy management

Canopy management is the manipulation of tree canopies to optimize the production of quality fruits. The canopy management, particularly its components like tree training and pruning, affects the quantity of sunlight intercepted by trees, as tree shape determines the presentation of leaf area to incoming radiation. An ideal training strategy centers around the arrangement of plant parts, especially, to develop a better plant architecture that optimizes the utilization of sunlight and promotes productivity. Light is critical to growth and development of trees and their fruits. Strong bearing branches tend to produce larger fruits. The problem of a fruit grower is initially to build up a strong and balanced framework of the trees, then equip them with the appropriate fruiting. Obviously, pruning in the early years has to be of a training type to provide strong and stocky framework with well-spaced limbs or any other desired shape.

4.6.8. Soilless culture

Soilless culture is one of the best techniques to overcome local water shortages, while also producing high quality produce, even in areas with poor soil structure and problematic conditions. Soilless culture is a method of growing plants without soil. The application of these systems is likely to increase close to existing cities as well as in mega-cities worldwide in the near future.

In order to meet out the growing demand for soilless culture technology, ICAR-IIHR has initiated a project on development and standardization of soilless cultivation of vegetables on Arka Fermented Cocopeat under open and protected conditions during the year 2015. Under this project, the production technology including nutrient formulations for open and polyhouse soilless cultivation of most commonly consumed vegetables, viz. tomato, chilli, cabbage, cucumber, brinjal, cowpea, dolichos, French bean, garden peas, ridge gourd and leafy vegetables and few exotic vegetables like zucchini, broccoli and color cabbage using Arka Fermented Cocopeat as substrate has been standardized. The results of most of the experiments conducted with different vegetable crops in grow bags under open-field and polyhouse soilless culture indicated that the plants grown in soilless culture recorded higher yield and better quality particularly in mineral nutrient content compared to soil grown plants.

4.6.9. Vertical farming

Vertical farming is the practice of growing crops on a smaller land area, by the utilization of vertical space, which is usually left unutilized in traditional agriculture. Though vertical farming has been in vogue since ancient times as evinced by the Hanging Gardens of Babylon, the modern concept of vertical farming involves the union of plant biology and suitable engineering know-how. It can be considered as an extension of indoor farming that evolved in the 1700s with the advent of greenhouses, with the primary objective of harnessing the off-season crop cultivation potential during unfavorable seasons. This involves the stacking of crops growing platforms in a vertical fashion and providing the necessary nutrition and lighting by artificial means in order to cultivate crops all-round the year. Most commercial vertical farm ventures in developed countries operate from existing warehouses or abandoned factories/buildings that have been suitably converted to hydroponic/aeroponic facilities with LED based illumination systems. Further, utilizing vertical space, is highly energy intensive since artificial illumination needs to be provided for crop production in multi-tiered structures. Therefore, uninterrupted power supply would be a limiting factor for vertical farming under Indian conditions. Apart from this, the high initial costs of the infrastructure and the operating costs may act as deterrents for large scale vertical farming, therefore, there is a need to develop country specific infrastructure and technology before vertical farming can be adopted on a large scale in India. Other limitations of the vertical farming include the cultivation of crops without sufficient scientific validation, lack of varieties/hybrids that have been exclusively bred for the purpose of vertical farming, lack of good agricultural practices for vertical farming situations and the design of existing high-rise structures that are not amenable for vertical farming. The above limitations could be taken up as researchable issues to promote vertical farming in India.

4.6.10. Quality planting material production

Availability of planting material of good quality is one of the most important elements of successful horticultural production. Planting material available to small-scale farmers in different areas is often of insufficient quality, which undermines potential yield and performance of crop production. There has been a significant increase in the production of horticultural crops in recent years in India. At present in India more than 4,409 fruit nurseries including 1,575 under government sector and 2,834 under private sector are functioning which have an annual target of producing 1,387 million fruit plants. This accounts for 30-40% of the demand of planting materials of fruit sector.

In order to produce true to type planting materials, most of the fruit crops are propagated by asexual or vegetatively except the crops like papaya, phalsa and mangosteen.

4.6.11. Biotechnological approaches

Biotechnology offers vast potential for improving the efficiency of crop production, thereby lowering the cost and increasing the quality of food. The tools of biotechnology can provide scientists with new approaches to develop higher yielding and more nutritious crop varieties, to improve resistance to pests including weeds and adverse conditions, or to reduce the need for fertilizers and other expensive agricultural chemicals.

In the first 20 years (1996 to 2015), biotech crops were planted by up to 18 million farmers (up to 90% were small/poor farmers) in 28 countries annually. With an increase of 100-fold from 1.7 million hectares in 1996 to 179.7 million hectares in 2015, this makes biotech crops the fastest adopted crop technology in recent times – the reason – biotech crops have the trust of millions of farmers because they deliver significant and multiple benefits. Accordingly, the number of biotech countries has more than quadrupled from 6 in 1996, to 16 in 2002 and 28 in 2015.

4.6.11.1. *Bt* brinjal in India: India continues to debate the relevance and need of biotech crops since it imposed the moratorium on *Bt* brinjal on 9th February 2010. *Bt* brinjal is an important biotech crop, which is ready to be commercialized, would trigger a new phase of growth and momentum in the horticultural crop biotech sector in the country:

The approval of country's first vegetable crop *Bt* brinjal (eggplant) by revisiting its 5-year old moratorium in the context of the large-scale commercial planting of Bt brinjal in the neighboring country of Bangladesh; it is noteworthy that a noticeable increase in pesticide residues is occurring in important vegetables and fruits; biotech crops could help reduce the use of pesticides on food crops.

The management of brinjal fruit borer, *Leucinodes. orbonalis* is extremely difficult. Often per cent borer damage due to *L. orbonalis* in spite of weekly pesticide sprays exceeds 25%. Further, the pattern of pesticide use has resulted in resurgence of secondary pests such as whiteflies, mites, aphids and thrips.

Because of the ease with which pesticide crystal protein genes from *Bacillus thuringiensis* can be transferred into crops, this avenue of research was exploited to develop sustainable control measures to combat *L. orbonalis*. This will help to reduce pest population build up prior to fruiting.

Bt Brinjal is the first genetically modified food crop in India that has reached the approval stage for commercialization. It was developed by inserting a gene cry1Ac from a soil bacterium called *Bacillus thuringiensis* through an *Agrobacterium*-mediated gene transfer. It is a genetically modified brinjal developed by the Maharashtra Hybrid Seed Company Ltd. (Mahyco), a leading Indian seed company (Fig. 4.2). *Bt* Brinjal event EE1 has been developed in a Public Private Partnership mode under the aegis of the Agriculture Biotechnology Support Project (ABSP) from Cornell University where the *Bt* technology available with M/s Mahyco has been transferred (free of cost) to the Tamil Nadu Agricultural University (TNAU), Coimbatore; the University of Agricultural Sciences (UAS), Dharwad; and the Indian Institute of Vegetable Research (IIVR), Varanasi. The Event EE 1 was introgressed by plant breeding into various local varieties.

The Genetic Engineering Approval Committee (GEAC) cleared *Bt* brinjal for commercialization on 14 October 2009. Following concerns raised by some scientists, farmers and anti-GM activists; the Government of India officially announced on 9 February 2010 that it needed more time before releasing *Bt* brinjal, with Indian Environment Minister Jairam Ramesh saying that there is no overriding urgency to introduce *Bt* brinjal in India. On 17 February 2010, Jairam Ramesh reiterated that the Centre had only imposed a moratorium on the release of transgenic brinjal hybrid, and not a permanent ban, saying that "until we arrive at a political, scientific and societal consensus, this moratorium will remain".

Fig. 4.2. Damage caused by brinjal fruit and shoot borer (left) and healthy *Bt* brinjal developed by MAHYCO (right)

The imposed moratorium has been criticized by some scientists as not being based on any compelling scientific evidence and potentially setting Indian biotechnology decades back (Shantharam, 2010).

4.6.11.2. *Bt* brinjal in Bangladesh: The Bt brinjal project in Bangladesh may lay claim to be the first crop biotechnology transfer project to deliver a product to farmers. Bt brinjal was developed as an international public private partnership, between an Indian seed company Mahyco generously donating technology to the Bangladesh Agricultural Research Institute (BARI), facilitated by Cornell University led project ABSP-II and funded by USAID. Bangladesh approved Bt brinjal for commercial cultivation on 30th October 2013 and in record time – less than 100 days – on 22 January 2014, a group of small farmers (120) planted the first four commercial Bt brinjal varieties (Uttara, Kajla, Nayantara, and Iswardi/ISD 006) on 12 hectares. In 2015, a total of 250 hectares of Bt brinjal were planted by 250 farmers and the area is expected to increase substantially in 2016. Five additional Bt brinjal varieties (Dohazari, Shingnath, Khatkati, Chaga, and Islampuri). – three of them received NCB approval in 2015, will be available in 2016 and remaining two in the near future.

Bt brinjal increases marketable yield by at least 30% and reduces the number of insecticide applications by 70-90%, reduced cost of production, increases marketable yield and improves fruit quality; with a net economic benefit of US$ 1,868 per hectare, equivalent to a gain of up to US$ 200 million per annum. Agronomic performance data released by BARI from 19 *Bt* brinjal demonstration plots, established in 108 farmers in 19 districts showed close to zero pest infestation, increased yield up to 100% compared to non *Bt* variety. Farmers have successfully sold Bt brinjal fruits in the open market labelled as "BARI *Bt* Begun #, no pesticide used".

4.6.12. Protected cultivation

Among the productivity enhancing technologies, protected cultivation has a tremendous potential to increase the yield of vegetables and flower crops by several folds.

Protected cultivation technology, using different kinds of structures, offers several advantages to produce vegetables and flowers of high quality with high yields and minimum risks due to uncertainty of weather (Fig. 4.3). It also ensures chances for efficient use of land and other natural resources. The main purpose of protected cultivation is to create a favorable environment for the sustained growth of plant so as to realize its maximum potential even under adverse climatic conditions. There is good opportunity for greenhouse cultivation of vegetables, selected fruits and flowers by economic methods of water and fertilizer use such as fertigation. Low cost greenhouses along with micro-irrigation and fertigation techniques may be popularized in areas where evaporation exceeds precipitation during several months in a year. This becomes relevant to the growers in various parts of the country who have small holdings, say less than 2 ha. They would be interested in a technology, which helps them to produce more crops each year from their land, particularly during off-season when prices are higher in nearby markets. Vegetable and flower production under location-specific and cost-effective structures may give substantial benefits even to small growers.

4.6.12.1. Increase in productivity: The yield under polyhouse cultivation can be achieved to the level of 5-8 times as compared to the open crop cultivation. Various trials conducted at Agricultural Research Centers in northern India indicate that bell pepper (planted in mid-September), cucumber (planting in mid-October) and tomato (November planting) under polyhouse produced 106, 146, and 153 tons per ha. The duration of these crops were 4- 9 months and more than 90% of total yield was obtained during off-season (during winter before the start of summer) which fetches significantly higher market price (2-4 times than normal season). Further, the crop duration can be extended up to July–August with the application of micro-irrigation and fertigation and yield can be achieved to the level of 20-25 kg/m². Therefore, it is possible to harvest a single crop round the year with minimum additional inputs and higher income can be generated. Further, cut flowers like carnations, gerbera, lily, rose, orchids, anthuriums, *etc.* can be grown under polyhouses/ net houses giving high returns and top-quality produce. The potential of floriculture under protected cultivation is huge for Indian and global markets.

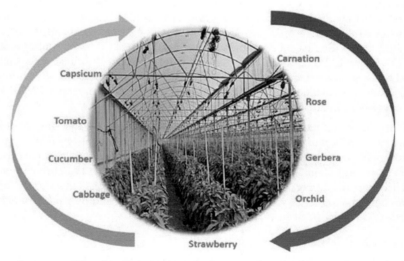

Fig. 4.3. Crops grown under protected cultivation

The protected cultivation of high value crops has become irreplaceable both from economic and environment points of view. It offers several advantages to grow high value crops with improved quality even under unfavorable and marginal environments. However, due to high training needs of the greenhouse growers and some poor-quality produce with pesticide residues has been a matter of great concern. These issues can easily be addressed by integrating various production and protection practices including location specific designing and construction of the polyhouses for efficient input use. Creating awareness among the greenhouse growers for judicious use of pesticides for safe production can be instrumental in providing quality products without polluting the environment.

4.6.12.2. Advantages and limitations

i) **Advantages:** Protected production can reduce the amount of water and chemicals used in production of high value crops compared to open field conditions. The comparative advantages are:

» Productivity in greenhouse is increased manifold in comparison to open field.
» Year round production of crops.
» Adverse climate for crop production can be overcome by different systems of protected cultivation.
» Multiple cropping on the same piece of land is possible.
» Off-season crop production to get better returns to growers.
» Production of high quality and healthy seedlings for transplanting in open field supporting early crop, strong, and resistant crop stands.
» Increase production as well as productivity per unit of land, water, energy and labor.
» Supports the production of high quality and clean products for export.
» Makes cultivation of crops possible in areas where it is not possible in open conditions such as high altitudes and deserts.
» Makes vertical cultivation of crops possible using technologies like hydroponics, aeroponics, *etc.* and use of vertical beds for production.
» Disease-free seed production of hi-tech crops becomes easy.
» Meets the potential demand for producing good nutrition, healthy foods, and quality vegetables free from pesticides can be fully exploited.
» Controlled environmental conditions are used for early raising of nurseries, off-season production of vegetables, seed production, and protecting the valuable germplasm.
» Management and control of insect-pests, diseases, and weeds is easier.
» Maintenance of stock plants, cultivating grafted plantlets, and micro propagated plants is easy.
» Early maturity.
» Tropical vegetables like cucurbits, capsicum, brinjal, okra *etc.* are rare in hilly region but these crops can be grown in greenhouse.
» Ideally suited for Indian farmers having small holdings.

ii) **Limitations**

» Manual or hand pollination in cross pollinated vegetables like cucurbits or development of their parthenocarpic hybrids/varieties.
» Expensive, short life and non-availability of cladding materials.
» Lack of appropriate tools and machinery.
» Structure cost initially looks unaffordable. Farmers with zero risk affordability do not come forward to adopt it.

4.6.13. Post-harvest management

There is a considerable gap between the production and net availability of the horticultural produce due to post-harvest losses due to lack of elaborate harvesting equipment, collection centers in major producing areas, suitable containers, commercial storage plants and lack of cold chain in entire post-harvest operation.

The total post-harvest losses in fruits are estimated to be 17 to 35% amounting to nearly Rs. 2000 crores annually. In a systematic study on post-harvest losses in fruits have been estimated between 17.1 to 36.7% in mango, 12 to 19% in banana, 8.3 to 30.7% in orange, 10 to 25% in apple, and 23 to 90% in grapes. Considerable losses occur during transport. Minimizing these losses has, therefore, been recognized as an effective means of increasing fruit and vegetable availability without bringing additional areas under these crops. In fact, the reduction in post-harvest losses is a complementary means of increasing the productivity.

4.6.13.1. Pre-harvest treatments

i) Fruit crops

- » Pre-harvest application of GA delays ripening at market stage and improves storage life in mango and guava and color development in citrus.
- » Pre-harvest sprays of Topsin-M (0.05%) or Bavistin (0.1%) at 15 days before harvest can control anthracnose and stem-end rot of mango.
- » Application of thiophanate methyl (0.05%) effectively controls post-harvest losses in mango by controlling maturity and delaying ripening.
- » Post-harvest decay of Nagpur mandarin can be controlled by 3 pre-harvest sprays of Benlate or Topsin-M (0.1%) at 15 days interval before harvest.
- » Pre-harvest sprays of 0.6% calcium chloride 10 to 12 days prior to harvest improved shelf-life and reduces physiological loss in weight in Thompson seedless grapes.
- » Application of 2.5% calcium chloride at 50 ml/sucker in the core of pineapple plant about 300 days after planting showed better ripening and improved overall quality.

ii) Vegetable crops

- » Pre-harvest application of maleic hydrazide (MH) reduces sprouting of onion and potato during storage. In onions, application of 1500 to 2000 ppm MH at 75 to 90 days after transplanting reduced sprouting during 4 to 5 months storage in ventilated structures.
- » Post-harvest diseases of tomato and onion can be controlled by 3 pre-harvest sprays of 0.2% Difolatan at an interval of 10 days before harvest.
- » Pre-harvest application of N-benzyladenine (growth promoter) at 10 to 20 ppm prolongs shelf-life of leafy vegetables.
- » Stoppage of irrigation 10 to 15 days before harvesting onion results in enhanced storage life of bulbs.

4.6.13.2. Post-harvest treatments

i. **Use of chemicals:** Post-harvest losses can be reduced by use of chemicals, fungicides, waxing etc. Post-harvest application of 0.1% Bavistin or 0.1% Topsin controls storage diseases in mango. Hot water treatment with 0.1% Bavistin or 0.1% Imazil reduced storage losses of Nagpur mandarins. Spraying with aqueous emulsion of CIPC at 50 and 100 mg/kg of tubers resulted in complete inhibition of sprouting of potatoes in evaporative cool chamber for 4 and 5 months, respectively.

Protective skin coating with wax increases storage life of fruits and vegetables at ambient temperatures. This treatment protects fruits and vegetables against excessive moisture loss and higher rate of respiration. To protect against microbial spoilage, fungicides can be added to the wax emulsion.

ii. **Disinfection**: Vapor heat treatment of tropical fruits at 48°C for 1 hr eliminates all stages of fruit flies.

iii. **Irradiation**: Sprouting of onion can be checked by gamma irradiation at a dose of 0.06 to 0.1 KGY. In potato, gamma irradiation at 0.1 KGY can inhibit sprouting completely and also inhibits the light induced synthesis of chlorophyll and the toxic alkaloid solanin. The irradiated potatoes could be stored successfully for 6 months at 15°C with 10% weight loss.

Fruits of climacteric class such as banana, guava, mango and papaya where irradiation in the mature but unripe pre-climacteric stage is given at dosage of 0.25 to 0.75 KGY showed improved shelf-life due to delay in the rate of ripening and senescence.

iv. **Cold storage:** Storage at low temperatures immediately after harvest reduces the rate of respiration resulting in reduction in build-up of heat, thermal decomposition, microbial spoilage and also helps in retention of color and quality and freshness for a longer period. At present, there is a cold storage capacity of 30 million cubic meters in India of which 90% capacity is utilized for storage of potatoes. Fruits and vegetables together utilize only about 8.1% of total cold storage capacity.

Different methods employed in cold storage include Controlled atmospheric (CA) storage and Zero energy cool chamber.

a. **Controlled atmospheric (CA) storage:** CA storage involves modification of atmosphere i.e. reduction in O_2 and/or elevation of CO_2 concentration combined with low temperature. It markedly retards respiration and delays softening, senescence and changes in color and quality of the stored fruits. Most fruits tolerate up to 5% CO_2 and minimum of 2% O_2. This technique can be used for long term storage and for long distance transport.

b. **Zero energy cool chamber**: Indian Agricultural Research Institute, New Delhi, has developed a zero energy cool chamber for short-term storage of fruits and vegetables based on evaporative cooling system. It is a double walled brick structure with sand in between which is watered at regular intervals. During summer when the outside temperature goes beyond 44°C, the temperature inside the chamber never goes beyond 28°C. The RH is maintained at 90% in the structure. A commercial size structure of 12 x 12 x 12 m has also been successfully tried.

4.6.13.3. Value addition: Processing of fruits and vegetables for value addition is an indispensable part of horticulture industry. In countries like USA and Brazil, nearly 60 to 70% of fruits and vegetables are used for processing as against 1 to 2% in our country. Steps should be initiated for achieving at least 10 % level of processing by 2010 which would generate an investment of Rs. 1 lakh crores and create almost half a million direct employment and another 2 million indirect employment. Hence, attempts to promote processing industries in this country is very essential. In mango, banana and onion, our share is nearly 50, 15 and 8% of the total world production and hence a great scope exists for development of newer products from these commodities. Dehydrated onion and tomato products have great potential as export commodities.

4.6.14. Integrated pest management

4.6.14.1. Weeds: Integrated weed management system (IWMS) involves co-ordinated efforts towards development of a long lasting and environment friendly practice of weed management. IWMS includes all preventive, crop husbandry and direct control measures aimed at reducing and maintaining weed population at sub-economic level. In this, all weed control methods like preventive, cultural, mechanical and biological are included in different permutation and combinations are tried to achieve maximum weed control which is not possible in any one single method. IWMS is well suited for adoption in different cropping systems of fruit crops.

4.6.14.2. Insect pests, diseases pathogens and nematodes: Integrated Pest Management is an important principle on which sustainable crop protection can be based. IPM allows farmers to manage pests in a cost effective, environmentally sound and socially acceptable way. According to FAO, IPM is defined as "A pest management system that in the context of the associated environment and the population dynamics of the pest species, utilizes all suitable techniques and methods, in a compatible manner as possible and maintains the pest populations at levels below those causing economic injury". IPM strategies are based on 3 main components/tools like prevention, monitoring/ observation/ regulation, and intervention.

An important difference between conventional and biointensive IPM is that the emphasis of the latter is on proactive measures to redesign the agricultural ecosystem to the disadvantage of a pest and to the advantage of its parasite and predator complex.

Biointensive integrated pest management is a systems approach to pest management based on an understanding of pest ecology. It begins with steps to accurately diagnose the nature and source of pest problems, and then relies on a range of preventive tactics and biological controls to keep pest populations within acceptable limits. Reduced-risk pesticides are used if other tactics have not been adequately effective, as a last resort, and with care to minimize risks.

BIPM options may be considered as proactive or reactive.

i.   **Proactive options:** Proactive options, such as crop rotations and creation of habitat for beneficial organisms, permanently lower the carrying capacity of the farm for the pest. The carrying capacity is determined by the factors like food, shelter, natural enemy complex and weather, which affect the reproduction and survival of a pest species. Cultural control practices are generally considered to be proactive strategies. Proactive practices include crop rotation, resistant crop cultivars including transgenic plants, disease-free seed and plants, crop sanitation, spacing of plants, altering planting dates, mulches, etc.

The proactive strategies (cultural controls) include:

» Healthy, biologically active soils (increasing below-ground diversity).
» Habitat for beneficial organisms (increasing above-ground diversity).
» Appropriate plant cultivars.

ii. **Reactive options:** The reactive options mean that the grower responds to a situation, such as an economically damaging population of pests, with some type of short-term suppressive action. Reactive methods generally include inundative releases of biological control agents, mechanical and physical controls, botanical pesticides and chemical controls.

4.7. POTENTIAL OF HORTICULTURAL CROPS FOR ENSURING FOOD AND NUTRITIONAL SECURITY

The fruit and vegetable sectors of horticulture are most potent to fulfil food requirements of the country and elevate the nutritional and economic status of people (Rana, 2010). In general, antioxidants, vitamins, minerals, and fibers are considered as main nutrients contributed by horticultural crops to a balanced diet, and thus special attention should be addressed to this group of nutrients (Siddiqui *et al.*, 2013a,b).

Fruits and vegetables provide substantial amount of nutrients, important for human health and can play significant role for improving the nutritional intake especially of predominantly vegetarian population. They are important source of micronutrients, vitamins, minerals and folic acid (Siddiqui and Dhua, 2010). Fruits have proved to be essential for a balanced diet as a rich source of vitamins, protein, minerals, organic acids, carbohydrates, and antioxidants (Siddiqui *et al.*, 2013b). Whereas, the nutritional value of vegetables as a vital source of essential minerals, vitamins and dietary fibers has well recognized. Vegetables play an important role in human nutrition as a neutralizing agent for acid substances produces during digestion of high-energy foods (Siddiqui *et al.*, 2013a). Due to their high water content and fiber, leafy vegetables and roots probably aid in digestion and utilization of more concentrated food in human diet (Chopra, 2010). Man needs a wide range of nutrients to perform different metabolic functions and to lead a healthy life and their deficiency leads to a number of diseases in children and adults. Fruits and vegetables contain nutraceutical substances that provides medical and health benefits and their daily consumption have been strongly associated with reduced risk for some forms of cancer, heart disease, stroke, and other chronic diseases (Siddiqui and Dhua, 2010; Siddiqui *et al.*, 2013a,b). Some components of fruits and vegetables (phytochemicals) are strong antioxidants and function to modify the metabolic activation and detoxification/disposition of carcinogens, or even influence processes that alter the course of the tumor cell. It is better to consume a variety of commodities rather than limiting consumption to a few with the highest antioxidant capacity as it varies greatly among fruits and vegetables (Kalt, 2002). Nutritive constituents of fruits and vegetables that have a positive impact on human health and their sources are present in Table 4.9.

Table 4.9. Bioactive substances present in some horticultural crops

Crop	Bioactive substances	Crop	Bioactive substances
Amla	Vit. C, Polyphenols	Beans	Genistein, isoflavonoid
Apple	Flavonoids, polyphenols	Broccoli	Glucosinolate, selenium
Citrus	Vit C, bioflavonoids, Chalcone	Carrot	α- carotene, lycopene
Grapes	Flavonoid, phenols	Chilli	Capsaicin, capsinthin
Strawberry	Flavonoid, phenols	Ginger	Gingerol
Tomato	Lycopene	Turmeric	Curcumin
Onion	Quercetin	Radish	Iso-thiocyanate

Source: Rana, 2010; Ayala-Zavala *et al.*, 2011; Siddiqui *et al.*, 2013b.

4.8. FUTURE THRUSTS

Our future line of action should focus on strengthening of integrated cold chains at farmgate pack houses; expanding the harvesting window by mainstreaming of horticultural crops even during offseason to increase availability; developing nutrient dense biofortified varieties and last but not the least, considering perennial fruit crops as a source of promoting clean environment. It goes without saying that our last mile, more than 700 Krishi Vigyan Kendras (KVKs), need to be aggressively plugged into farmgate supply and value chains to advise on increasing productivity, yield and market reach of horticultural crops. India can achieve the goal of bringing to the plate horticultural crops of the poorest of the poor, not as a special food but as a daily necessity.

4.9. CONCLUSION

The horticulture sector in India possesses unique advantage because of availability of varied climatic conditions for production of various horticultural crops and several ICAR institutes and SAUs are working for the benefit of farmers and other stakeholders. Besides, there is a huge varietal diversity in most of the cultivated fruits, vegetables, flowers, plantation and spices, which increase the harvesting season and more revenues to the farmers. The production and supply of quality planting materials, implementation of INM, IDM and IPM, multi cropping system, integrated farming system (IFS), pollinators, pre- and post-harvest management, forward and backward linkages in marketing are very important for successful cultivation of horticultural crops. In spite of higher production, the quantity of processed products from the raw materials are poor and very low in quantity produce, which is due to the lack of GAP for most of the horticultural crops and poor post-harvest handling. Several prediction models have suggested that the area under horticultural crops is going to increase and cereal crops will be decreasing, hence, it is time to breed the climate resilient varieties to withstand the climatic aberrations. Hence, a comprehensive plan covering R and D aspects for all horticultural crops is a must and convergence of scientists, policy makers and public is the need of the hour for marching ahead and to take the horticulture to newer heights. However, there are certain cause of concerns like-horticulture does not enjoy a safety net like the Minimum Support Price (MSP) for food grains; lack of good cold chain storage and transport networks to extend the life of perishable products; very less or limited input by machinery and equipment so it is tough to minimize the time restraints; higher input costs than food grains make it a difficult set up, especially when there is no support from the local governments to the smaller farmers; and limited availability of market intelligence, mainly for exports–make it a tougher option to choose.

Development of horticulture in India needs some critical management inputs particularly that of supply chain management - collaboration among various stakeholders along with efficient vertical and horizontal integration. The horticulture sector in particular has to prioritize development of research in the issues of genetics, biotechnology, integrated and sustainable production systems, post-harvest handling, storage, marketing and consumer education. Diversification offers an attractive option and a major source of pushing up growth of agricultural sector. While technological up-gradation and associated institutional changes are identified as thrust areas for future development of the horticulture sector, exports are considered to be most important for the growth of the sector. India can look forward to emerge as a major producer of horticultural products and thus secure reasonable market access for its agro exports, which are largely dependent on the competitive technologies that will help in enhancing export potential. This development will also help in overall growth of the economy through generation of extra foreign exchange, creating employment opportunities and also upliftment of the small and marginal farmers, with definite positive implications on income and employment. The government should create a positive environment that will ensure a mutually beneficial relationship between farmers and organized sector.

Horticultural crop diversification should be encouraged by intercropping horticultural with non-horticultural crops. This will yield more food, more income and better soil health. To increase the production and productivity of fruits and vegetables, introduction of vegetables in the crop rotation and adoption of recommended practices is very important. The use of vegetables in intercropping also helps in increasing the incomes of farmers during the period when the fruit orchard has not become commercially viable. The diversification plan of the horticulture sector needs to identify potential crops area-wise and the area under low yielding vegetables and fruits should be shifted to more productive and profitable one. There is a strong need to strengthen the research on horticultural crops to develop demand-driven technology by improved variety, pest management, etc. The horticultural development requires a minimum set of basic production factors, an optimal crop management infrastructure, post-harvest infrastructure, entrepreneurial management and horticultural expertise, logistical infrastructure and supporting financial infrastructure.

Development of horticultural sector should be accompanied by the growth of the agro processing industry. The opportunity exists to promote the industry by intensifying production of a required, appropriate variety of tomatoes, cucumbers, mangoes, pineapples, lemons, *etc.*, for the products like ketchup, sauce, juice and pickles. Thus, the production strategy should target not only meeting the domestic and export demand of fresh products but also of the processed products. There is the need to improve post-harvest operations related to handling, storage and marketing of fresh and processed produce. Volumes saved in post-harvest losses are actually the surpluses generated, without additional cost.

The horticulture sector has an immense potential of generating employment. Additional employment can be generated by developing the horticulture based agro-processing units. Empirical analytical evidence has shown that the horticultural export has increased the opportunity for higher earnings for smallholders and that the much higher land sizes owned by horticultural smallholders are indeed a cause or consequence for their participation in the sector. This sector needs to be developed as an organized industry and has to be run collectively by all the stakeholders with farmers as entrepreneurs and private sectors.

Chapter - 5

Yellow Revolution
(Oil Seed Production)

5.1. INTRODUCTION

Oilseed crops consist of a wide variety of plants, the seeds of which are utilized primarily for extracting oil. Important edible oilseed crops grown in India include groundnut (*Arachis hypogaea*), rape and mustard (*Brassica* spp.), sesame (*Sesamum indicum*), safflower (*Carthamus tinctorius*), sunflower (*Helianthus annuus*), soybean (*Glycine max*) and niger (*Guizotia abyssinica*) and non-edible oilseeds include linseed (*Linum usitatissimum*) and castor (*Ricinus communis*). Oilseed crops are the second most important determinant of agricultural economy, next only to cereals within the segment of field crops. These oilseeds possess about 20–60% oil, which is chiefly consumed as food and energy source. They contain essential fatty acids, carbohydrates, and vitamins (A, D, E, and K).

Edible oil is an important constituent of the Indian diet. Besides being a source of energy, they add a special flavor and palatability to food. The annual per capita consumption is 11.1 kg against the world average of 14.5 kg and the average of 26 kg in developed countries. Edible oil consumption is likely to increase with rising of per capita income.

However, the daily in-take of fat should not contribute more than 15-20 per cent calories. There is potential to produce about 2.5 million tons of oil from non-conventional sources, but hardly about 0.8 million tons are being utilized. It is important to work out a strategy to exploit maximum potential from these sources.

Despite being the fifth largest oilseed crop producing country in the world, India is also one of the largest importers of vegetable oils today. There is a spurt in the vegetable oil consumption in recent years. The demand-supply gap in the edible oils has necessitated huge imports accounting for 60 per cent of the country's requirement (2016-17: import 14.01 million tons; cost Rs. 73,048 crores). Despite commendable performance of domestic oilseeds production of the nine annual crops (Compound Annual Growth Rate of 3.89%), it could not match with the galloping rate of per capita demand (~6%) due to enhanced per capita consumption (18 kg oil per annum) driven by increase in population and enhanced per capita income.

The productivity of oilseeds continues to be as low as 944 kg per hectare when compared to the world level at 1,632 kg per hectare. At present, there is not much scope to expand the cultivable area under oilseeds.

5.2. SOURCES OF VEGETABLE OILS

5.2.1. Primary sources of vegetable oil

Nine oilseeds are the primary source of vegetable oils in the country, which are largely grown under rainfed condition over an area of about 26 million ha. Among these, soybean (34%), groundnut (27%), rapeseed and mustard (27%) contributes to more than 88% of total oilseeds production and >80% of vegetable oil with major share of mustard (35%), soybean (23%) and groundnut (25%).

Andhra Pradesh (groundnut), Gujarat (groundnut), Haryana (Mustard), Karnataka (Groundnut), M.P. (Soybean), Maharashtra (Soybean), Rajasthan (Mustard and Soybean), Tamil Nadu (Groundnut), U.P (Mustard), West Bengal (Mustard) contributing more than 95% of total oilseed production in the country. India is producing about 7-8 million tons of vegetable oils from primary sources.

5.2.2. Secondary sources of vegetable oil

In addition to nine oilseeds, 3 million tons of vegetable oil is being harnessed from secondary sources like cottonseed, rice bran, coconut, Tree Borne Oilseeds (TBOs) and oil palm. Oil palm which is categorized as secondary sources of oils should be included as primary source as it gives the highest per ha oil yield (4-5 t/ha).

5.3. AREA, PRODUCTION AND YIELD

In India, annual oilseeds are cultivated over 26.67 million hectares of area producing 30.06 million tons annually (quinquennium ending 2016-17) (Fig. 5.1). Majority of the oilseeds are cultivated under rainfed ecosystem (70%). The area under oilseeds has experienced a deceleration in general, and this is due to their relative lower profitability against competing crops like maize, cotton, chickpea *etc.*, under the prevailing crop growing and marketing situations.

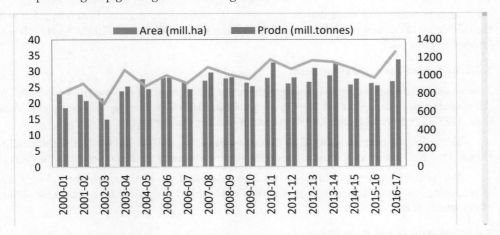

Fig 5.1: Trends in area, production and yield of annual oilseeds in India (2000-2017) (Source: DFI Committee Estimates based on data compiled from DACNET)

The annual production of oilseed crops was virtually stagnating at about 10 million tons over a span of more than 15 years in spite of a considerable increase in area under oilseed crops from 10.73 m ha in 1950-51 to 19.02 m ha in 1985-86. Till mid-eighties, the growth in output also lagged far behind the growth in demand, thus forcing the government to resort to large scale import of edible oils (to the tune of US $1,100 million during 1981-86) to bridge the demand-supply gap.

India attained an average productivity of 1,087 kg per hectare for the triennium ending 2012-13. The average yields of most of the oilseeds are invariably low (Table 5.1).

Table 5.1. Oilseeds productivity (kg/ha) in India *vis-a-vis* World (2012)

Crop	India	World	Country with highest productivity*
Groundnut	1179	1676	4699 (USA)
Rapeseed-Mustard	1140	1873	3690 (Germany)
Soybean	1208	2374	2783 (Paraguay)
Sunflower	706	1482	2494 (China)
Sesame	426	518	1315 (Egypt)
Safflower	654	961	1489 (Mexico)
Castor	1455	1162	1455 (India)
Linseed	260	752	1358 (Canada)
Oil palm fruit	12380	14323	21901 (Malaysia)

Source: FAOSTAT, 2012

* from among the countries with >80% global contribution

5.4. CAUSES OF LOW PRODUCTIVITY

5.4.1. Energy-starved conditions

As oilseeds are energy-rich crops, they must be grown in energy-rich conditions. However, unfortunately, the cultivation of oilseeds in India is mostly done under energy-starved conditions, where soil is both "thirsty and hungry."

5.4.2. Low or no use of plant nutrients

Low or no use of plant nutrients is another important factor affecting the low yield of oilseed crops. All oilseed crops require plant nutrients in adequate quantities for achieving their high yields (Aulakh *et al.*, 1985 ; Tandon and Sekhon, 1988 ; Pasricha and Tandon, 1993) .

The estimated nutrient removal by oilseed crops during 2007–2008 was 3.358 thousand tons $(N + P_2O_5 + K_2O)$, whereas the contribution in nutrient uptake by fertilizers was only less than 15%.

The nutrient requirement of oilseeds is high, in general, for all the nutrients, and for P and S in particular, and hence these nutrients require to be furnished with adequate amounts for obtaining higher yields of oilseed crops.

5.4.3. Imbalanced use of fertilizers

Planting of high-yielding varieties of oilseed crops without concomitant use of fertilizers leads to nutrient mining and nutrient imbalance in soils. This results in low soil fertility, hidden hunger, and low factor productivity. Recently, the response of oilseeds for N, P_2O_5 and K_2O fertilization has decreased, mainly due to the deficiency of secondary and micronutrients (Hegde and Sudhakara Babu, 2009) .

5.4.4. Lack of irrigation facilities

There is no doubt that many Indian farmers have shifted to growing oilseed crops from rainfed conditions to irrigated situations. As such, there is a change from about 89% oilseed rainfed cultivated area to nearly 72% (Ameta *et al.*, 2001; Hegde and Sudhakara Babu, 2009). However, this 72% of the oilseed rainfed area is subject to uncertainties of moisture availability. Absence of rain at crucial stages of crop growth, especially prior to maturity, causes significant loss in oilseed crops. The problem is further accentuated by the fact that a substantial portion of the rainfed area comprises marginal lands, with the result that a certain degree of instability is inherent in the oilseed crops production process.

It is worthwhile to mention that, with the exception of rapeseed, Indian mustard and castor, the irrigated area in other oilseed crops varies from less than 1% in safflower and niger to 19% in groundnut.

5.4.5. Lack of high-yielding varieties of oilseeds

Although the Genetics and Plant Breeding Departments of State Agricultural Universities and those under the Indian Council of Agricultural Research, have contributed a lot in developing a number of varieties of oilseed crops, yet the evolution and introduction of high yielding improved variety hybrids with in-built resistance to biotic and abiotic stresses is lacking (Sharma, 2000). This is more in case of rapeseed and mustard.

5.4.6. Poor adoption of other improved technology

The majority of farmers still continue to adopt traditional practices, which result in low yields of oilseed crops. Many a time, crops are neither sown in time nor due emphasis is laid on weed control and plant protection measures. This is more so in the case of rapeseed and mustard production, especially in Rajasthan, where these crops contribute to major Rabi oilseed crops.

5.4.7. Small and marginal farmers

The size of farm holdings is fast decreasing, with the national average holding size hardly 1.4 ha. As a matter of fact, in India, per capita cultivable land holding has declined from 0.5 ha (1951–1952), 0.35 ha (1970–1980), and to 0.14 ha in 2,000. By the end of 2020, it is further expected to decline from 0.14 to 0.10 ha. Merely 15% of Indian farmers can be called large agriculturalists having 2 ha or more land, and the remaining 85% are known as small farmers having a land holding of 1 ha or less.

No bank would finance individual heavy machinery, such as tractors and other implements, if a farmer's land holding is below 3.2 ha. The rural poor, as such, still meet 84% of their credit needs from nonformal sources; levels of debts and unusual stresses intensifying, especially in rainfed areas (Singh, 2009) . In this context, many agricultural experts are of the firm view that there is a need to revise existing laws and land reform measures in order to discourage the division of land below the economically viable area of 2.0 to 3.2 ha per individual family. They also recommend the formation of clusters of farming community, in line with a cooperative pattern of collective farming, and sharing the cost and earning.

5.4.8. Less priority to grow oilseed crops

The progress in respect of oilseeds has not been substantial mainly because the growth of food grains is given first priority by peasants. Scientists too give more preference in research and development programs to food grain production.

5.4.9. Improved practices

Improved agricultural practices, compiled and formulated with respect to oilseed crops by the scientists of State Agricultural Universities as well as by the scientists of ICAR, are not fully followed by oilseeds growers. This results in low yield of oilseed crops.

5.4.10. Short supply of essential inputs

Many times, yield increasing inputs such as improved seeds, fertilizers, pesticides, biocontrol agents, *etc.*, are inadequate in the market at a critical period during the farming season or their cost goes beyond the reach of small and marginal farmers.

5.4.11. Poor post-harvest technology and storage facilities

Even after securing good yields in the fields, heavy losses may occur due to poor harvesting technology and inadequate storage facilities.

5.4.12. Inadequate plant protection measures

Pests and diseases cause huge loss to oilseed crops. Mustard aphid causes about 65 and 28% losses in rapeseed and mustard, respectively (Sharma, 2000) . But most farmers do not use pesticides to protect their crops from the hazards of pests and diseases. It may be due to lack of funds or latest technical knowhow.

5.4.13. Lack of credit facilities to farmers

Adequate and timely loans at a low interest rate for the purchase of yield increasing inputs are not available to farmers. As a result, the crop is grown mostly under poor management conditions, resulting in low yields.

5.5. BRIDGING YIELD GAP

There exists a tremendous potential for enhancing the yield of nine oilseed crops by adopting the technologies already available. This contention is based on the results of 23,118 frontline demonstrations (FLDs) (2010-2015) conducted on nine oilseeds crops under real farm situations in different agro-ecological conditions of India over a period of five years. The productivity (yield) gap between improved technology and farmers' practices ranged from 21 % in sesame to 149 % in sunflower (Table 5.2).

Table 5.2. Productivity potential (kg/ha) of improved technology of oilseeds.

Crop	Improved technology Yield (kg/ha) (IT)*	National Average Yield (kg/ha) (NAY)**	Increase in IT over NAY (%)
Groundnut	2264	1439	57
Soybean	1603	1182	36
Rape-Mustard	1692	1181	43
Sunflower	1742	700	149
Sesame	536	441	21.5
Safflower	1061	567	87.1
Niger	406	313	29.7

Castor	2032	1647	23.4
Linseed	1090	484	125.2
Mean	1541	1019	51.3

Bridging yield gap across oilseeds can increase oilseeds production significantly that would concomitantly reduce the dependence on imports of vegetable oil besides realizing higher profitability to oilseed farmers.

5.6. THE TECHNOLOGY MISSION ON OILSEEDS

Realizing the demand-supply gap, the Government of India appointed the Technology Mission on Oilseeds in May 1986, with the objective to create/manage conditions that would harness the best of production, processing and storage technologies to attain self-reliance in edible oils in the foreseeable future. A target of producing 16-18 million tons of the nine annual oilseeds (groundnut, rapeseed-mustard, soybean, sunflower, safflower, sesame, niger, linseed and castor) was fixed to substantially cut down the imports by the year 1990. This approach envisaged developing and taking modern technological inputs to the peasants. Such technological inputs consisted of high-yielding varieties of various oilseed crops having a higher oil content, better production stability, resistant to diseases and pests, and good agricultural practices.

The Mission started functioning as a consortium of concerned Govt. departments, namely, Agricultural Research and Education (DARE), Agriculture and Cooperation (DoAC), Civil Supplies (DoCS), Commerce (DoC), Science and Technology (DST), Biotechnology (DBT), Planning, Health, Irrigation and Economic Affairs. The Mission adopted a four-pronged strategy under the following Mini-Missions:

5.6.1. Mini-Mission-1

Improvement of crop production and protection technologies for realizing higher yields and profit to farmers.

5.6.2. Mini-Mission-ll

Improvement of processing and post-harvest technology to minimize the losses and increase the oil yield from both traditional and non-traditional sources of oil.

5.6.3. Mini-Mission-IIl

Strengthening the input support system to ensure availability of right kind of seed, fertilizers, pesticides, irrigation, credit, etc. and to bring awareness among farmers about the potential of the farm worthy technology through massive transfer of technology programs.

5.6.4. Mini-Mission- IV

Improvement of post-harvest operations for effective procurement, handling, disposal including price support system to farmers and financial and other supports to processing industry.

The constitution of Technology Mission on Oilseeds (TMO) in 1986, spearheaded by Dr M.V. Rao, resulted in the country's oilseed production surpassing the target of 18 million tons, fixed for the Seventh Five-Year Plan with an impressive annual growth rate of nearly 6 per cent in the short-run. Thanks to TMO, the import got reduced to almost negligible. Hence, India achieved near self-sufficiency in edible oils during early 1990s, which was popularly referred to as 'Yellow Revolution". The growth rate in the per capita edible oil consumption during this period was 3.66 per cent with an average per capita consumption of 6 kg per annum. The increase in the per capita consumption vis-à-vis the previous period was 54 per cent. As a result of concerted efforts under TMO, a quantum jump in oilseeds production from 10.83 million tons (1985-86) to 24.75 million tons (1998-99) was made possible through effective coordination among different Ministries, Departments and Organizations like ICAR and SAUs under the able leadership of Dr M.V. Rao.

The ICAR Institutes and SAUs developed high yielding varieties with disease and insect resistance suited to various agro-climatic conditions. Ministry of Agriculture provided needed support for timely supply of inputs like seed and propagation of production technology through State Extension Services, arranging credit facilities, marketing and processing, storage and price support, *etc*. It is overwhelming to record that the area under oilseed cultivation increased from 19 million ha (1985-86) to 26 million ha (1996-97) and production increased from 10.83 to 24.38 million tons during just one decade, registering an increase of 36 per cent in area and 125 per cent in production. Similarly, productivity of all the annual oilseed crops, on an average, increased from 570 to 926 kg/ha, being an increase of 62 per cent during this period. This golden era witnessed the release of 200 varieties and hybrids and performance of improved crop technologies under real farm situations, leading to significant improvements both in yield and profits to the farmers. As a result, India achieved a status of 'self-sufficient and net exporter' during early nineties, rising from the 'net importer' state. At the same time, the imports declined from Rs. 700 crore in 1985-86 to Rs.300 crore in 1995-96.

5.7. TECHNOLOGICAL INTERVENTIONS

5.7.1. Research set up and breeders' seed production

India has the largest research network for the development of location specific oilseeds crop production technologies. It includes the All India Coordinated Research Project on Oilseeds (AICRPO) and the Directorate of Oilseeds Research (DOR) at Hyderabad (Andhra Pradesh), three National Research Centres, one each for soybean at Indore (Madhya Pradesh) groundnut at Junagadh (Gujarat) and rapeseed-mustard at Bharatpur (Rajasthan).

Besides, the State Agricultural Universities, other ICAR institutes and private/corporate sector have also been involved in oilseeds research and development programs. A multitude of improved technologies have been generated and transferred on to farmers' fields over the time through this network.

The oilseed research set up in the public sector has about 500 well trained scientists engaged in research and technology generation purposes. Realizing the fact that quality seed is the most crucial, critical and vital input to enhance productivity of oilseed crops, separate breeders' seed production units in these crops were also established. Consequently the breeder seed production, as a basic input to quality seed production, increased 3.7 folds.

The following factors played their crucial role in bringing success to the oilseed mission and thereby the yellow revolution in India.

» Strong socio-economic and political will to become self-sufficient in vegetable oils,

» Biodiversity and the matching diversity in agro-ecology and farming situations for various annual oilseed crops, in spite of the paucity of more prospective oil crops such as the oil palm.

» Economically viable and sustainable improved oilseeds production technologies generated with the help of strong and vibrant oilseeds research network coupled with encouraging financial and policy supports to research.

» Attractive incentives to the farmers in terms of minimum support prices and input subsidies.

» Institutional support for the overall oilseeds research and development by public, corporate and private sectors, particularly the setting up of the Technology Mission on Oilseeds by the Government of India.

» Effective implementation, monitoring and periodical evaluation of the technology transfer programs, especially the "Frontline Demonstrations in Oilseeds" Project.

» Integrated, effective, efficient and transparent functional farmer-research-industry- policy interface.

5.7.2. Crop improvement

During the past two decades, over 240 improved varieties/hybrids have been developed in annual oilseeds, which have shown 9 to 38 per cent yield superiority over the local cultivars. A further scope is envisaged to capitalize on this potential area with the development 26 of varieties/hybrids which can out yield 50 per cent more or even higher than the existing varieties under farmers' field.

5.7.2.1. Plant genetic resources: Sustained utilization of the genetic resources in different oilseeds augmented through collection/introduction of native/exotic plant species and the genetic stocks developed by crop specific breeding programs contributed effectively towards breeding of improved varieties. The exotic oilseed crops/varieties, particularly soybean and sunflower, have also helped produce good results.

5.7.2.2. Hybrids and improved varieties: Development of two safflower hybrids in India is recorded as the 'first' in the world. The safflower hybrid DSH-129, developed by the Directorate of Oilseeds Research, Hyderabad, and released for commercial cultivation for all safflower growing areas in the country, has an average yield potential of 1750 kg/ha. It offered 22 per cent higher seed yield and 29 per cent higher oil yield over the presently grown varieties during rabi/summer season in black soils.

The development and release of hybrids in castor namely, DCH-32 and GCH-4 and a high yielding variety DCS-9 (Jyoti) revolutionized castor production in the country.

New hybrids in sunflower and castor that were brought out in series have provided new opportunities. A few more hybrids in rapeseed- mustard are around the corner. Efforts are also on to evolve hybrid in sesame. Breeding efforts in oilseed crops to meet the objectives such as the oil content in the seed, resistance to biotic and abiotic stresses, reduced crop duration *etc.* have been quite successful.

5.7.3. Crop production

5.7.3.1. Improved packages of cultivation practices: To maximize yield gains from improved technology in terms of varieties/hybrids, agronomic packages for efficient crop management have been developed for different crops and situations, which assured better results.

5.7.3.2. Plant nutrients: Nitrogen (N) is an integral part of chlorophyll and all proteins. It is partly responsible for the dark green color of leaves, vigorous growth, branching/tillering, and enlargement of the leaf surface. Indeed, "Nitrogen is to a plant but petrol is to a car." It plays a pivotal role in the growth of oilseed crops

Phosphorus (P_2O_5) is essential for growth, cell division, root growth and elongation, seed and fruit development, and for counteracting the bad effects of excessive use of nitrogen such as lodging and late ripening. It is a constituent of ADP and ATP (adenosine diphosphate and adenosine triphosphate) which act as energy-rich bonds, and also becomes a constituent of phospholipids and amino acids.

Potassium (K) is involved in the activation of a number of enzymes which assist in the production and translocation of photosynthates to storage organs, in membrane permeability, regulation of transpiration, and providing resistance against a number of pests and diseases.

The use of high analysis Sulfur (S)-free fertilizers has led to more widespread and more intense S-deficiencies in Indian soils (Jena *et al.*, 2006). Hitherto, S-deficiency was observed only in 130 districts of Indian soil but now about 45% of districts in the country have shown more than 40% S-deficiency in their soils. In Orissa, visual symptoms of S deficiency in groundnut have been reported in a number of locations (Desh, 2002) .

Since S is one of the essential plant nutrients, which is directly involved in the biosynthesis of oil, being constituent of the sulfur containing amino acids such as methionine, cystine, and cysteine, the application of S is a must to increase the yield of oilseed crops and their oil content (Singh *et al.*, 2006) .

The deficiency of Ca can be a problem in acidic, coarse textured, and calcareous soils having neutral to alkaline reaction. An application of 40 kg Ca ha^{-1} to Ca-deficient soils is desirable for getting satisfactory yields of various crops, especially in case of oilseed crops.

Iron (Fe), manganese (Mn), zinc (Zn), copper (Cu), boron (B), molybdenum (Mo), and chlorine (Cl) are listed as micronutrients because they are used by plants in very small amounts. These plant nutrients may upset plant growth either due to their low content in soils or because some conditions may reduce their availability in soils.

5.7.3.3. Mixed/ intercropping: The system of mixed/intercropping renders a very important contribution toward checking the lack of balanced diet of the people. For instance, growing cereals provides carbohydrates, pulses proteins, edible oilseeds and vegetable fats in the food.

In South India, 24 crops are grown in mixture with finger millet (ragi), of which 7 are cereals, 6 are pulses, and 11 are other crops, including oilseeds. In North India and Peninsular India (north of the river Krishna), 12 crops enter in mixture with wheat, of which cereals are 4, pulses are 2, and others are 6 which principally consist of oilseeds.

On an average, the yield under crop mixture, including oilseed crop, was 9.92 q ha^{-1} against 5.62 q ha^{-1} under single cropping, indicating a bonus yield of 4.30 q ha^{-1}. The second highest additional yield was seen in the mixed crop of safflower + coriander.

All the crop mixtures proved profitable, offering an average net profit of Rs. 451 ha^{-1} and net income was higher under safflower + coriander combination (Rs. 1,278 ha^{-1}).

Based on average values, crop mixtures provided 57.83 men days of employment per hectare as compared to 45.13 men days of employment per hectare under single cropping system, indicating 28% increase in employment generation over single cropping system. Mixed crop of safflower + coriander gave 50.19 men days ha^{-1} as against 39.11 men days ha^{-1} by single crop of coriander.

Maize–potato–sunflower is another intercropping system which has been found to be the most beneficial to all soil types, and adopted by all classes of farmers, irrespective of size of holdings in Farrukhabad and Kannauj districts of Uttar Pradesh (Singh, 2009). Apart from this, castor (crowd bunch farming) is also grown as intercropping in cluster beans + red gram. The castor is grown in the system to provide support to the bean and also for seed purpose.

Cotton and groundnut are the important crops in Gujarat, especially in Saurashtra region. Both these crops are grown either alone or as intercrops (Raheja, 1973; Vekaria *et al.*, 2000). Highest returns were obtained by planting cotton alone at 180 cm spacing, followed by the treatment when two rows of groundnut alternated with one row of cotton. At Junagadh, cotton alone and cotton + groundnut in alternate rows gave almost equal money value than mixtures of groundnut + castor and cotton + castor. Similarly, an intercrop of cotton and groundnut gave the highest money return under Chennai soil conditions in the first year. In the subsequent 2 years, a mixture of finger millet (ragi) and groundnut was found more profitable.

In North East Hill Region, groundnut can be successfully grown as an intercrop in upland rice (3:1 or 4:1 row ratio) or in maize (1:1 or paired row). This intercropping system has been found to increase total productivity per hectare (Panwar *et al.*, 2001) .

In rice, sowing of groundnut as intercrop can be done during May/June, while with maize, it can be sown in the last week of April, or sown after 20–25 days of maize sowing.

Based on research findings, the package of practices for gobhi sarson (*B. napus*) as an intercrop in "Autumn Sugarcane" have been recommended (Toor *et al.*, 2000) .

Intercropping of sorghum + green gram/ black gram/soybean/ groundnut (2:1) 60 cm row-to-row distance has been found profitable for eastern parts of Madhya Pradesh (Shrivastava *et al.*, 2000) .

Intercropping of sunflower + soybean (1:1) produced highest net returns (Rs 14,865 ha^{-1} and benefit cost ratio of 1.41) under temperate soil conditions of Kashmir (Khan *et al.*, 2006).

5.7.4. Minimum support price policy

Considering the fact that oilseed crops being considerably influenced by market forces are prone to wide fluctuations in prices of oilseeds, a positive view was taken by the government, which resulted in increased minimum support price of oilseed as compared with other crops over the years.

The ruling open market prices had been about 20 per cent higher than the support prices during the harvest, times and about 50 per cent higher during the lean period. The announcement of support prices before sowing of crops provided the guaranteed market clearance and thereby geared up to switch over from cultivation of staple cereal crop(s) to market oriented non-food crops.

As a part of the Market Intervention Operations (MIO) by the National Dairy Development Board (NDDB), the imported oil (other than that released through the public distribution system) was channelized into the market and the vanaspati industry, and buying, stocking and selling of oilseeds/oils continued to be undertaken in the domestic market.

The NDDB also introduced vegetable oil in the consumer packs under the brand name of 'Dhara', (meaning 'The Flow'), which not only narrowed the range of prices of different edible oils (thereby discouraging adulteration of costlier oils with cheaper ones, being an economically unattractive proposition) but also helped in popularizing blended edible oils.

5.7.5. Extension and training in oilseeds technology

5.7.5.1. Frontline demonstrations - technology transfer: With a view to demonstrate under real farm situations the productivity potentials and profitability of a spectrum of improved oilseeds crop production technologies, evolved by the oilseeds research network in the country from time to time, on-farm demonstrations were organized through various technology transfer programs of the Central and State Governments, State Agricultural Universities and voluntary organizations.

Among the most successful of such programs figured the "Frontline Demonstrations in Oilseeds" Project, a component of the Oilseeds Production and Development Program (OPDP) of the Government of India that supplemented the Oilseed Technology Mission. During 1990-91 to 1996-97, more than 8,000 demonstrations were organized in different oilseed crops across various agro-ecological and crop growing situations.

It has been unequivocally proved over years that the improved technologies offered yield advantage ranging from 24 per cent to 107 per cent over the prevailing farming practices, with the benefit – cost ratio varying from 1.35 to 3.33 across different crops, regions and situations. These demonstrations tremendously convinced the farmers of the efficacy of improved technologies, thereby promoting their on-farm adoption.

5.7.5.2. Training and education: Several short- and medium-term training programs were organized and thousands of extension functionaries have been trained over the time by oilseeds research network involving ICAR, SAUs and other development organizations.

The trainees included officials from the State Departments of Agriculture of different states, such as subject matter specialists, Joint/Deputy/Assistant Directors of Agriculture, Project Officers of OPDP of State/GOI, scientists from SAUs, extension personnel of other voluntary organizations and progressive farmers. The untiring and sustained efforts by the oilseeds research and extension machinery to disseminate the knowledge on improved technology included:

» Publication and distribution of package of cultivation practices for increasing production for each of the oilseed crops in English, Hindi and regional languages.

» Publication and distribution of technical bulletins on specific weed/insect/disease management technologies, seed production and frontline demonstrations.

» Use of mass media such as video films on improved crop cultivation practices, radio talks and television programs by scientists.

» Organization of field days/farmers' rallies *etc.*, among others.

5.7.6. Institutional support and linkages

Besides linking the Commission for Agricultural Costs and Prices (CACP), the National Agricultural Cooperative Marketing Federation (NAFED), the National Dairy Development Board (NDDB), Oilseeds Growers' Societies (village level) and Oilseeds Growers' Federations (state level) for implementing the support price policy, projects to evolve and perfect new technologies for oilseeds and transfer these on to farmers' fields were initiated.

A development project for groundnut was launched in 1980-81 and that for soybean in 1981-82. Also, a program for distribution of mini kits of improved seeds and fertilizers for oilseeds was started in 1980-81. Indigenous annual oilseed crops like sesame have got adequate attention for varietal development and seed production.

Exotic soybeans have also drawn particular attention of breeders and seed producers in India. In 1985-86, the National Oilseeds Development Project (NODP) was launched with a view to accelerating the production of four major oilseeds, namely, groundnut, rapeseed-mustard, soybean and sunflower.

Under this program, seeds of improved varieties, plant protection chemicals, fertilizers and Rhizobium culture were made available to the growers at subsidized rates. The efforts of State Departments of Agriculture, ICAR institutes and SAUs were devoted towards demonstrating the potentials of improved technologies on farmers' fields.

Initially, the efforts were concentrated in potential areas of 12 states, but later the project was extended to 180 districts of 17 states in the country. While the interaction of remunerative prices and new technology started showing its positive impact on the production of oilseeds, the need for integrated efforts on harnessing the best of production, processing and management technologies of the oilseed economy was strongly felt.

Towards this end, the Technology Mission on Oilseeds (TMO) launched in May 1986, "Oilseeds Production Thrust Project (OPTP)" initiated in 1987-88, covering 246 districts of 17 states and later extended to more areas, helped in attaining the developmental goals. To provide further operational teeth, in 1990-91, the NODP and OPTP were merged under one program, namely, the Oilseeds Production Program (OPP).

The achievement of self-sufficiency in vegetable oils in India that nullified the foreign exchange drain witnessed during eighties is a sure success. This could be an example to follow in the Asia Pacific region whereby other countries may achieve similar results. The holistic approach in the form of mission mode adopted by the Government of India to tackle the problem of burgeoning import bill on edible oil front is worth commendation.

5.8. POST MISSION PERIOD

The increasing per capita income led to enhanced consumption of edible oils. The gap between the domestic production and the requirement became widened at an alarming rate. This completely eroded the gains that the country had achieved during the TMO period. In addition, the increasing biotic and abiotic stresses, strong intervention of market and non-market forces led to a sticky domestic oilseeds production and profitability.

Although enhanced growth rates of 5.94 per cent were evidenced on the domestic availability of edible oils for the period ending 2011-12, it could not match the rate of growth of imports of edible oils which was 6.99 per cent. The per capita consumption of edible oils grew at a rate of 5.65 per cent. Growth analysis of individual oilseed crops during the decade 2000- 01 to 2010-11 suggests that there has been acceleration in area under soybean, rapeseed-mustard and sesame while stagnation/ deceleration has been observed in groundnut, sunflower, niger, safflower and linseed. The growth in area under castor crop, although marginal, resulted in production enhancement through considerable productivity improvement. The trend of vegetable oils production over the years did help to a considerable extent in reducing imports (Table 5.3). On the contrary, the Government policy allowed greater freedom to open market and encouraged healthy competition rather than protection or control.

Table 5.3. Domestic production, imports, per capita consumption and self-sufficiency in edible oils in India

Year	Prodn. of edible oil (million tons)	Imports (million tons)	Per capita consumption (kg/yr)	Self-sufficiency (%)
1986–87	3.87	1.47	6.2	72
1990-91	6.37	0.53	6.5	92
1994-95	7.19	0.35	7.3	95
1998-99	7.00	2.62	9.8	73
2002-03	5.15	4.36	8.8	53

2006-07	8.00	4.27	11.2	65
2010-11	9.78	7.24	13.6	57
2011-12	9.02	9.94	13.8	48

The country is meeting now more than 50 per cent of its oil requirement through imports resulting in huge drain on our foreign exchange. The current import bill is around Rs. 56,000 crores annually, which is indeed phenomenally higher than in the past.

However, since the oilseed crops are sensitive to market forces and are mostly grown in moisture and nutrition scarce conditions, sustaining the success in the long run is not so easy a task. Sufficient rethinking has to be done in several spheres to sustain and flourish. Research funding by the Government needs to be continued/strengthened for oilseeds research and development in the era of liberalized world trade.

All the research priorities/activities need to be organized/refined in a matrix mode of operation. The support price policy needs to be continued and strengthened. The market mechanism of offering higher price for quality oilseeds with higher oil content and of better quality may have to be searched for.

Quality consideration of oilseeds is desired as a matter of principle. This would encourage and provide a sense of direction to the oilseed growers and researchers alike. The "seasonally variable import duty" may have to be continuously followed to safeguard the overall interests of oilseed sector.

Owing to wide intra and inter seasonal fluctuations in commodity prices of oilseeds and vegetable oils, the move to allow "futures trading" in these commodities needs to be hastened. Since oilseeds generally require less water as compared to many other crops, irrigation water charges should be commensurate with the water used rather than on flat area basis. India should strive to export value added products instead of exporting direct items like oilseeds, oil and oil cakes.

Castor oil which forms the basis for many oleo-chemicals has a great potential. Hydrogenated castor oil, dehydrated castor oil, sebacic acid, undecylenic acid heptaldehyde are some important oleo-chemicals with high value and can earn larger foreign exchange. Similarly, mustard oil is rich in erucic acid which is a useful industrial raw material.

Cultivation of varieties with low/zero levels of toxic constituents or adoption of some detoxification techniques can improve the export of some seeds and cakes. The existing farmer-research-industry-policy sectoral interface needs to be strengthened for achieving the overall development of the oilseed sector in terms of commercial exploitation of untapped yield reservoir, value addition to oilseeds and their products/by-products, demand driven research agenda and congenial public policy environment.

5.9. FUTURE ROAD MAP

5.9.1. Policy issues needing perspective changes

All options for risk mitigation like timely availability of inputs and credit, MSP and procurement crop insurance long-term policy, linking farmers to market, buffer stock options, and other commodity price stabilization schemes, *etc.* need to be put in place for oilseeds sector as a matter of priority.

5.9.2. Minimum support price

Given the fact that input prices across crops remained the same, the relative prices should have been accordingly adjusted. On the contrary, the MSP index analysis clearly indicated that it mainly favored wheat and paddy against pulses, coarse cereals and oilseeds. During TMO period, there was effective implementation of MSP through NDDB that gave confidence to farmers about the minimum expected returns. It, therefore, fortifies revival of an institutional mechanism to implement MSP effectively for reaping the benefits by oilseed growers.

5.9.3. Need for institutional linkages

It must be recognized that the core strength for the success of technology mission on oilseeds was due to effective dovetailing and coordination among institutions linked with production, processing, input supplies, trade and pricing like National Dairy Development Board (NDDB), National Agricultural Marketing Federation (NAFED) and the flagship program of the government in oilseed sector; Integrated Scheme on Oilseeds, Pulses, Oil palm and Maize (ISOPOM)]. Some systems need to be revisited again to give much needed push to oilseeds sector.

5.9.4. Eco-regional approach for productivity enhancement

The concept of eco-regional approach refers to the practice of delineating efficient zones for specific crops for realizing potential yields with high input-use efficiency. Supporting services like input supply, marketing and processing have to be linked to these ecological zones besides strengthening research and extension systems and infrastructural facilities.

The classical examples in high productivity of spring season sunflower in Indo-Gangetic Region of Punjab, Haryana, Western Uttar Pradesh and Bihar; high productivity of safflower in Malwa region (Madhya Pradesh) and Gujarat, high productivity of sesame in West Bengal in summer season, high productivity potential of soybean in Punjab, Haryana and Eastern UP, *etc.* are mainly due to the optimum ecological conditions which are beyond input and management. Hence, providing necessary input supply, technology, market and extraction facilities in these areas can help realize quantum jump in productivity with ease.

5.9.5. Natural resource management

Currently only 28 per cent of area is irrigated under oilseeds. Water requirement in oilseeds is, therefore, a key factor for ensuring higher yields. Correcting the present limitation and imbalance in soil nutrients can provide rich dividends. Cultivation of oil seed crops in fertile soils result in high productivity. Watershed management with appropriate rainwater harvesting both *in situ* with proper disposal and storage farm ponds provide excellent opportunity to mitigate the expected dual problems of long droughts and floods with advantage. Site specific land configuration and management for effective soil and moisture conservation and its economic use can operationalize the drought mitigation strategy. Enhancing drought tolerance in oilseed crops is therefore, a priority with associated practices to improve profitability through achieving 'more crop (oil) per drop' of water, resource use efficiency and preferential edge over other competing crops.

Improving nutrient use efficiency of fertilizers through better product development and method of application should now be a priority for achieving profitable oilseeds production. Exploiting nutrient interactions as per the soil test and crop response results in higher efficiency and reduced cost. Organic manures are central in the integrated nutrient management (INM) of oilseeds under rainfed situation along with other components such as secondary and micronutrients, like use of sulfur bioinoculants, crop residues, *etc.* Precision crop management with conservation agricultural practices and customized fertilizer application schedules would usher higher efficiency and profitability. Emphasis on integrated natural resource management in oilseeds should, therefore, be our high priority.

5.9.6. Crop Improvement strategy

Conventional breeding coupled with modern tools such as biotechnology should now be the primary focus in crop improvement programs. Heterosis breeding should be the major focus in crops like sunflower, castor, rapeseed-mustard, safflower, and sesame. Research emphasis needs to be on (i) augmentation/ identification of trait specific germplasm; (ii) pre-breeding and genetic enhancement; (iii) allele mining, (iv) functional genomics, proteomics, metabolomics, and interactomics; (v) marker assisted breeding and gene pyramiding; and (vi) trait improvement through genetic engineering.

5.9.6.1. Role of Biotechnology: The two main options of biotechnological approaches for crop improvement include molecular marker-based selection and transgene manipulations. Some of the crop-specific needs that are to be addressed through biotechnological interventions include: pests, like bud fly in linseed; Antigastra and phyllody in sesame; necrosis, leaf spot and powdery mildew in sunflower; wilt and Alternaria in safflower; and Botrytis and lepidopteran pests in castor; quality aspects such as presence of anti-nutritional compounds (oxalic acid and phytates) in sesame; oil quality in mustard; toxic proteins (ricin and *Ricinus communis* agglutin) in castor; and herbicide tolerance in soybean. Apart from these crop-specific issues, there are research areas of generic nature such as abiotic stress (drought, salinity, cold) tolerance, increased oil content, altered fatty acid profiles to suit different industrial and human consumption requirements, *etc*. Understanding the molecular basis of trait manifestations such as stress tolerance, oil accumulation, and interactions among different metabolic pathways under varied environmental conditions and at crop growth stages are expected to pave the way for development of designer oilseed crops to meet both domestic and industrial requirements. The success of 'doubled haploids' in developing superior inbreds is a potential area for immediate gain in oilseed crops limited by availability of superior inbred development. The required infrastructure and support need to be ensured for operationalization.

5.9.6.2. Transgenic approach: Transgenics are a reality in crops like canola and soybean and what the introgressed traits can do in sustaining and increasing the productivity of crops is already well demonstrated. Modifying the fatty acid profile of the oil to suit industrial, pharmaceutical, nutritional, cosmetological requirements using genetic engineering approaches has been a priority in application of biotechnology in oilseed crops. Similarly, imparting biotic and abiotic stress tolerance, improved resource use efficiencies through transgenic approaches have been the areas of focused attention for sustained productivity levels under changing as well as challenged environmental situations.

In spite of the progress made in the use of transgenic technology in the oilseed crops, current policy controversy, is a setback, delaying the fruits of results to reach stakeholders.

5.9.7. Exploring frontier sciences

Significant innovations in frontier science and technologies such as nanotechnology, genetic engineering and biotechnology, synthetic lipid science and technology, information science and modeling, simulation and forecasting and the recent developments in related sciences such as hydroponics, vertical farming and protected agriculture; precision agriculture systems; biosecurity and biodiversity management provide unlimited opportunities for supporting higher production and product development to meet the changing requirements through precision farming and protection/conservation practices. Post production, developments in dynamic integration of production, processing, quality with global trade would make production of vegetable oils profitable and competitive. These frontier sciences will have to be harnessed and integrated into ongoing research programs for productivity improvement; increasing the resource use efficiency, improving processing, value addition, diversified uses, improved access to stakeholders through ICT, enhanced delivery systems, better targeting of technologies for yet better production and marketing environments, including supply chain mechanisms. The traditional knowledge should also be valued for its wisdom for technology generation, refinement and adoption.

5.9.8. Public-private partnership (PPP) and linkages

The potential of public-private partnership (PPP) through linkages in all aspects of oilseeds production and marketing needs to be harnessed for a win-win situation. The grey areas for PPP in oilseeds include incentives for seed production, forward/backward linkages for processing, value addition, contract research in niche areas, contract farming, joint ventures for higher order derivatives and speciality products, etc. The shift in consumer preference for branded edible oils has resulted in the corporate sector targeting on packaged edible oil segment in the last few years. Hence, PPP mode for R and D efforts towards value addition emerges as new priority to move forward.

5.9.9. Diversification and value addition

Profitability of oilseeds solely from the primary products like seed and oil will not be sustainable. Besides the primary product oil, oilseed crops provide immense scope for diversified uses with high value speciality products and derivatives. Designer oils with requisite blends can meet the expectation and to that extent individual oilseed crop's potential would be seen for the yield of oil or the desired fatty acid and not as oil from specific crop. Thus, the present wide diversity of oilseed crops may narrow down to a few high oil yielding crops.

Major opportunities for oilseed crop diversification and value addition include introduction as catch crop in paddy fallows to utilize residual moisture and fertility; component crop in major wide spaced field crops such as sugarcane, pigeon pea, cotton, maize, *etc.* for sunflower; and as main crop with groundnut, soybean, finger millet, pigeon pea, cluster bean, short duration pulses, *etc.* for castor and sunflower; with chickpea and coriander for safflower; rabi castor under limited irrigation protection and sunflower for Indo-Gangetic plains of Punjab, Haryana, Western Uttar Pradesh in spring, and Bihar, Odisha and West Bengal in rabi/summer. Soybean also offers opportunity for rice wheat cropping system in the north.

5.9.10. Adaptation to climate change

Oilseeds need higher inputs for increased productivity. Adaptation strategy for drought, high temperature and rainfall variations must, therefore, be put in place as a matter of priority. Oilseeds production is constrained by several biotic stresses like insect pests and diseases that are being further aggravated by changing climatic conditions. Botrytis, root rot and capsule borer have emerged as major threats to castor production. Sunflower production is limited by diseases like Alternaria leaf blight, sunflower necrosis, downy mildew and powdery mildew while mealy bug is an emerging pest. The foliar diseases, Alternaria and Cercospora leaf spots and Macrophomina root rot are becoming increasingly important while wilt and aphid continue to challenge safflower production. Global warming induced climate change is expected to trigger major changes in population dynamics of pests, their biotypes, activity and abundance of natural enemies and efficacy of crop protection technologies. Studies on the epidemiology of plant diseases including variation in pathogen population in the light of climatic change are necessary to develop integrated disease management (IDM) modules. Studies on wilt disease etiology in the context of reniform nematode in castor and sunflower and root-knot nematode in sunflower coupled with identification of sources of resistance deserve attention. There is a need to generate information on the likely effects of climate change on pests so as to develop robust technologies that will be effective. The approach to pest management has seen a significant change over the years from chemical control to IPM with emphasis currently on bio-intensive integrated pest management (BIPM) involving use of pest-resistant varieties, bio-agents, bio-pesticides and natural products like botanical pesticides and pheromones. Several eco-friendly products of biological origin have been developed at the Directorate of Oilseed Research, Hyderabad, for management of important pests of oilseed crops like castor semi- looper, sunflower head borer, tobacco caterpillar as well as wilt of castor and safflower. However, the relative efficacy of many of these pest control measures is likely to change as a result of global warming necessitating identification of temperature tolerant strains.

5.9.11. Transfer of technology

Concerted efforts are urgently needed for the dissemination of technologies and new approaches on a participatory mode to be strengthened for effective delivery mechanism by show-casing the potential technologies/products. The Farmer-Institution-Industry linkage mechanism should be strengthened besides the existing formal delivery mechanisms so that the gap between the potentially attainable yield and the yield realized on the farmers' fields is reduced and it makes the industry more vibrant and profitable on account of assured quality supply, reduced obstacles in supply chain, enhanced capacity utilization and increased economic surplus with benefits to both the producer and the consumer. The potential Information and Communication Technology (ICT) tools should be harnessed on a dynamic and interactive mode. This can minimize the dissemination loss while sharing information and provide benefits to all the stakeholders involved in oilseeds. Also a dedicated TV channel on agriculture will help in faster dissemination of knowledge. Creation of agriclinics with provision of outsourcing through involvement of new breed of young well-trained technology agents would go a long way in out-scaling innovation for greater impact.

5.10. FUTURE PROSPECTS

In order to increase the production and productivity of rapeseed, mustard, and other oilseed crops per unit area per unit time, the following strategies/approaches need to be adopted:

1. Replacement of high-yielding varieties of rapeseed and mustard, linseed, soybean, groundnut, safflower, sunflower, sesame, *etc.*, use of quality seed, mass scale plant protection measures, especially botanical/biological ones, application of organic and inorganic fertilizers and other improved technologies require to be given top priority.

2. Expansion of irrigated area under rapeseed and mustard and other oilseed crops is the need of the day.

3. Organizing front line demonstration of improved package of practices on farmers' fields' is the basic concept of demonstration. It is because "Seeing and Believing" is considered the most powerful instrument to motivate and convince farmers to adopt beneficial innovations to them.

4. Timely and adequate supply of inputs such as seeds, fertilizers, plant protection, biochemical, and credit facilities need to be strengthened.

5. Marketing facilities should be improved to safeguard the interest of the oilseed crop growers.

6. There is a need to arrange mobile training camps to train the farmers at their own locations. Arranging of such camps will help in quick dissemination of technical knowledge. These camps should be of short duration, preferably 1 or 2 days. Trained personnel must then be entrusted with the responsibility of arranging more camps in their villages to answer the queries of fellow farmers.

7. Arranging of exhibitions is another suitable method for teaching all types of people, including

peasants. Exhibits should display major ideas of oilseed production technologies with supportive information. The exhibitions should be held at the village, block, tahsil, district, state, national, and international levels.

8. Effort should be made to show films relating to various aspects of oilseed crops. Rural people reveal keen interest in such shows whenever these are organized in villages.

9. Publications including leaflets, pamphlets, folder, bulletins, popular articles in local languages such as Dogri, Urdu, Gojri, Gujarati, Tamil, Rajasthani, *etc.*, of scientific information with respect to cultivation of various oilseed crops should be published and distributed among potential users.

10. Radio talks and television programs should be arranged from time to time in regional languages, concerning cultural practices of oilseed production.

11. Extent of the knowledge of groundnut cultivation, regarding improved crop technology such as primary tillage, sowing time and seed rate, use of FYM and chemical fertilizers, weeding and intercultural, *etc.*, generated in the Saurashtra region by State Agricultural Universities and other departments, must be utilized by groundnut growers. They should also use high-yielding varieties of this oilseed crop. While sowing the crop, they are required to maintain adequate space.

5.11. SPECIFIC RECOMMENDATIONS

Following research, development and policy strategies would be needed urgently for increasing both oilseeds production and vegetable oils availability in the country.

5.11.1. Research

» Greater emphasis on innovation to achieve quantum jump in productivity using new science and translational research.

» Integration of all oilseed research institutes under NARS for holistic research approach on systems mode.

» Develop short duration, high yielding genotypes for better adaptation to climate change through integration of modern biotechnological tools like MAS and transgenic breeding, supplementary to conventional breeding and develop cultivars with in-built resistance to biotic and abiotic (specially drought and heat) stresses. In this context, greater use of germplasm through pre-breeding will be highly desirable.

» Develop small farm machinery for different operations specific to each crop so as to ensure timely farm operations and efficient use of costly inputs.

» Increased emphasis needs to be given on post-harvest technology and value addition for diversified uses in order to ensure higher profitability.

5.11.2. Development

» Establishing strong linkages for successful operation of 'seed village concept' with producers, technocrats, certifying agencies and concerned State Departments of Agriculture for timely procurement and distribution to ensure higher seed replacement by improved varieties/hybrids.

» Promote oilseeds cultivation in new and non-traditional areas and seasons for ensuring crop diversification and additional area for expansion. Eastern region offers option for potential area expansion especially in paddy fallows. Similarly, soybean offers great opportunity for diversification of rice-wheat cropping system in Northern India.

» Adopt location specific efficient dry farming technologies for drought proofing and sustainable oilseeds production. Integrate oilseeds production with watershed programmes for holistic development and to ensure life support irrigation for assured harvest.

» Increase area under protective irrigation and promote efficient irrigation methods, especially micro-irrigation, for achieving higher production and stability.

» Promote adequate and balanced fertilization with emphasis on use of sulfur and limiting micronutrients through proper soil amendments, based on soil testing.

» Effective transfer of technology with assured input, market and technological backstopping by both public and private sector agencies.

» Promote intercropping systems involving oilseeds for achieving higher efficiency of resources, profitability and risk minimization.

» Adopt need-based plant protection measures through effective and bio-intensive integrated pest management (IPM).

» Largescale production of promising small farm equipment through involvement of state governments that will help in improving efficiency in farm operations. Also provision of credit and incentives for manufacturing of small farm equipment and machinery by small-scale industries and promotion of custom hiring to ensure resilience in farming.

» Greater thrust on use of soybean as food rather than only as oil and feed will help the nation in addressing current major concerns for protein malnourishment, while ensuring nutrition security.

» Exploit additional features of crops like high value safflower petals and fiber from linseed for realizing additional profits. Also, there is a need to accelerate area expansion of oil palm plantations and extend assured irrigation, power, local processing facility and competitive prices for realizing higher production of vegetable oil per unit area per unit time.

» Avoid use of rice bran directly as feed in order to promote greater extraction and use of rice bran oil.

» Promote scientific processing of cotton seed for higher oil recovery and to get high protein retention (42%) compared to traditional processing practices (22%).

» Improve efficiency of extraction of oil through solvent extraction for hard seeds (<20% oil) and expeller extraction for soft seeds (35 to 40% oil).

5.11.3. Policy

» Regulate import of vegetable oils through adoption of appropriate import policy aiming at increased domestic production. Hence, vegetable oils should be viewed beyond export – import balance with the goal of achieving self-sufficiency to a greater extent. The need for achieving self-sufficiency in vegetable oil should be seen in the context of improved livelihood, higher profitability of oilseed farmers and for processing industry.

» Ensure market intervention for effective implementation of MSP through needed procurement of oilseeds, being a major national priority.

» Appropriate regulations to amend the Agricultural Produce Marketing Act for making it pro oilseeds producers and enhance proper trading and fair pricing to both producer and consumer.

» Encourage establishment of large scale 'captive plantations' and specialized 'seed gardens' of oil palm by declaring oil palm as a plantation crop and also ensure proper pricing policy for profitability.

» Creation of enabling environment to strengthen private participation in collaborative research, development, extension and marketing operations.

» To avoid diversion especially of edible oils for biodiesel production and other industrial uses.

» Similar to sugarcane model, oil expeller industry should promote local/regional oilseeds production for assured and adequate supply of raw material as per pre-determined assured prices. The industry should also be involved in supporting technology development and extension activities.

» Revival of Oilseed Mission, through a special purpose vehicle, with greater thrust on 5 Ps: Priorities, Policies, Productivity, Profitability and Private sector participation, with emphasis on increased oilseed production in the country be the highest priority of the Government.

» Greater emphasis and investments on public awareness about rationalization of vegetable oil consumption for proper health becomes our national priority.

5.12. CONCLUSION

Since the oilseed crops are sensitive to market forces and are mostly grown in moisture and nutrition scarce conditions, sustaining the success in the long run is not so easy a task. Sufficient rethinking has to be done in several spheres to sustain and flourish. All the research priorities/activities need to be organized/ refined in a matrix mode of operation. The support price policy needs to be continued and strengthened. India should strive to export value added products instead of exporting direct items like oilseeds, oil and oil cakes. Castor oil which forms the basis for many oleo-chemicals has a great potential. Hydrogenated castor oil, dehydrated castor oil, sebacic acid, undecylenic acid heptaldehyde are some important oleo-chemicals with high value and can earn larger foreign exchange. Similarly, mustard oil which is rich in erucic acid is a useful industrial raw material.

Increased availability of vegetable oils would involve greater commitment of various stakeholders (farmers, scientists, policy makers, NGOs, private sector industry, *etc.*) Assessing the problems and prospects of all these stakeholders and the establishment of strong as well as viable linkages among them towards the goal of improving vegetable oils situation in the country is indeed a challenging task. The success of 'Yellow Revolution', achieved through mission mode approach of TMOP during eighties, fully justifies revival of Oilseeds Mission approach with greater zeal and commitment of all to tide over the present crisis of large-scale import of edible oils. We must have clear national policy of bridging the yield gaps and increased oilseeds production with specific aim to reduce our vegetable oil imports, as was achieved during earlier TMOP. No doubt to achieve this, we would need clear policy directions and also missionary zeal and commitment of all concerned.

Chapter - 6

Pulse Revolution

6.1. INTRODUCTION

Pulses are an important group of food crops that can play a vital role to address national food and nutritional security and also tackle environmental challenges. The share of pulses to total food grain basket is around 9-10 per cent and is a critical and inexpensive source of plant-based proteins, vitamins and minerals. Pulses are critical in food basket (dal-roti, dal-chawal), are a rich source of protein (at 20-25 per cent, it is double the protein content of wheat and thrice that of rice) and help address obesity, diabetes, and malnutrition. The United Nations declared 2016 as "International Year of Pulses" (IYP) to heighten public awareness of the nutritional benefits of pulses as part of sustainable food production aimed at food security and nutrition.

Pulses are the chief sources of proteins, vitamins and minerals and are popularly known as "Poor man's meat" and "rich man's vegetable", contribute significantly to the nutritional security of the country. For the average Indian household, dal has to be part of the daily menu. Per capita net availability of pulses in India, however, has reduced from 51.1 g/day (1971) to 41.9 g/day (2013) as against WHO recommendation of 80 g/day. For the majority, dal is the only source of protein.

More than a dozen pulses crops are grown in different parts of India. Among them, chickpea (Bengal gram), pigeon pea (tur), (mung bean), black gram (urad bean), lentil (masoor) and field pea (matar) are most common ones. Pulses like pigeon pea, black gram, green gram and French bean (Rajmah) are grown as *Kharif* crops, while chickpea, Khesari and lentil are generally *Rabi* crops.

In India, pulses are generally produced in poor soils not suited to other crops, with a minimum use of resources and have a very low water footprint. They are vital constituent of cropping and consumption pattern. Of the total net sown area of 141.40 million hectares, 52 per cent i.e. 73.20 million hectares is rainfed.

Pulses play a greater role in sustaining the economy of the rainfed farming community in a variety of ways. Besides improving soil fertility and physical structure, pulses fit well in mixed/inter-cropping systems, crop rotation and dry farming, provide green vegetable (pods/beans) and nutritious fodder for cattle as well, thereby contributing to a more sustainable food system. Cultivation of pulses build-up a mechanism to fix atmospheric nitrogen in their root nodules (72 to 350 kg N per ha per year), thereby meeting their own nitrogen requirements to a great extent. The cultivation of the pulses under irrigation is only about 20% of their cropped area, with remaining 80% are being grown under rainfed conditions.

Important reasons for their cultivation include:

» Suitability for human and animal consumption.

» Adaptability for inter- or mixed cropping.

» Agronomic management of legumes is relatively easy.

» Legumes are relatively hardy crops and grown in some of poorer soils and harsh growing conditions and face lower incidence of pests and diseases.

» Input (especially nitrogen fertilizer) requirement is lower compared to other crops.

» Legumes are also considered as cash crops.

Though legume cultivation has several advantages they also suffer from some limitations restricting their cultivation, especially limited availability of quality seeds of improved varieties, harvesting is tedious, in addition to labor requirement for value addition, and most importantly volatility of markets.

Historically India is the largest producer, consumer and importer of pulses. Although it is the world's largest pulses producer, India has been importing 3-4 million tons of pulses every year to meet its domestic demand. The productivity of pulses in India (694 kg/ha) is lower than most of the major pulse producing countries and yield potential attained at research stations and on-farm demonstrations. However, during the last decade, growth in pulses production has increased significantly.

6.2. AREA, PRODUCTION AND PRODUCTIVITY

The total world acreage under pulses is about 85.40 million ha with production of 87.40 million tons at 1023 kg/ha yields level. India is the largest producer and consumer of pulses in the world accounting for about 29 per cent of the world area and 19 per cent of the world's production. India, with >29 million ha pulses cultivation area, is the largest pulse producing country in the world. It ranks first in area and production with 34 per cent and 26 per cent respectively. During 2017-18, the country's productivity at 835 kg/ha, is a significant increase over Eleventh (662 kg/ha) and Twelfth plans (745 kg/ha). Among various pulse crops, chickpea dominates with over 40 percent share of total pulse production followed by pigeon pea (18-20%), green gram (11%), black gram (10-12%), lentil (8-9%) and other legumes (20%) (IIPR Vision, 2030).

In India, total pulse area and production during 2017-18 has been >29.3 million hectares and 24.5 million tons, respectively. Out of the total area, >7.3 million ha is in Madhya Pradesh alone, earning a prime status in pulse production registering a remarkable 25% of the country's pulse area with 33% production, thereby ranking first both in area and production. This is followed by Rajasthan in respect of area (16 per cent) and Maharashtra in case of total production (13 per cent).

More than 90 per cent of total pulse production has been the contribution of 10 states namely, Madhya Pradesh, Maharashtra, Rajasthan, Uttar Pradesh, Karnataka, Andhra Pradesh, Gujarat, Jharkhand, Tamil Nadu and Telangana Tables 6.1 to 6.3.

Table 6.1. Major state-wise area, production and yield of *Rabi* pulses (2017-18)

States	Area (million ha)	Production (million tons)	Yield (kg/ha)
Madhya Pradesh	4.643	5.756	1239
Maharashtra	2.103	1.812	861
Karnataka	1.512	0.886	586
Uttar Pradesh	1.469	1.630	1105
Rajasthan	1.410	1.524	1080
Andhra Pradesh	1.003	1.055	1051
Tamil Nadu	0.593	0.396	667
Bihar	0.439	0.371	845
Jharkhand	0.038	0.384	1005
Others	2.046	1.685	823
Total rabi pulses	15.600	15.500	993

Source: DES, Ministry of Agri. & FW (DAC&FW), Govt. of India; 2017-18*- Third advance estimates.

Table 6.2. Major state-wise area, production and yield of *kharif* pulses (2017-18)

States	Area (Lakh ha)	Production (Lakh tons)	Yield (kg/ha)
Rajasthan	3.410	1.455	426
Madhya Pradesh	2.680	2.282	851
Maharashtra	2.247	1.453	646
Karnataka	1.601	0.966	603
Uttar Pradesh	0.892	0.604	677
Gujarat	0.560	0.508	907
Telangana	0.457	0.351	852
Jharkhand	0.412	0.351	614
Odisha	0.386	0.237	614
Andhra Pradesh	0.366	0.181	494
Others	0.756	0.560	740
Total kharif pulses	13.766	9.007	654

Source: DES, Ministry of Agri. & FW (DAC&FW), Govt. of India; 2017-18*- Third advance estimates.

Table 6.3. Major crop-wise area, production and yield of pulses (2017-18)

States	Area (million ha)	Production (million tons)	Yield (kg/ha)
Chickpea	10.573	1.158	1056
Pigeon pea	4.459	4.180	937
Black gram	5.031	3.284	653
Green gram	4.070	1.901	467
Other *kharif* pulses	1.874	0.799	426
Other *rabi* pulses	3.353	3.183	903
Total rabi pulses	15.600	15.500	993
Total kharif pulses	13.760	9.006	654
Total pulses	29.360	24.506	835

Source: DES, Ministry of Agri. & FW (DAC&FW), Govt. of India; 2017-18*- Third advance estimates. India imported 11.281 million tons of pulses valued at Rs. 22160 crores.

6.3. CONSTRAINTS

6.3.1. Inputs

Quality and timely availability of critical inputs seeds, varieties, bio-fertilizers, micronutrient and critical irrigation were identified across the states and felt necessary to be addressed as one of the major strategies under this category.

6.3.2. Marketing

Distress sale, lower minimum support prices compared to cost of production, unfavorable Exim policy, non-accessibility to market, post-harvest losses *etc.*, were identified as major market related constraints, especially in major pulse producing states. The policies to provide remunerative prices to the farmers including the procurement facilities were felt vital by the government.

6.3.3. Extension and their interventions

Lack of guidance in respect of certified seed production/variety identification, insect pests/diseases identification and management phases, importance and procedure of seed treatment/rhizobium inoculation, lack of information/knowledge on current advances in production, management technology, and also poor or no knowledge about organizing seed production and its protection for succeeding crop.

Poor knowledge base on nutrient use efficiency (NUE), IPM, method of preparation of spray solutions and multiplicity of extension system on IPM, especially, pesticide dealers *etc.*, identified as technology transfer or extension related constraints.

6.4. OVERCOMING SOCIO-ECONOMIC CONSTRAINTS

6.4.1. Semi-formal seed systems

Even though there are good HYVs released for all major chickpea and pigeon pea locations in India, and there is enough Breeder seed and Foundation seed produced, at farmers level there is a shortage of Certified/Truthful seed. Both public and private agencies have failed in the supply of enough quality seed and the seed replacement ratio is very low. There has been some success in establishing semi-formal seed systems to produce Truthfully Labelled seed, in which linkages were established between the formal and informal seed sectors through supply of basic quality seed by the NARS, and quality of seed production is monitored by university/non-governmental organizations/ farmers' associations. In this way, there will be enough quantity of seed production at the local level. It should also be coupled with farmer participatory varietal selection (FPVS), which gives farmers an opportunity to select from a range of improved varieties (Abate, 2012).

6.4.2. Awareness about new technologies

Farmers' awareness on improved varieties, seed availability of improved varieties and other good agricultural practices are the key factors in the spread of improved varieties and technologies. The television would be the most popular media for increasing awareness and also FPVS will also be helpful.

6.4.3. Cash and credit

Cash is a key element for enabling smallholder farmers to shift from low input-low output to high input – high output agriculture. But access to credit by these farmers is low because of their low asset base, low risk bearing ability and high-risk environments. This can be effectively tackled by the insurance-linked credit to pulse crops without any collateral security. The scale of finance should be sufficient enough to cover all the costs of the recommended practices.

6.4.4. Marketing

Markets for legumes are thin and fragmented due to scattered production and consumption across states. Farmers/village traders sell their marketed surplus immediately after harvest, while some large traders/wholesalers trade between major markets and hoard pulses to take advantage of speculative gains in the off-season. Due to this, farmers do not benefit from the higher market prices of pulses. Investments in market infrastructure, cold storage, warehouses, market information systems both in public and private sector through PPP models and viability gap funding models need to be encouraged in SAT India.

6.4.5. Farm mechanization and land lease market

One of the reasons for success of expansion of area under chickpea in Andhra Pradesh is the development of suitable varieties for farm mechanization. Hence, farm mechanization in peak season activities like harvesting and threshing needs to be encouraged through the distribution of subsidized farm machinery to cope with labor shortage and higher wage rates. With the expansion of irrigation facilities through ground water and also through canal irrigation systems, there is a scope for expansion of irrigated area under pulse crops, especially summer, *rabi* and spring season crops, as yield response is higher.

In short, to increase area and production of pulse crops we need crop specific and region-specific approaches. Already ICAR and ICRISAT, with the support of state and central governments, are involved in the development of short duration, photo-thermo insensitive varieties for different agro-ecology, development of hybrids in pigeon pea, development of efficient plant architecture in major pulses crops, development of bio-intensive integrated pest management modules, design of improved machines to cope with labor shortage, production of breeder seed of latest released varieties and in organizing frontline demonstrations in farmers' fields. The efforts under NFSM-Pulses and R and D under NARS needs to be strengthened with the major thrust on:

» Replacing cereal crops in the prevailing rice-wheat cropping systems with high yielding varieties of pulses.

» Encouraging R and D on extra early maturity pigeon pea suited to multiple cropping and improved crop management.

» Developing pigeon pea genotype suitable for *rabi/* spring and summer seasons.

» Including short duration varieties of pulses as catch crop; introducing black/green gram (spring) will utilize unutilized land and water in the spring/summer season with high returns.

» Using genomics and biotechnology tools for development of multiple disease and pest resistant varieties to reduce yield loss of standing crop and to increase yields.

» Reducing storage losses and improving market information and infrastructure.

» Technology dissemination and input delivery mechanisms were too weak for pulses. Coordinating research, extension and farmers to encourage farmer-participatory research.

» Linking MSP to market prices can bridge the gap between demand and supply.

6.5. INTERVENTIONS FOR PULSES REVOLUTION

Interventions required for increased production of pulse crops include:

» Strengthening seed delivery system by encouraging high-quality seed production of pulses.

» Developing short-duration and pest- and disease-resistant cultivars, with adequate funding support for R and D.

» Popularization of extra early and stable dwarf type suitable for multiple cropping and improved crop management in sequence with wheat under irrigated conditions in the states of Uttar Pradesh (UP), Haryana, Punjab and northern parts of Madhya Pradesh (MP)

» Replacement of other dryland crops like cotton in states with less water availability like Gujarat, Karnataka, Andhra Pradesh, Maharashtra and Tamil Nadu.

» Popularization of *rabi* pigeon pea in the states of Orissa, Gujarat, West Bengal, Bihar and eastern UP.

» Increasing area through inter-cropping of pigeon pea with soybean in MP, Maharashtra and Rajasthan; and with cotton, sorghum, pearl millet and groundnut in the states of Andhra Pradesh, Maharashtra, Karnataka, Gujarat, MP and UP, which is expected to get additional coverage under pigeon pea by at least 1 million hectares by the turn of the century,

» Management of pod borer, fusarium wilt and sterility mosaic.

» Ensuring remunerative prices for farmers by judicious consideration of Minimum Support Price (MSP).

» Effective procurement by arranging procurement centers close to producers.

» Skilling of pulse growers on modern production practices with help from Krishi Vigyan Kendras.

» Efficient crop insurance mechanism focused on pulse growers.

» Expansion of area under pulses by utilizing fallow lands and reclaimed wastelands for pulses production.

» Forming Farmer-Producer Organizations (FPOs) for value addition through processing of pulses and shortening of the value chain.

» Customization and development of farm equipment, including app-based hiring.

» Setting up of storage and warehousing in rural areas.

» Foresight for international trade, with tools to predict market demand/supply.

> » Integrating pulses into the public distribution system (PDS) to ensure minimum consumption by poor households even during scarcity.

6.5.1. Expanding *rabi* pulses in targeted rice-fallow areas (TRFA) of eastern India

Vast areas of rice-fallows (about 10 million ha) available in eastern India (Jharkhand, Bihar, Chhattisgarh, Odisha and West Bengal) offer opportunities for expanding area under *rabi* pulses. Under area expansion of *rabi* pulses in rice fallows, TRFA program has been initiated since 2016-17 in six Eastern States of Assam, Bihar, Chhattisgarh, Jharkhand, Odisha and West Bengal. Against a target of 4.5 million ha rice fallow under pulses, an area of > 2.6 million ha has so far been achieved until 2017-18. Most of the rice fallows were covered under pulses viz. pea, lentil, black gram, green gram, chickpea, pigeon pea and lathyrus.

The earlier experiments clearly demonstrated that chickpea, lentil and field pea are suitable pulse crops for rice-fallows, provided suitable varieties and technologies for crop establishment in rainfed rice-fallows are available. The most important traits required in the varieties for rice-fallows include early growth vigor, early to extra-early maturity, and tolerance to reproductive stage heat stress. For example, an early maturing and heat tolerant chickpea variety JG 14 (ICCV 92944) released for late sown conditions of Madhya is already becoming popular in Eastern India. This variety and other heat tolerant varieties can be promoted along with suitable sowing equipment and technologies for ensuring better crop establishment and plant stand. Similarly, early maturing varieties of lentil like Pant L 6, HUL 57, DPL 62, Moitree *etc.* are getting popularity among farmers of eastern Uttar Pradesh and Bihar as these varieties are suitable for late sown and rice fallow conditions. Recently, short duration field pea varieties viz., Vikas and Prakash are being accepted by the framers of North East Hill states of India as farmers are getting more price when they sell immature pods in market.

Recommendations for improving pulses productivity under rice fallow include:

6.5.1.1. Mechanization of field operations:
Residual soil moisture in surface layer at the time of planting *rabi* crops is the major constraint in rice fallows. Relay cropping in standing rice is often practiced but with use of combine for rice harvesting, the option is now shifting for direct seeding using zero-till drill or turbo type Happy Seed drill which need to be designed for different situations. For harvesting and threshing, appropriate machines need to be designed and developed.

6.5.1.2. Scaling-up crop management practices:
Tillage and plant population management, application of nutrients and weed management in *rabi* crops pose serious challenges in rice fallows. Early-maturing crop varieties, relay cropping, higher seed rate, seed priming, seed inoculation with Rhizobium culture, seed pelleting, mulching, foliar spray of nutrients *etc.* are recommended practices which need to be further refined and standardized for different ecosystems. Work on development of short-duration, high-yielding varieties, appropriate seeding techniques, water harvesting and recycling, post-emergence herbicides, biotic and abiotic stresses *etc.* need to be strengthened.

6.5.1.3. Creation of community water reservoirs:
Despite heavy rains during kharif season, soil moisture becomes the most critical limiting factor for raising second crop during winter as most of the runoff is wasted. It is, therefore, necessary to create farm pond and community water reservoirs in the area well supported by Government. This will serve as important source for life-saving and supplemental irrigation. Further, the loss of soil and plant nutrients from productive lands will be reduced.

6.5.1.4. Quality seeds:
Timely availability of quality seeds is often a major constraint for delayed planting and poor yields. Hence, community-based seed production programs need to be launched with appropriate processing and storage facilities. The national and state seed Corporations should strengthen their activities in these areas.

6.5.1.5. Ensuring timely availability of other critical inputs: Traditionally, the winter crops on residual soil moisture are grown using local varieties without application of plant nutrients, bio-fertilizers, pesticides and other agro-chemicals due to their non-availability. Since crop productivity is the driver for area expansion, which in turn is influenced by better crop management, emphasis needs to be placed on timely availability of all critical inputs.

6.5.1.6. Seed priming and optimum seed rate: Overnight soaking of seeds (seed priming), hastens seed germination and crop establishment under relay cropping. Adoption of 20-25% higher seed rate over the recommended rate is recommended ensures desired plant stand.

6.5.1.7. Foliar nutrition: Since application of fertilizers under relay cropping is not feasible, seed pelleting and foliar application of nutrients should be practiced. Foliar application of 2% urea at flowering and pod formation significantly improves yields of chickpea under rainfed conditions by increasing leaf N content and making them photo synthetically more active. Seed pelleting with micronutrients like Zn and Mo is also recommended as a part of nutrient management strategy in rice fallows.

6.5.1.8. Planting strategy: In rice fallows, planting is generally delayed. Under relay planting, seeds should be broadcasted 2-5 days before harvest of rice. Zero-till seed-cum-fertilizer drill should be used wherever feasible when planting is done after harvest of rice. It is necessary to use short to medium maturing varieties of rice for timely planting of *rabi* crops

6.5.1.9 Plant protection: Since post-emergence herbicides are not commercially available specially for crops like chickpea and lentil and inter-cultivation is difficult due to hard soil, hand pulling of weeds is the only option which should be done at an early stage. Post-emergence herbicide (Imazethapyr at 50 g/ha) has been found quite effective against seasonal grassy weeds in crops like groundnut, black gram and green gram. It should be applied at 3-4 leaf stage. Similarly, quizalofop can be used to check ratooning of rice stubbles which cause substantial moisture loss. Insect-pests and diseases should be promptly controlled. Seed dressing with fungicides like carbendazim should be done.

6.5.2. Promoting early-maturing, drought and heat tolerant and disease resistant varieties for central and southern India

Drought and heat stresses during the reproductive phase and with increasing severity towards the end of the crop season are the major abiotic stresses of chickpea and other *rabi* pulses as these crops are generally grown rainfed (68%) on residual soil moisture and experiences progressively receding soil moisture conditions and increasing atmospheric temperatures towards end of the crop season. Early maturity is an important trait for escaping these terminal stresses. In addition, we need cultivars with enhanced tolerances to these stresses. For example, some of the promising varieties of chickpea possessing these traits include JG 11, JG 130, JAKI 9218, KAK 2 and Vihar. The adoption of such varieties needs to be enhanced in central and southern India. Farmers of central India usually have preference for large seeded lentil varieties. Considering the demand, several early maturing varieties having large seeds were released for cultivation. Out of these varieties, DPL 62, JL 3, IPL 316 and IPL 526 are getting popularity in Bundelkhand tracts of Uttar Pradesh and Madhya Pradesh. Government of Karnataka has also taken initiative to promote early maturing lentil varieties in state. Similarly, early maturing field pea varieties like Adarsh, DDR 23, Ambika, Vikas, Indra, Shikha and Prakash are very popular and have helped in enhancing field pea productivity (1100 kg ha⁻¹) in India. During 2013, an early maturing green seeded variety 'IPFD 10-12' has been released for cultivation in central India. This variety has potential to replace some of the area of vegetable type pea. In southern India, field pea is a less known crop but have vast potential during *kharif* season. For example, many farmers in adjoining areas of Dharwad in Karnataka have already started cultivation of pea during kharif season. These varieties offer ample scope in central and southern India.

6.5.3. Extra-short and short-duration pigeon pea in high elevation and rice-fallow cropping system

Early duration pigeon pea varieties have a potential to grow in new niches considering its photo and thermo insensitivity. It can be grown in diverse range of latitudes (30° N) and altitudes (1250 MSL) like in Uttarakhand, Rajasthan, Odisha and Punjab. For instance, ICPL 88039, a short duration (140-150 days) pigeon pea variety, can enhance pigeon pea production in the states of Rajasthan Uttarakhand and Odisha. This variety provides an opportunity to increase crop intensity by growing a post-rainy season crop after harvesting pigeon pea. Since its cultivation does not require any additional inputs and the grains have good market value, its adoption by the farmers will be quick. Similarly, super-early lines (ICPL 11300, ICPL 11285 and ICPL 20325) maturing in 90 days have potential to be adaptive and productive in the rice-fallow cropping system and rainfed hilly areas of India.

6.5.4. Expanding pigeon pea hybrid production

During the past 5 decades, pigeon pea productivity in India has remained almost stagnant around 700 kg ha^{-1}. In this context, the hybrids can produce more biomass (more than 50%) and productivity (more than 30-40%) than improved varieties. To break the yield barrier in pigeon pea, ICRISAT and its partners (Department of Agriculture, State Agriculture Universities, public and private organizations, and farmer organizations) have developed several CMS-based medium maturity pigeon pea hybrids (ICPH 2671, ICPH 2740, and ICPH 3762) which enhanced productivity (30-40%) and adaptability in Maharashtra, Andhra Pradesh, Odisha, Telangana State, Madhya Pradesh, Karnataka and Jharkhand. ICRISAT and partner institutions can expand commercial hybrids substantially.

6.5.5. Creation of infrastructure under Enhancing Breeder Seed Production (EBSP)

Breeder seed infrastructure strengthening comprises of several need-based items at different centers such as works (threshing floors), Seed Processing Plants (SPP), farm implements (tractors, sprinkler systems, power sprayers, rotavator, BBF planter, ridge planter, storage godown, irrigation channel, fencing, bore-well, power thresher, seed-cum fertilizer-drill, pick-up van, irrigation pipes, hydraulic trolley, gravity separator, weighing machines, combine harvester).

6.5.6. Seed village program

To address critical input, the seed village program has been operationalized to improve the quality and stock of farm saved seeds of pulses enhancing production/productivity. To upgrade the quality and varieties of farm-saved seeds which is about 80-85% of the total seed used, of cluster of 50 farmers @ 1 acre is provided with 60% financial assistance towards foundation/certified seed of pulses for production of certified/quality seeds. The farmers are also imparted training on seed production technology.

During 2017-18, the seed production program was conducted in > 9 lakh ha area by distribution of 7 lakh qtls. of seed. In all, about 28 lakh farmers were benefitted in > 1 lakh villages of the country.

6.5.7. Creation of seed-hub

To enhance the quality and quantity of pulses seed in the country, a project on creation of seed-hubs (2016-17 to 2018-19) has been initiated under (NFSM) with the mandated objectives and targeted seed production of latest varieties (150 locations in 24 states, at ICAR Institutes 07, ICAR-AICRPs centers - 46, KVKs-97) across 24 states in the country. Each seed hub has to produce 100 tons of seed per year.

6.5.8. Distribution of pulses seed mini-kits

For introduction and popularization of latest released/ pre-released HYVs of pulse crops within 10 years of release, seed mini-kits distribution program has been initiated since 2016-17, including those belonging to below poverty line, to encourage farmers for seed multiplication of various crops at grass root level.

A total of 7.85 lakh mini-kits of different sizes of pulse crop have been distributed during *kharif, rabi* and spring/summer of 2020. Agencies involved in supply of seed mini-kits are NSC, NAFED, IFFDC, HIL, and KRIBHCO. Looking to the success of this intervention, >15% increase of seed mini-kits >09 lakh nos.(8,31,484 nos. have been distributed against the allocation of 9,15,900 numbers) are to be distributed during 2017-18.

6.5.9. Area expansion

In order to bring additional production from additional area coverage in rice fallows, crop-wise *Rabi* pulses strategies were adopted viz., Gram - Chhattisgarh, West Bengal, Bihar, Jharkhand, Odisha, Assam, Andhra Pradesh, Tamil Nadu; Lentil - Chhattisgarh, West Bengal, Bihar, Jharkhand, Assam; and green/black gram in rice fallow in coastal region. Intercropping gram with barley, mustard and linseed in Rajasthan, UP, Bihar, Vidarbha (Maharashtra); intercropping gram/lentil with autumn planted ratoon sugarcane in UP, Maharashtra, and Bihar.

In order to bring additional production from additional area coverage (diversion to other crops like cotton, oilseeds, coarse cereals), cultivation of *kharif* pulses as intercrop, planting of red gram on rice bunds, and cultivation of minor pulses in niche areas was undertaken.

Spring/summer pulses in Punjab, Haryana, Madhya Pradesh, Uttar Pradesh, Bihar, West Bengal, Gujarat, Jharkhand, TN and AP were chosen.

6.5.10. Promotion of pulses in intercropping systems and non-traditional areas

Pulse crops are grown as intercrops in many parts of the country during all three crop seasons (*rabi, kharif* and spring/summer) and forms integral part of rainfed agriculture. Vast potential exists for promotion of pulse crops in intercropping system. For example, chickpea/ lentil + autumn planted sugarcane in western Uttar Pradesh, Terai region of Uttar Pradesh, Maharashtra and Karnataka; mung bean + long/medium duration pigeon pea in Uttar Pradesh, Bihar, Jharkhand, Madhya Pradesh, Maharashtra and Gujarat; chickpea/lentil + mustard in Rajasthan, southern Madhya Pradesh and Uttar Pradesh; pigeon pea + soybean; and pigeon pea + sorghum etc. offers not only scope to enhanced pulses production but also to ensure sustainable agricultural production base.

6.5.11. Integrated crop management technologies

These includes standardization for basal dose of fertilizers/nutrients, seed rate, seed priming to ensure better germination, planting methods (raised bed, ridge and furrow, broad bed and furrow, dry sowing followed by light irrigation) for better crop establishment, pre-emergence application of weedicides, irrigation scheduling and micro-irrigation *etc.* Application of sulfur @ 20-25 kg/ha can enhance yield by 15-20 % irrespective of pulse crops. Similarly, soil application of 1 kg Ammonium Molybdate has helped in achieving higher productivity of chickpea in soybean belt of Madhya Pradesh (Gupta and Gangwar, 2012). Supplemental irrigation with a limited amount of water, if applied to rainfed crops during critical stages can result in substantial improvement in yield and water productivity.

6.5.12. Integrated nutrient management

The low productivity of pulses is mainly due to their cultivation on poor fertile soils with inadequate and imbalanced nutrient application without the application of organic manures and micronutrients like boron, zinc and molybdenum. The use of phosphate solubilizing bacteria (strains from the genera of *Pseudomonas*, *Bacillus* and *Rhizobium* are among the most powerful P solubilizers) as inoculants simultaneously increases P uptake by the plant and thus crop yields (Khan *et al.*, 2009). INM involves optimum use of indigenous nutrient components i.e. crop residues, organic manure, biological N fixation as well as chemical fertilizer and their balancing interactions rises N and P recovery. Pulses are generally fix atmospheric nitrogen so treat seed before sowing for enhancing N-fixation.

6.5.13. Productivity enhancement

6.5.13.1. Promotion of Sulphur and Zinc: Wide spread deficiency of sulphur and zinc noticed in pulse growing regions constraints the productivity of pulses. In major pulse growing areas, 44 districts have shown 40- 60% sulphur deficiency and 82 districts with 50-60% zinc deficiency. In view of encouraging response to application of S and Zn with cost benefit ratio of 10-21%, their application was vigorously pursued after 2014-15.

6.5.13.2. Promotion of biofertilizers [Rhizobium and Phosphate Solubilizing Bacteria (PSB)]: About 40% pulse growing regions have low to medium population of native Rhizobium. Seed inoculation with bio-fertilizer (Rhizobium and PSB); low cost inputs; are known to increase pulse productivity by 10- 12%. Rhizobium + PSB provided under NFSM.

By the utilization of biofertilizers like Rhizobium strains, phosphate soluble bacteria, Cyanobacteria, VAM and many other useful microorganism fertilizers can help in uptake of nutrients, formation of nodules which can help in the mineralization process of converting unavailable form of nutrients into the available form of nutrients for easy uptake. Mostly biofertilizers plays a symbiotic relationship with host plant to increase in their growth and production.

i.   **Symbiotic nitrogen fixing bacteria:** It includes the species of Rhizobium which have a symbiotic relationship whit host which helps in fixation of nitrogen. It includes:

   » *Rhizobium leguminosarum:* It is a species of Rhizobium which helps in the fixation of nitrogen in Pea groups (includes all types of pea, lentil, and some beans)

   » *Rhizobium japonicum:* This type of Rhizobium is observed in the soybean crop which helps in the nitrogen fixation.

   » *Rhizobium phaseoli:* This species of Rhizobium helps in the fixation of nitrogen in the crops like kidney beans and garden beans.

   » *Rhizobium meliloti:* The species of Rhizobium is observed in the alfalfa groups of pulses (mainly in Lucerne).

ii. **Non-symbiotic nitrogen fixing bacteria**

   » **Azatobacter:** These are aerobic microbes which helps in fixation of Nitrogen works well in neutral soils and sensitive to inadequacy of phosphate.

   » **Clostridium:** These are anaerobic bacteria helps in fixation of nitrogen but fixes less as compared to Azatobacter.

   » **Mycorrhizae:** It is a symbiotic association of fungi and roots of higher plants. It is of two types:

     • **Endomycorrhizae:** These groups are called arbuscular mycorrhizal fungi [AMF] which improves uptake of phosphorus, imparts resistance against drought and certain root infecting fungi.

- **Ectomycorrhizae**: This group grows on surface layers of roots.
- **Frankia:** It is an association between actinomycetes and plants. Actinomycetes are interim between bacteria and fungi. These are responsible for the smell of earth when it rains which is caused by the breakdown of compounds like gaosmin.

iii. **Phosphorus solubilizing bacteria (PSB):** Most of the phosphorus sources are gets fixed in soil and becomes unavailable to plants so availability and absorption of phosphorus is induced by the utilization of phosphorus solubilizing microbes such as *Aspergillus*, *Pseudomonas*, *Bacillus*, and mycorrhizal fungi.

6.5.14. Integrated pest and disease management

In absence of access to resistant/tolerant varieties, quality bio-pesticides (*Trichoderma* spp., *Pseudomonas fluorescens* and *Bacillus subtilis*) and fungicides (carbadenzim or thiram) for seed treatment, seed as well as soil-borne diseases dominated by Fusarium and collar rot alone causes 10% seedling mortality in pigeon pea, chickpea and lentil (Sharma *et al.*, 2015). Treat seeds with bio-fertilizers and organic pesticides or fungicides. Soil application of *Trichoderma harzianum* @10 g/kg FYM for controlling the wilt or seed treatment @10g/kg seeds in pigeon pea for wilt in pigeon pea needs promotion. Besides, insecticides use, in situation of chickpea and pigeon pea crop, fixing of 4-5 pheromone traps per ha and erection of 20-30 bird poles per ha and neem oil spraying decreases larval population of *Helicoverpa*. Foliar spray of Dimethoate @ 0.03% for managing the aphids in lentil; foliar spray of Imidacloprid 17.8 SL @ 2.5 ml/10 liter of water or Thiamethoxam 25WG @ 2-3g/10 liter of water (first spray after 15 DAS and subsequent sprays at 15 days intervals as per need) to control insect pests in green gram and black gram during *kharif*.

6.5.15. Integrated weed management

Integrated weed management (IWM) became an accepted and frequently used term by weed scientist in the early1970's. In IWM system use two or more than two methods in combination for efficient weed management. Non-chemical methods like preventive measures, mechanical and cultural methods including stale-seed bed practice are effective for weed management. Pre-emergence and pre-plant herbicides provide effective control of weeds during initial crop growth at least one month. In some pulses, when the canopy is not adequately developed, one hand weeding or application of post-emergence herbicides is required at 20–30 days of growth. Post-emergence herbicidal control of weeds in pulses is limited. Herbicides like Imazethapyr and Quizalofop-p-ethyl can be used in specific crops for controlling specific weed flora.

6.5.16. Post-harvest technology

Storage of pulse grains is prone to loss due to stored grain pests causing huge economic losses in comparison to split ones (in form of dal). Value addition certainly ensures more money from farm produce to the farmers. IIPR, Kanpur has designed and developed a low capacity dal mill "IIPR Dal Mill", however, other Dal Mills developed by CFTRI and CIAE are also available. Still, there is need to divert sincere efforts for improvement in milling.

6.5.17. Processing and value addition

There is very little value addition for pulses. Pulses are mostly consumed whole or split, apart from desi chickpea which is usually consumed in the form of flour/besan and has growing demand. Most of the processing units are in production regions mainly to minimize the transportation cost for procuring raw materials and use traditional technology. However, the growing health consciousness, preference for quality packaged products and shortage of labor drives the processors to use modern technology. Due to inconsistent policies and lacklustre support, the players in the value chain of pulses are hesitant to come forward to make investment decisions including those related to R and D, marketing and input supplies.

6.5.18. Making seeds and other inputs available to farmers

There has been slow adoption of improved cultivars and production technologies by farmers. The major reasons include unawareness of farmers about improved cultivars and technologies or unavailability of seeds and other required inputs. Concerted efforts on training and other awareness activities for farmers, strengthening formal and informal seed systems and increasing access to other inputs are needed for enhancing adoption of improved cultivars and technologies. Considering the huge requirement of quality seed of improved varieties, and limited interest of private players in the pulses seed sector, there is need to encourage 'Seed Village' concept through involvement of farmers in quality seed production. We will also need to develop good linkages between the formal and informal seed system so that the entire seed system chain can be strengthened. It is also a known fact that many tribal and poor farmers of Jharkhand, Chattisgarh, Assam and North Eastern Hill region usually consume or sell green immature pods of chickpea and field pea, therefore proper training and awareness among farmers need to be created for production of quality seeds, and policy support from government is inevitable.

6.5.19. Technology demonstrations

The frontline demonstrations were conducted in different agro-climatic regions on important pulse crops with a view to demonstrate and assess the benefits of new varieties and technologies under diverse cropping systems have revealed the existing potential of productivity to be exploited through technological interventions.

A package technology like improved cultivar, Rhizobium inoculation, use of Sulfur, INM, application of weedicide, foliar spray of urea, IPM *etc.* were vigorously pursued.

For good crop establishment, seed priming (soaking the seeds overnight in water surface, drying and sowing next day), seed treatment with effective Rhizobium strain, sowing of seed into deeper moist soil (in case of chickpea), lime pelleting for acidic soil and gypsum in saline areas was encouraged under NFSM pulses/CFLDs.

Government focused on key areas like seeds of improved varieties, irrigation tailored to pulses (especially micro-irrigation), bringing new niche areas under pulse cultivation, attractive minimum support price (MSP) and market that allow farmers to increase their profitability aligned to improved farmers welfare.

6.5.20. Technology transfer through CFLDs.

Government has initiated National Level Cluster Frontline Demonstrations on pulses, through Krishi Vigyan Kendras under 11 Agriculture Technology Application Research Institute (ATARIs) to demonstrate the production potential of new varieties and the related technologies; increasing production through area expansion and productivity enhancement in a sustainable manner in the identified districts of the country; restoring soil fertility and productivity at the individual farm level;

and enhancing farm level economy (i.e. farm profits) to restore confidence amongst the farmers.

The CFLD program has been started from 2015-16. The total targets during 2015-16 in 15382 hectares have been increased at 126% during 2018-19.

The transfer of technology through CFLDs have increased yield levels up to 42% and 54% over local check and normal yield, respectively.

6.5.21. Interventions related to marketing - phenomenal increase in MSP of pulses

To address price security, distress sale, access/connectivity to markets, farmers' exploitation and heavy storage loss (20-30%) *etc.*, initiatives to link all APMCs under e-NAM and procurement of pulses under PSS and PSF on MSP by National Agricultural Marketing Federations Ltd. (NAFED) / FCI, SFAC *etc.*, also paid dividends.

6.5.22. Remunerative price to pulse grower

6.5.22.1. Enhanced procurement of pulses (2014-15 to 2017-18): NAFED has done record procurement of pulses jointly with SFAC and FCI during the year 2017-18 under Price Support Scheme (PSS) and Price Stabilization Funds (PSF), funded by Ministry of Agri. & FW. It procured more than 2008.52 thousand tons of pulses (Gram - 188.59 thousand tons, Masoor - 27.07 thousand tons, green gram - 407.74 thousand tons, black gram - 290.62 thousand tons and pigeon pea 1094.49 thousand tons).

NAFED has been involved in creating buffer stocks and stabilizing the prices of pulses. A substantial quantity from buffer stock has been proposed to be supplied to Para-military and Defense forces. Also the supplies will be made to state governments as per their requirements under PDS and other such schemes. Consequently, the procurement agency has positioned itself as the 'Pulse Arm' of the government.

6.6. CROP-WISE STRATEGIES FOR INCREASING PULSES PRODUCTION AND PRODUCTIVITY

6.6.1. Chickpea

» Development of input responsive and non-lodging varieties.
» Abiotic stress tolerance - Drought and heat.
» Biotic stress tolerance - Pod borer (*Helicoverpa armigera*), dry root rot and Fusarium wilt
» Extra-large kabuli varieties for domestic and international market.
» Super-early varieties for green grains - 60-70 days for green pods used as a vegetable or snack.
» Machine harvestable and herbicide tolerant varieties - with 30 to 40% more height with semi-erect to erect growth habit.
» Nutritionally enhanced varieties - protein content of 20 – 25% with higher beta-carotene (precursor of Vitamin A) levels and micronutrient (iron and zinc) contents.

6.6.2. Pigeon pea

» Development of varieties/hybrids tolerant to Fusarium wilt, sterility mosaic disease, pod borer and Phytophthora blight.
» Development of extra-short duration genotypes (< 120 days maturity) to different cropping systems in north western plain zone.

» Development of genotypes (> 180 days maturity) with frost resistance for north eastern plain zone.

» Development of super-early genotypes (90-100 days maturity) for different cropping systems.

» Integrated pest and disease management - avoidance of virus inoculum, vector control (for sterility mosaic disease), management of pod borer.

6.6.3. Lentil

» Germplasm enhancement and pre-breeding - identification of new sources of resistance/ tolerance to diseases, insect pests, nematodes and post-emergence herbicides.

» Early maturing varieties for rice fallow areas – with high biomass; resistance to diseases like Stemphylium blight, rust and wilt; tolerance to low temperature at vegetative stage and high temperature at reproductive stage, and terminal soil moisture stress.

» Nutritionally dense varieties for culinary and export purposes - for enhancing total production of protein, by enhanced yield and increasing seed protein content, bio-fortification to enhance availability of minerals like Fe and Zn.

» Restructuring existing plant type - tall, erect, non-lodging and non-shattering varieties suitable for mechanical harvesting.

» Climate smart varieties - tolerance against different stresses (heat, drought, increased water use efficiency, water-logging, salinity)

6.6.4. Green gram

» Germplasm enhancement and pre-breeding - quality traits, yield contributing traits, and resistant sources against biotic (MYMV and bruchids) and abiotic stresses.

» Climate resilient varieties - photo- and thermo-insensitive varieties

» Biotic stress resistance - to major insect pests (thrips, jassids and pod borer) and diseases (yellow mosaic virus, anthracnose, powdery mildew, Cercospora leaf spot), pyramiding of useful genes to develop multiple stress resistant varieties.

» Short duration varieties for crop diversification - incorporation of seed dormancy or pre-harvest sprouting, reduce maturity duration at least by another 8-10 days to fit for wheat-rice dominated cropping system.

» Restructuring plant type - future varieties with determinate growth habit, photo- and thermo-insensitivity, early maturity, high harvest index and high yield (>2.0 tons ha^{-1}) and resistant to lodging.

6.6.5. Black gram

» Germplasm enhancement and pre-breeding - requirement for diverse germplasm lines for development of biotic and abiotic stresses resistant/tolerant varieties (Chaturvedi, 2009).

» Short-duration varieties for sustainability of cereal-based cropping system.

» Varieties with multiple diseases resistance.

» Development of efficient plant type - efficient plant type would be photo-thermo-insensitive, with determinate growth, early maturing, high yielding (>2000 kg ha^{-1}) with high harvest index.

» Developing herbicide tolerant varieties - high seedling vigor and post-emergence herbicide tolerance.

6.6.6. Field pea (Dry pea)

» Restructuring plant types for lodging resistance and mechanical harvest - plant type with strong and thick stem can keep plant standing and erect till maturity.

» Dual purpose, short duration varieties – animal feed and forage varieties possessing high biomass and early maturity, development of dwarf, non-lodging and disease resistant varieties.

6.7. WAY FORWARD

It is an established fact that a human body requires a daily intake of about 55 to 60 g of protein. The protein malnutrition and other health indicators like maternal and infants mortality rates, anaemia *etc.*, is a matter of concern for the government. Nutritious food has direct bearing on health and affects work performance of the people. Out of the 22 amino acids required in the human diet, the body supplies 14. The remaining eight have to come from food. If all the eight amino acids are present in a single food item, it is called a complete protein food. Since all proteins from animal sources are complete proteins, it is easy to meet the dietary protein requirements of people with non-vegetarian food as the main diet. However, for vegetarian population the main sources of protein are leguminous plants to which the pulses belong. However, in general, pulses have lower concentrations of protein than animal sources. Besides, none of the pulses, except soybeans, are complete proteins. Therefore, combinations of two or more pulses are needed in a vegetarian diet. Dairy products, which are complete proteins, may also be used to supplement pulse proteins in vegetarian diets.

Keeping in view the production issues and recent pulse scenario in India, it is now established that the domestic supply of pulses is able to meet the growing demand of our consumers. Availability, price and the dietary preference for specific types of pulses in different parts of the country is largely responsible for this. About 29 million hectares of land is under pulses producing about 24 million tons annually. Our population is increasing as well as demand will also increase which we have to resort to increase the production. It is a challenge for us to ensure supply of pulses as pulse crops are primarily taken up for cultivation under rainfed condition in poor soils and are prone to production losses due to moisture stress.

Two major issues have emerged in respect to pulses production in the country. First, the limited genetic potential for high yields and second their vulnerability to pests and diseases. Compared to other food grain crops, yield potential of the pulses has been rather low. Newer varieties of pulses need to be developed so that the crop cycle fits well into cropping systems that the farmers adopt. Another important issue is limited mechanization potential, especially for planting and harvesting of the crop. Suitable plant types need to be developed for mechanical harvesting with pods above the canopy and sturdier plants.

Due to the high protein content in pulses, the crop is highly vulnerable to pests and diseases. It is estimated that about 30% of pulses crops are lost on account of pest attacks and diseases every year. Attack by pod borer and pod fly is so severe that the entire standing crop is devastated at several times. Research efforts need to be prioritized in this direction by employing modern biotechnological tools for developing pest resistant varieties of pulses.

Weed is one of the problematic areas for limiting yield both in *kharif* and *rabi* pulses. Manual weeding appears to be difficult in future to manage this weed menace, therefore development of herbicide tolerant pulses as well as varieties suitable for mechanical harvesting is the necessity of the time to minimize the labor-intensive crops.

Another problem peculiar to the Indo-Gangetic plains is the menace of large-scale grazing by blue bulls. We need to support the efforts of the farmers for higher acreage under pulses crops without contravening the legal provisions of the Wild Life Protection Act which prohibits killing of these animals. There is thus a huge possibility and potential of bringing innovative solutions to save the pulses crops and encourage more intensive promotion of production technologies.

Besides varietal research, we need to address the issues relating to farmers' preference for the competing crops to pulses through development and promotion of crop production and crop protection technologies. Under National Food Security Mission, BGREI and CDP, these aspects have been taken up for more aggressive promotion of available technologies through cluster demonstration to ensure that the farmers are able to harvest better crops. The Government significantly increased the Minimum Support Price of the pulses and strengthened pulses procurement mechanism by designating additional central agencies to support the farmers. In fact, the minimum support price has been doubled in last three year with quantum jump given this year with an increase of more than 150% of cost of production during *Kharif* 2018. Use of drip irrigation in pigeon pea and agronomic practices like transplantation and nipping of branches are showing very encouraging results. Our import and export policies are linked to our ability for better crop forecasts and the principle of balancing the farmer's interest in a manner that the prices are not distorted and the Indian farmer continues to get a good return for their produce. This will hopefully help in better planning and management of supply chain. With better yields through development of pest and disease resistant varieties, increased MSP support and lessons learnt from exchange of knowledge, the Indian farmer will definitely adopt pulse based cropping systems to produce more pulses with increased acreage. This will significantly ease the supply side constraints in the years to come.

To meet the challenges faced by the pulses sector, government has given emphasis on research efforts for developing biotic stress resistant and stress tolerant varieties, to be encouraged along with public-private initiatives for better logistics planning and handling of pulses. The crop cycle should be such as to fit into the overall cropping system that the farmer takes during the year.

Aggressive promotion of available technologies under the cluster front line demonstration has been taken up. The pulse procurement mechanism has been strengthened by designating additional central agencies and the minimum support price (MSP) has been significantly increased to more than 150% of the production cost. Thus, by balancing the farmer's interest in such a way that the domestic prices are not distorted and the Indian farmer continues to get a good return for his produce, the use of new production technologies and agronomic practices, and government support put together could maintain to self-sufficiency.

Given the important role that pulses play in the human diet, their availability needs to be increased indigenously. The possibility of improving pulse productivity two to three times through existing varieties and available package of technologies is well demonstrated in scientific experiments.

In order to sustain the growth of pulses at various levels i.e. among the states, districts, within districts and to bridge the yield gap between FLDs and farmers' practice, DAC&FW has envisioned a road map which include:

6.7.1. Area expansion

Pulses have tremendous scope for area expansion. As per estimation, about 2.5 million ha additional area can be brought under different pulses through cropping system improvisation like green gram and black gram as catch crops in summer/spring under cereal-based cropping systems, intercropping short-duration pulses (green gram, black gram, cowpea) in sugarcane, millets, cotton, *etc.* advocating new cropping systems such as pigeon pea–wheat in the north, rice– lentil in the east and black gram –rice in the southern peninsula.

6.7.1.1. Horizontal expansion

> » Horizontal expansion through bringing additional area under pulses, and diversification of rice-wheat system in Indo-Gangetic plains (IGP) through popularization of short duration varieties of pigeon pea, kabuli chickpea, field pea and summer green gram.

> » Bringing additional area under pulses through promoting black gram/ green gram cultivation in rice fallow in peninsular India and chickpea lentil in NEPZ and Chhattisgarh.

> » Promotion of pulses in intercropping viz., short duration thermo-insensitive varieties of green gram/ black gram with spring sugarcane.

> » Pre-*rabi* chickpea with mustard/linseed; pigeon pea with groundnut/ soybean/millets, etc.

> » Development and popularization of black gram/ green gram for late planting (mid Aug-early Sept in north India).

6.7.1.2. Vertical expansion

> » Vertical expansion through increasing productivity and bridging the yield gaps.

> » Development of high yielding short duration varieties having multiple and multiracial resistance to diseases.

> » Development of new and efficient plant types; development of input use efficient genotypes.

> » Exploitation of hybrid vigor in pigeon pea.

> » Popularization of improved crop management practices and bridging yield gaps.

6.7.2. Development of resilient pulse crops to climate adversaries

> » Development of resilient pulse crop varieties to mitigate the impacts of climate change;

> » Critical monitoring of diseases and pest dynamics with reference to climate change;

> » Production and supply of quality seeds through seed hub and ensuring seed production accountability to SAUs/ ICAR to organize location specific recommended latest varieties of pulses in their jurisdiction;

> » Active involvement of private sector, NGOs, and farmers' help groups in production of quality seeds;

> » Mandatory target to public sector seed corporations; popularization of seed village concept with buyback system;

> » More incentive on production of seeds of new varieties; promotion of farmer to farmer exchange of seeds and seed village program.

> » Reducing post-harvest losses, refinement and popularization of harvesters, threshers and graders;

> » Development of stored grain pest resistant varieties;

> » Popularisation of storage bins and mini dal mills;

> » Strengthening of FPOs and establishment of processing units;

> » Development and popularization of low-cost safe storage structures.

6.7.3. Advance seed plan

» The DAC&FW has made advance planning for quality seed production of pulses for three years (2018-19 to 2020-21) in order to ensure sustainability in growth of pulses production in the country.

6.7.4. Ensuring timely availability of critical inputs and advisory

» Promotion of IPM technologies against *Helicoverpa*.
» Ensuring timely availability of quality bio-pesticides- HaNPV, Trichoderma & herbicides e.g. Pendimethalin *etc*.
» Seed dressing of fungicides for controlling seed borne diseases.
» Providing safe storage structures like Pusa Bin and warehouse facility.
» Creation and maintain/ sustain production units of quality biofertilizers and bio-pesticides.
» Fortification of fertilizers with specific nutrients like S, Fe, Zn, B *etc*., in specific regions.
» Popularization of sprinklers and micro irrigation techniques in rainfed areas.
» Establishment of single window input supply centres for cluster of villages.
» Advanced forewarning and forecasting systems for pest and disease outbreaks.

6.7.5. Improving yield stability

» Development of transgenics against drought and gram pod borer.
» Efficient water management in rainfed areas.
» Rainwater harvesting and recycling through farm ponds and community reservoirs.
» Promoting short duration varieties in drought prone areas.
» Promoting micro irrigation system.
» Adoption of moisture conservation practices.
» Development of location specific suitable varieties and ensure availability of quality seeds.

6.7.6. Product diversification

» Research on creating novel innovative products using pulses as ingredients need to be emphasized. In countries like Canada, research is already at advanced levels to explore opportunities in blending pulses protein with other flour such as wheat flour. But in India, the private and the public sector R and D are yet to fully explore these opportunities and others such as fortification and/or blending pulses with other ingredients that may accelerate consumer demand.

» Pulses have captured the attention of the United Nations and General Assembly of the UN has declared 2016 as the 'International Year of Pulses'. The Food and Agriculture Organization of the United Nations (FAO) has been nominated to facilitate the implementation of the Year in collaboration with Governments, relevant organizations, non-governmental organizations and all other relevant stakeholders. The Year provides an unique opportunity to encourage connections throughout the food chain that would better utilize pulse-based proteins, further global production of pulses, better utilize crop rotations and address the challenges in the trade of pulses.

» The need of the hour is to forge a holistic understanding of the issues affecting the pulses value chain, major reforms in agri-food policies, increased need for more R and D on the input side and food processing innovations, increase awareness as well as interest of consumers, policy makers, food industry and NGOs in pulses and their health, nutrition and environmentally sustainable benefits.

6.7.7. Efficient transfer of technology

» Organizing farmers training and exposure visits; popularization of improved technology through mass media.

» Close interaction of research organizations, state departments of agriculture and private agencies.

» Market led initiatives for organized village level seed production to exploit the high demand for improved varieties of pulse crops as well as branding of local germplasm e.g. Baigani pigeon pea in tribal belts of MP.

» Exploiting the commercial pulse processing units at village level.

» Promotion of pulse production as cash crops in unconventional areas like hills, coastal and tribal belts of the country.

» Promotion of farmer information and communication technologies-based pulse knowledge management to increase production and productivity of smallholder farmers.

6.7.8. Renewed emphasis on research and broad basing extension

High priority needs to be accorded to research in pulses, with emphasis on varietal development to suit the local specific conditions, collection and characterization of pulse germplasm to identify climate resilient ones, development of short duration variety to facilitate intercropping and mixed cropping without affecting the yield of cereals and other crops. The public sector extension services though having widespread network has not been able to reach the smallholders whereas the private sector is yet to rise up to the mark in case of pulses which are less remunerative. The awareness about correct improved agronomic practices, such as timely sowing, regular mechanical weeding, timely harvesting, post-harvest handling could go a long way in bridging yield gap and also enhancing income of smallholders. Thus, there is a need to involve public institutions (ICAR, SAUs), NGOs, seed companies, farmers' associations and private entrepreneurs in quality seed production, transfer of technology, processing and value addition and supply of critical inputs.

6.7.9. Efficient marketing system

One crucial missing link that can incentivise smallholders and address the price fluctuation in the pulses is the efficient marketing system. In order to facilitate better price discovery and transparency, facilities have been created for the electronic trading of pulses. A few APMCs in Karnataka, Andhra Pradesh and Telangana have been modernised and facilitated trading of pulses. In the backdrop of the move towards National Common Market for agricultural commodities, a lot of such initiatives have to be taken to build an efficient marketing system for pulses.

6.7.10. Announcement of MSP well in advance

» Assured procurement and creation of procurement centers in production zones.

» Development of organized markets for pulses.

» Linking farmers with FPOs, aggregations and e-NAM (markets).

» Promotion of export of pulses like lentil and kabuli chickpea and arid legumes.

  » Production of value-added products and use of by-products.
  » Branding of produce and promotion of organic pulse production.

6.7.11. Value chain approach

Value chain approach starting from the production at farm level and encompassing post-harvest processing, packaging, transportation and marketing has the potential to link the smallholders with the market and improve their incomes. Though are instances of successful value chains in other commercial crops they are yet to be developed in case of pulses. At present, post-harvest processing of pulses is mainly handled by the private sector. Installation of efficient, small dal mills /processing units in villages will reduce the cost of processing and ensure their ready availability at cheaper rates. Distribution of seed-storage bins to the farmers and mass awareness campaign for adoption of scientific methods of storage of pulses at the village level are likely to reduce losses from stored grain pests in pulses.

6.8. CONCLUSION

It is evident that the agricultural policy in India till now have not paid much attention to the long run sustainability of pulse production in India. Till very recently, agricultural strategy albeit with an intention of ensuring grain sufficiency continued to favor only rice and wheat. This only led to burgeoning stocks and increasing subsidy bills. Simultaneously such a biased policy pushed out other crops like pulses, thereby having a strong impact on soil quality, groundwater usage and the domestic insufficiency. The Government has to move beyond increasing MSP. Pulses need to receive its due share of attention. This will require measures for ensuring adequate procurement, creating a buffer stock and give attention to research and technological innovations. With a better seed – fertilizer – irrigation support higher yields will be achieved and greater area devoted to pulses. There is also a need to be region specific in the long-term policy on pulses. If green revolution made India self-sufficient in food grains and the white revolution made the country the largest producer of milk, it would not be long before suitable policies can make the nutrition deficiency in India get addressed by the rainbow revolution, where pulses play an important part.

Global supply of pulses is inadequate, as India occurs to be the largest producer and consumer of pulses. Hence, India wants to produce the required quantity, but also remain competitive to protect indigenous pulses production. There is need to improve the extension activities so that the farmers can be made aware of the nutritional deficiencies in the soil hindering the production as well as to disseminate the enhanced production technologies. The much-talked schemes "*Pradhan Mantri Krishi Sinchai Yojana*" and "*Pradhan Mantri Fasal Bhima Yojana*" of the Government of India have capacity to bring sea changes in pulses cultivation.

The National Food Security Mission (NFSM), launched in 2007, outlines policy packages involving field demonstrations of best farming practices, incentives for adoption of modern technologies, and resource conservation and management practices. In the past six years, the government has continued to increase the MSP of *kharif* pulses by over 45% and that of *rabi* pulses by 50–60%. However, it will take much more than these incentives for India to achieve self-sufficiency in pulses. Drawing on the above recommendations can prove to be invaluable for policy makers while formulating strategic plans to increase and improve pulse production in India.

Chapter - 7

Silver Fiber
(Cotton Production) Revolution

7.1. COTTON INDUSTRY

With a market value of $4.5 billion, cotton is the India's fifth largest crop (Monsanto). Cotton is the most important commercial crop of India contributing up to 75% of total raw material needs of textile industry and provides employment to about 60 million people. India has the largest area under cotton cultivation with relatively low productivity. India ranks first in world area wise, whereas, it ranks second in production next to China. Cotton is grown in developed and developing countries but differences in input intensity and degree of mechanization are huge. The main GE cotton growing countries in terms of area planted are India, China and the USA (Table 7.1).

Table 7.1. GE cotton area of major producers (2011)

Country	GE cotton area (million ha)	Total cotton production (million bales of 480 lbs.)
India	10.6 (88% area of 12.1 million ha)	27.0
United States	4.0	18.0
China	3.9	33.0
Pakistan	2.6	10.3
Australia	(95% area)	4.2
World	25.0	---

The extent of achievement can be drawn from the fact that about 11.7 million hectares (93.6 percent) of total 12.5 million hectares under cotton cultivation in 2019-20 was planted with Bt cotton seeds.

Over the last decade, cotton yields in the country increased by more than 300%; pesticide consumption reduced by ~ 50%; acreage increased by 150% and production increased by 400%. This technological advancement enabled over 7 million farmers earn an additional farm income of US$ 16.69 billion. While yield increase was not a claim for the Bt technology, excellent control of bollworms led to reduced stress on the crops that resulted in higher yields. Before Bt cotton was used, the total income per hectares was ₹7,558 in the rain-fed area. However, it went up to ₹16,000 in rainfed areas and even up to ₹25,000 in irrigated areas within a decade.

A common feature for cotton production around the globe is the high level of chemical pesticides that are used to protect this crop from damage inflicted by pests and diseases. In industrialized and developing countries alike, a proportionally much greater share of pesticides is used in cotton than in most other crops and negative externalities, for example impact on the environment and human health, are widely reported. Moreover, this high level of pesticides applied mainly to control caterpillar pests of the Lepidoptera family, results in high production costs and hence reduces the net revenue for farmers. Unable to face the consequences of crop failures and mounting debts, thousands of cotton farmers across India ended their lives in the last five years. For this reason, cotton was among the first crops for which research on genetically engineered insect resistance resulted in commercially approved varieties.

The cotton production reached an all-time high of 27 million bales and surpassed the production of US to rank second in the world. One of the reasons attributed for this growth is introduction of Bt cotton in India. Currently Bt cotton occupies an area of 88% of total cotton area in India. Bt cotton was commercialized in India in the year 2002.

7.2. YIELD GAP

The average productivity of cotton in our country is still lowest among the major cotton growing countries of the world. As against world average of 616 kg lint/ha, India's productivity was only 278 kg lint /ha during 2000-01.

But from the year 2005-06 the productivity increased substantially having as high as 563 kg lint per ha during 2007-08 (Table 7.2). Therefore, the gap between world and India in productivity is reducing. The gap which was 338 kg lint per ha during 2000-01 reduced to 212 kg lint per ha during 2007-08 (Table 7.3). The average yield of Gujarat, Tamil Nadu, Andhra Pradesh and Punjab are at par with world average, ranging from 650 to 743 kg lint per ha (Table 7.4) as per CAB estimates.

Table 7.2. Yield gap of Indian cotton

Year	Yield (kg lint per ha)		
	India	World	Gap
2000-01	278	616	338
2001-02	308	647	339
2002-03	302	649	347
2003-04	399	647	648
2004-05	470	751	281
2005-06	472	734	262
2006-07	521	754	233
2007-08	563	765	212

Source : CAB

Table 7.3. States with more productivity of cotton yield

States	Yield (kg lint)/ha
Gujarat	743
Tamil Nadu	691
Andhra Pradesh	667
Punjab	630
World	765

Source : CAB

7.3. TECHNOLOGY MISSION ON COTTON (TMC)

Considering the poor cotton production, productivity and quality in India, the Government of India launched the Technology Mission on Cotton (TMC) in February 2000. The TMC consists of four Mini Missions MM-I, MM-II, MM-III and MM-IV.

7.3.1. Objectives

» To increase the income of the cotton growers by reducing the cost of cultivation as well as by increasing the yield per hectare through proper technology dissemination.

» To improve the quality of cotton, particularly in respect of trash, contamination, *etc*. by improving the infrastructure in the market yards and by upgrading / modernizing the existing ginning and pressing factories.

7.3.2. Structure

The structure of Technology Mission on Cotton is presented in Table 7.4.

Table 7.4. The structure of Technology Mission on Cotton

Mini mission	Objective	Nodal agency	Nodal institution
MM-I	Research and technology generation	Indian Council of Agricultural Research, New Delhi	Central Institute of Cotton Research, Nagpur
MM-II	Transfer of technology	Ministry of Agriculture	Directorate of Cotton Development, Mumbai
MM-III	Improvement of market infrastructure	Ministry of Textiles	Cotton Corporation of India, Mumbai
MM-IV	Modernization of ginning and pressing factories	Ministry of Textiles	Cotton Corporation of India, Mumbai

7.3.3. Mini Mission I: Advancement of research

The Central Institute for Cotton Research (CICR), Nagpur is the nodal agency for MM- I. It envisages improving productivity and quality of Indian cotton with reduced cost of cultivation to make cotton profitable to cotton growers and ensure abundant supply of quality cotton to end users so as to compete globally in the free trade regime in future was its vision. The overall objective of this Mini Mission is to develop farm worthy production and protection technologies with potential for enhancing cotton productivity by 15-20% on a sustainable basis in five years.

7.3.3.1. Genetic improvement: The period between 1980 to 1996 can be called an era of germplasm augmentation and today the total collection touched to more than 10,000 accessions at CICR, Nagpur.

India is the pioneer country in the world for developing first hybrid H-4 in 1970 from GAU, Surat. Two strains viz., CINA-316 (from Nagpur) and PA-402 (from Parbhani) have been released through AICCIP and PA-402 has already been released by MAU, Parbhani during 2003-04 with the name 'Vinayak'. Considering the beneficial features like tolerance to drought, resistant to pests and diseases of *desi* cotton by improving fiber properties, a variety PA 255 (Parbhani Turab) was evolved at Parbhani and released during 2000.

7.3.3.2. Good agricultural practices: The following good agricultural practices have been developed:

- » Integrated nutrient management
- » Integrated water management
- » Refining regional level prediction of yield
- » Farm mechanization
- » Biotic stress management
- » Diagnostic tools for pathogens and insect pests
- » Technology interventions

7.3.4. Mini Mission II: Technology transfer

The Department of Agriculture & Cooperation, Ministry of Agriculture, Govt of India is the nodal agency for Mini Mission-II. Directorate of Cotton Development, Ministry of Agriculture, GOI, Mumbai has been designated to implement and monitor the MM-II of TMC.

The set objectives of this Mini Mission are to increase production and productivity of cotton, to made available quality seed to the farmers of improved varieties/hybrids, to transfer production technology to farmers through frontline demonstrations, and training of farmers/extension workers, to bring more area under irrigation and efficient use of water by popularizing drip and sprinkler irrigation, to minimize the losses to cotton crop by pests through popularizing the IPM module and IRM strategies, pest surveillance *etc.* and to promote the use of bioagents, biopesticides, and quality inputs in cotton through standard package of practices.

7.3.5. Mini Mission-III: Improvement of market infrastructure

By providing the required civil infrastructure in markets, the sources of contamination are being effectively plugged. Setting up of grading laboratory enables the farmer to get a price commensurate with cotton quality. FICs provide information for better crop management and price realization by which the profitability of cotton cultivation has improved. Till today 104 APMCs have reported completion of their market yards. With the anticipated completion of all the 250 market development projects about 90% Indian cotton would be transacted in clean environments.

Development of 51 market yards during IXth Plan and another 60 market yards during Xth Plan was the targets fixed. Subsequently, in June 2005, the target was revised to 250 for the entire 8-year period up to 2006-07. Against the target of 250 market yards, 251 market yards have already been approved.

7.3.6. Mini Mission-IV: Modernization of ginning and pressing factories

MM-IV has direct impact on the textile growth of the country in recent years. The number of mills which was 1569 increased to 1700 in 2006 in the country during the last decade with spindle capacity of 34.34 million. The rotors which were hardly 1.36 lakh increased to 3.95 lakh. The hand loom sector also showed increased growth.

Due to increased textile sector growth, the yarn, fabric and textile export increased manifold. Today, India is producing 3458 million kg spun yarn, 1179 million kg man-made fiber, and 49577 million kg fabrics which are much higher than 1996 period. As a result, the per capita fiber availability increased to 36.10 sq. m. in 2006 which was 27.99 sq. m. in 1996.

In the textile basket, out of the total fiber requirement, cotton fiber contribute 62% as raw material, out of total yarn requirement, 52% are cotton yarn and out of total fabric requirement, 47% contributed by cotton.

7.3.7. The impact

The impact of launching Technology Mission on Cotton (TMC) in the year 2000 is well documented in recent years in terms of increasing production, productivity, generating improved technologies, reducing contamination and improving quality. Practically TMC has brought new era in cotton research, development, marketing and textile sector. All the cotton stakeholders and farming community were appreciated the launching of TMC and its benefits. The farmers have been immensely benefitted through the TMC and now confident enough to grow better cotton with IPM/INM/IRM and other modern technologies with higher return. Cotton Sector has now started to create success stories. Our farmers are now growing better cotton with latest production technologies who are trained through Farmers Field Schools (FFSs). The FFSs made revolutionary change in technology adoption in an integrated manner. The direct and indirect impact of MM-II is described below.

- » Increase in area, production and yield
- » Increased growth rate
- » Enhanced seed production and distribution
- » Farmers Field Schools
- » Reduction in pesticide consumption and number of sprays
- » Export/import and demand /supply
- » Impact of insecticide resistance management (IRM) program
- » Increase in Bt cotton area
- » Enhanced availability of bio-agents/bio-pesticides
- » Human resource development

7.4. LOCATION SPECIFIC COTTON PRODUCTION TECHNOLOGIES

7.4.1. Transplanting

Bt seed is a costly input and crop is sown only after receipt of rains or release of canal water in the irrigated regions. Cotton planting is often delayed either due to a delay in release of canal water or delayed onset of monsoon. In such situations, transplanting is a viable option. Salakinkop (2011) observed that the transplanting method was superior to the cotton planted after the release of canal water.

7.4.2. Conservation tillage for soil management

All tillage operations are tractor based in north India. However, in central India, especially Maharashtra, cultivation is still done by the bullock drawn plow. Clean cultivation for weed control is a common practice in the entire cotton belt of India. Thus, frequent tillage operations are done. As a result, soil organic C is lost to the atmosphere. The soil also becomes more prone to erosion and ultimately leads to soil degradation. These could be reasons for the decline in crop productivity. Interest in conservation tillage practices has gained popularity all over the world. In the irrigated north zone, reduced tillage systems for cotton and minimum tillage system for wheat was found to enhance remuneration, as present tillage operations involve nearly 50% of the total production costs (Jalota *et al.*, 2008). Long-term tillage experiments conducted at CICR, Nagpur, indicated that significant yield increases with reduced tillage treatment (Blaise and Ravindran, 2003; Blaise, 2011).

7.4.3. Balanced nutrition and soil test fertilizer recommendations

Initially, single nutrient concept with focus on the three primary nutrients (N, P and K) was the main approach to meet crop nutrient demands. However, the focus shifted to the development of balanced fertilization as deficiencies of other nutrients surfaced (Blaise and Prasad, 2005). Data from long-term experiments clearly pointed out to the benefits of applying and following balanced fertilizer schedule (Blaise *et al.*, 2006).

Because of the vastness of the cotton acreage, variations in soil type, fertility status and also the soil water regime exists besides the climatic factors. Therefore, location specific crop production modules were developed. The most notable advancement that occurred was the soil-test based fertilizer recommendations which further refined to the site specific nutrient management (SSNM). The concept of the SSNM was proposed for cotton in 2006-07, but it is not fruitful (Blaise, 2007) due to practical limitations.

A further advancement is the development of nutrient deficiency is the advice given to the farmer when to apply fertilizer based on the nutrient sufficiency level of the cotton crop.

7.4.4. Cry toxin expression a function of N supply and timing of application

Cry protein, that is toxic to the bollworm, is dependent on N supply. With advancing crop age, toxin expression declines (Kranthi *et al.*, 2005). This also coincides with reduced uptake of nutrients. Therefore, foliar application or split application of N was found to be helpful (Hallikeri *et al.*, 2011).

7.4.5. Poly-mulching

In recent years, new concepts of using polythene mulch have emerging but not yet to become popular with the farmers. However, it has potential in the high value regions for example in the irrigated tracts. In the Bt cotton hybrids, which is spaced at a wide row spacing on drip irrigation, large areas are covered with polythene mulching by resourceful farmers. Nalayini *et al.* (2011) demonstrated an increase in seed cotton yield with the polythene mulch compared to the non-mulched plots, both in the non-Bt as well as Bt hybrids. Yield increases were greater with the Bt hybrids.

7.4.6. Organic mulch

Cotton crop provides very little residue and it is difficult to maintain a 30% of cover in monocrop situations prevalent in the rainfed areas (Blaise and Ravindran, 2003). In addition, there were concerns whether Bt hybrids could be recycled as the leaves would contain cry toxin. Research at CICR indicated no adverse effects of cultivating the Bt cotton hybrid on soil microflora and fauna (Velmourougane and Sahu, 2013). It also found that growing a green manure cover crop in between crop rows and place it as mulch is a feasible option (Blaise, 2011). There is a potential advantage of weeds being smothered and savings on fertilizer-N.

7.4.7. Integrated nutrient management (INM)

The concept of INM gained importance because large quantities of fertilizers are needed to meet projected demands (Blaise and Prasad, 2005). Relying on mineral fertilizers alone, would lead to soils that suffer from other nutrient deficiencies, low in organic matter and poor soil physical and chemical properties. Thus, a long-term studies conducted at CICR, Nagpur from 1985-86 to 2002-03 indicated the importance of INM wherein part of the fertilizer was substituted with organic manures such as FYM. INM practice not only increased the seed cotton yield but also imparted stability to the system (Blaise *et al.*, 2006).

7.4.8. Water management

In the irrigated cotton growing zones in north India, flood or check basin method of irrigation is the common practice. Recent studies at Bhatinda, Punjab, showed that the drip method of irrigation resulted in a saving of 50% irrigation water compared to the check basin method (Thind *et al.*, 2012). Further, 25% saving of fertilizer was also observed as the yields of the drip irrigated plots with 75% of fertilizer was equivalent to the check basin with 100% recommended dose of fertilizer.

Micro irrigation system (drip irrigation) can bring about a considerable saving in water ranging from 20 to 76% in Gujarat, up to 50% in Maharashtra and elsewhere (Kumar *et al.*, 2011). Field level survey data also clearly pointed out the benefits of drip irrigation in Bt cotton (Pawar *et al.*, 2013; Narayanamoorthy, 2008). Seed cotton yield was 4.65 t/ha with the drip systems compared to 2.16 t/ha with flood irrigation. This also led to increases in water productivity (kg/HP hour of water) (7.99 in the drip vs. 2.05 in normal). Because of these advantages, the drip irrigation systems are becoming popular and the State Governments are providing incentives.

7.4.9. High density planting system (HDPS)

High density planting system (HDPS) with straight varieties is a common practice followed in the major cotton growing countries such as USA, Australia, China, Brazil and Uzbekistan. The planting geometry is 8-10 cm distance between plants in a row with row to row distance spacing ranging from18 to 100 cm. The planting methods are referred as narrow row (NR) if the row-to-row spacing is less than 75 cm and ultra-narrow-row (UNR) if the spacing is less than 45 cm. Generally, wide row-to-row spacing is followed on deep soils and irrigated farms.

ICAR-Central Institute of Cotton Research, Nagpur came up with the HDPS technology to establish sustainable production systems. HDPS is the simplest of the technologies to achieve high productivity. However, HDPS can be taken up only with compact and early maturing cultivars. Several cultivars were identified that could be planted at high densities namely, PKV-081, NH-615, Suraj, KC3, Anjali, F2383 and ADB-39 at 45 or 60 cm row spacing depending upon the soil type. This technology was widely demonstrated on farmers' fields over the last three seasons (2013-16) across the country.

7.4.10. Bullock drawn cotton planter

Traditionally Bt cotton hybrid seeds are sown manually in check rows after cross wise marking the field with a marker in order to facilitate cross wise intercultural operations with a bullock drawn cultivator. Seed drills were used in north India but these seed drills do not ensure equal plant to plant spacing within the rows. Therefore, inclined plate planters were developed. A two-row bullock drawn cotton planter was developed at ICAR-CICR for small farmers, especially for vertisols. Vertical rotor type seed mechanism is used for metering of seeds. Similarly, 3-row, self-propelled check row planters with pneumatic metering mechanism are also available for cotton hybrids.

7.4.11. Weed management

Weed control is a labor intensive activity. Herbicides offer effective weed control. But the cost of the herbicide is high. To economize on the use of herbicide from wasteful spray, a wick applicator was developed.

7.4.12. Effective transfer of technology

Technologies are available. But these need to be communicated to the farmer. More than six million cotton growers are there in India residing in the nook and corner of the country. To provide rapid access to the technologies, e-kapas network technology was developed which provides advisories to the farmers. For enhancing cotton production and ensuring the livelihoods of the cotton cultivator, all technologies that are the best bet strategies need to be adopted. One has to carefully assess the situation and prepare a technology plan for the region as the saying goes 'one size does not fit all'.

7.5. PEST PROBLEMS IN COTTON

Cotton is attacked by several insect pests reducing the crop yield to a greater extent. The insect pests that attack cotton crop may be classified into sap sucking insects (aphids, jassids and white fly) or chewing insects (bollworms, leaf eating caterpillars *etc.*). Of the total pesticides used in Indian Agriculture, about 45 per cent is sprayed on cotton crop alone. To reduce pesticide usage in cotton, several strategies like use of genetic resistance to insect pests, integrated pest management (IPM), insecticide resistance management (IRM) *etc.* are advocated. In recent times, Bt cotton technology is found to be one of the best strategies to manage bollworms, the most important pest of cotton.

Among the main pests of cotton in most cotton producing areas are caterpillar pests, especially the cotton bollworm (*Helicoverpa armigera*) and the pink bollworm (*Pectinophora gossypiella*) (Fig. 7.1). The moths lay eggs on cotton (*Gossypium hirsutum*) and other host plant leaves and the hatching larvae feed on the plant tissue. The pest is especially difficult to control in cotton because larvae bore into cotton bolls and hence cannot easily be reached with chemical pesticides. The damage is huge, because most bolls with feeding damage are abscised by the plant.

Caterpillar insects are responsible for 60 to 70 percent of all insect damage to cotton plants (Monsanto). Efforts to control them chemically account for about 60 to 70 percent of a cotton grower's pesticide costs. Chemical sprays containing Bt are seldom practical in cotton, since Bt breakdown in sunlight, washes away in rain, and is nearly impossible to apply to the plant parts where insects feed. However, other chemical insecticides are used that may need to be applied as many as 10 times in one season. By planting insect-resistant cotton, researchers hope farmers can reduce the number of insecticide applications they must pay for.

Fig. 7.1. The cotton bollworm (*Helicoverpa armigera*) (*Left*), the pink bollworm (*Pectinophora gossypiella*) (*Middle*) and tobacco budworm damage (*Right*) on cotton bolls.

7.6. Bt COTTON

The genetic resistance, one of the important pest management strategies, is available in cotton gene pool against the sap sucking pests such as jassids, whitefly etc. and using this several resistant/tolerant varieties and hybrids have been developed and released in India. However, such kind of known resistance is not available against the bollworms. Hence, an alternate strategy is explored to circumvent this problem by cloning and transferring the genes encoding the toxic crystal δ - endotoxin protein from the soil bacterium *Bacillus thuringiensis*. The Bt transgenic cotton (Bollgard of Monsanto) has thus been developed successfully in USA, which has the ability to control the bollworms at the early stages of crop growth (up to 90 days) effectively.

The first commercial Bt cotton variety was released in USA by Monsanto (Bollgard), which contains Cry1Ac gene of *B. thuringiensis*. Globally nine countries are growing Bt cotton. Two developed countries (USA and Australia) and seven developing countries including three Asian countries (China, India and Indonesia) three from Latin America (Mexico, Argentina and Columbia) and South Africa. Worldwide the area under Bt cotton keep increasing year by year. Overall, about 12% of the world cotton is now planted with genetically modified varieties/hybrids (GMO) and ICAC has estimated that this may rise to 50% in 5-7 years. As of now, cotton is the most popular of the Bt crops: it was planted on about 25 million ha in 2011.

7.6.1. What is Bt cotton?

Bacillus thuringiensis or Bt is a naturally occurring soil bacterium used by farmers to control Lepidopteron insects because of a toxin it produces. Through genetic engineering, scientists have introduced the gene responsible for making the toxin into a range of crops, including cotton. Bt expresses the qualities of the insecticidal gene throughout the growing cycle of the plant. Cotton crops are very susceptible to pest attacks and use up more than 10 per cent of the world's pesticides and over 25 per cent of insecticides.

In India, while cotton is grown on 5 per cent of the total crop area, it uses up 45 per cent of all pesticides. Intensified chemical use has led to a dramatic rise in pest infestation as, over time, they have become resistant to insecticides. Given the high chemical dependence of the cotton crop, little wonder that cotton was one of the first crops to be genetically engineered by the US-based agrochemical multinational Monsanto, whose transgenic Bollgard (Bt) cottonseed varieties were a big draw among farmers the world over. Bt cotton, with its promise of reduced insecticide use and resistance to pest attacks - leading consequently to a rise in yields with lower costs - is being pushed by the multinational as an environmentally safe and cost-effective alternative to conventional cotton seeds.

But the Bt toxin targets only the bollworm complex, comprising the American bollworm, the Spotted bollworm, the Spiny bollworm and the Pink Bollworm. The Bt toxin Cry1Ac, approved for commercialization, is particularly specific to American Bollworm, which attacks the plant after 60 days of sowing. The Pink bollworm attacks the plant after 130 days of sowing - the time of the first pick. While Cry1Ac has only a moderate effect on the pink bollworm, none of the Mahyco hybrids has any impact on pests such as thrips, aphids and jassids, which attack the plant during its early phase. Thus, while the number of sprays against the bollworm could come down, there may not be a reduction in the use of pesticides against the other pests.

Using genetic engineering techniques, cotton varieties were developed to be resistant to major caterpillar pests (Figs. 7.2 and 7.3). Transgenic Bt-cotton varieties express a modified gene (mostly cry1Ac) that encodes an insecticidal crystalline delta endotoxin protein, derived from the common soil bacterium *B. thuringiensis*. The first plant transformation was *Agrobacterium tumefaciens*-mediated and conducted by scientists of the company Monsanto in the USA (Bollgard). The Bt-protein is toxic only for certain pests because it selectively binds to specific sites in the midgut epithelium of susceptible insects. Following binding to those receptors, cation-specific pores are formed that disrupt the midgut ion flow and thereby cause gut paralysis and eventual death due to bacterial sepsis. This way, Bt-toxins work highly selectively against a narrow range of Lepidopteron insects such as cotton bollworm, tobacco budworm and pink bollworm. Because there are no receptors for this protein on the surface of non-Lepidopteron insect guts or mammalian intestinal cells, non-target insects, livestock and humans are not susceptible to damage from these delta-endotoxin proteins (Betz *et al.*, 2000).

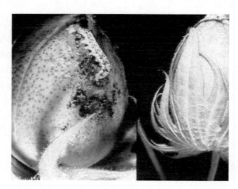

Fig. 7.2. On the left is a cotton boll being attacked by a cotton bollworm. The cotton plant that produced the boll on the right contains and expresses the gene for the Bt toxin.

Fig. 7.3. A higher number of cotton bolls were produced on field grown transgenic cotton producing NaPI and StPin1A (A) compared to Coker (B) the control non-transgenic parent

Bt-cotton varieties were first approved for commercial use in 1995 in the USA, and in 1997 in China. Since then Bt varieties have spread quickly and account for more than half of the cotton area in these two countries. Today India, China, USA, and Argentina are the main growers of Bt-cotton but several other countries have also approved Bt-cotton for commercial use.

7.6.2. Genetics of Bt cotton

The bacteria *B. thuringiensis* produce two types of toxins namely Cry (Crystal) toxin encoded by different cry genes and Kytolytic toxin. Over 50 genes have been noted to encode for cry toxin and they are sequenced for various studies. The various genes and their properties are tabulated here under (Table 7.5).

Table 7.5. The various cry genes and their properties

Gene	Crystal shape	Protein size (KD)	Insect susceptible
Cry I A(a), A(b), A(c), B, C, D, E, F, G	Bipyramidal	130 – 138	Lepidopteron larva
Cry II (Sub group) A, B, C	Cuboidal	69 - 71	Lepidoptera, Diptera
Cry III (Sub group) A, B, C	Flat irregular	73 – 74	Coleoptera
Cry IV (Sub group) A, B, C, D	Biopyrimidal	73 – 134	Diptera
Cry V – IX	Various	35 – 129	Various

The commercial Bt–cotton available today contain genes from the isolate *B. thuringiensis* ssp. *kurstaki* that produces Cry I A(a), Cry I A(b), Cry I A(c), Cry IIA.

7.6.3. First generation Bt-cotton

The most prevalent Bt-gene on a global basis Cry I A(c) was incorporated into Coker 312 cotton designated MON 531 by Monsanto and later named Bollgard cotton (first generation Bt-cotton). The high transformation efficiency was achieved in Coker 312 with *Agrobacterium tumefaciens*. The transformed Coker was than back crossed with lines from Delta and Pine land and other companies that had necessary agronomic qualities for commercial acceptance.

The advantages of the Cry IA (c) in Bollgard over the Bt-cotton spray are as follows:

» Active protein expressed in all plant parts.
» Active protein expressed throughout the season, hence timing of insecticide applications in relation to an infestation is not an issue.
» Less farmer exposure to insecticide.
» Labor saving technology due to elimination or reduction of insecticide sprays.
» Contribution to and provides the foundation for an IPM strategy.

7.6.4. Second generation Bt-cotton (Bollgard II cotton)

The Insect Resistance Management (IRM) strategy for Bt-cotton that Monsanto in conjunction with USDA developed second generation of improved Bt-cotton with two Bt-genes, now designated Bollgard II. The new product Bollgard II, Event 15985 was developed using particle acceleration plant transformation procedures to add the Cry IIA(b) gene to the cotton line DP 50B that already had the cry 1A(c) gene. Hence the Bollgard II cotton contains Cry 1A(c) and Cry IIA(b). The dual gene cultivars are expected to provide growers with a broader control over a wide variety of insects.

In 2002, Dow Agrosciences announced the development of new Bt-cotton with traits that confer broad spectrum resistance to Lepidopteron pests of cotton. The new Bt cotton product contains the dual genes Cry IA(c) and Cry IF, transformed with *Agrobacterium tumefaciens* and incorporated through back crossing into several high quality commercial varieties of cotton.

7.6.5. Performance of released Bt cotton hybrids

Comparative performance of Bt cotton hybrids developed by different private companies over the year (2001-2005) over different locations of three cotton growing zones of the country (India) indicated superiority of Bt hybrids over their non-Bt counterparts in terms of both yield and contributing characters. Apart from seed cotton yield, they were also superior in number of bolls/plant, boll weight, seed index, lint index and ginning outturn.

Large scale field trial results of four years from 1998 to 2002 indicated that Bt cotton was able to resist bollworm infestations thereby resulting in good boll retention and higher yields. Apart from the increase in yields there was a concomitant reduction in the use of insecticides due to Bt-cotton. Thus, it was concluded that Bt-cotton has potential to improve the lives of cotton farmers through the provision of favorable environmental and economic consequences (Table 7.6).

Table 7.6. Economics of Bt-cotton cultivation in ICAR trials 2001

Hybrid	Yield Q/ha	Gross income Rs./ha	Insecticide cost Rs./ha	Net income Rs./ha
MECH-12 Bt	11.67	21,006	1,727	16,854
MECH-12 Bt	13.67	24,606	1,413	20,768
MECH-12 Bt	14.00	25,200	1,413	21,362
Local check	8.37	15,066	2,845	12,221
National check	7.31	13,158	2,001	11,157

Results from extensive Bt cotton trials under farmer field conditions, conducted from 1998 to 2001 confirmed that Bt cotton with the Cry1 Ac gene provides effective and safe control of bollworm and related pests. Field trials have confirmed that, compared to conventional hybrids, Bt cotton can increase yields by up to at least 40%, reduce insecticide sprays by at least 50% or more (decrease from 7 to 2 or 3 sprays on average) (Table 7.7) equivalent to savings of Rs 2500/ha, and increase overall farmer income from Bt cotton from Rs 3500 to Rs 10,000 or more per hectare.

Table 7.7. Spray application reduction on Bt cotton during 2002-2003

Country	Year of introduction	Total cotton area (lakh ha)	Area under Bt-cotton		No. of insecticide sprays	
			Lakh ha	%	Non-Bt	Bt-cotton
USA	1996	62.0	20.0	33	5	2
Mexico	1996	0.8	0.3	35	4	2
China	1997	48.0	15.0	31	20	7
Australia	1997	4.0	1.5	36	11	6
Argentine	1998	1.7	0.1	5	5	2
S. Africa	1998	0.4	0.2	45	11	4
Indonesia	2001	0.2	0.1	18	9	3
Colombia	2002	0.4	0.1	10	6	2
India	2002	85.0	2.8	3	5	2

Mahyco commissioned a nationwide survey by AC Neilsen-ORG MARG in 2003. The survey covered 3,063 farmers from Maharashtra, Madhya Pradesh, Andhra Pradesh, Karnataka and Gujarat (Table 7.8). The data showed that a yield increase by about 29% (range 18 to 40%) due to effective control of bollworms, a reduction in chemical sprays by 60% (range 51 to 71%) and an increase in net profit by 78% (range 66 to 164%) as compared to non-Bt cotton. The net profit was estimated to an average of Rs.7,724 (range Rs. 5,900 to 12,696) per hectare. Mahyco conducted an independent survey during 2003 to assess the performance of Bt-cotton in fields of 3000 farmers. Results showed an average net profit of Rs 18,325 (range 15,854 to 20,196) per hectare.

Table 7.8. Results of the survey by A C Neilsen-ORG MARG, 2003

State	Pesticide reduction		Yield increase		Net profit	
	Rs/ha	%	Q/ha	%	Rs./ha	%
Andhra Pradesh	4,594	58	4.9	24	12,717	92
Karnataka	2,930	51	3.3	31	6,222	120
Maharashtra	2,591	71	3.6	26	5,910	66
Gujarat	3,445	70	2.9	18	8,564	164
Madhya Pradesh	2,200	52	5.4	40	9,594	68
Average	3,202	60	4.2	29	7,737	78

Field trials conducted by Mahyco during 2001, in 157 farms in 25 districts of Madhya Pradesh, Maharashtra and Tamil Nadu showed that there were no changes in the insecticide use for sucking pest control, but at least three sprayings meant for bollworm control were saved due to the Bt-technology. Thus, insecticide use of cotton bollworm was reported to have been reduced by 83% and yield increase by a staggering 80%. Global estimates show that, Bt cotton caused an average net income increase of $ US 50/hectare in the USA, $357/ha to $549/ha in China and $25-51/ha in South Africa.

Field trials have shown that farmers who grew the Bt variety obtained 25%–75% more cotton than those who grew the normal variety. Also, Bt cotton requires only two sprays of chemical pesticide against eight sprays for normal variety.

7.6.6. Benefits of Bt cotton

Bt cotton has several advantages over non-Bt cotton. Important advantages of Bt cotton include increase in yield, protection from bollworms, reduction in pesticide use, reduction in cost of cultivation, reduction in environmental pollution, genetic resistance, eco-friendly, no adverse effect on parasites, predators and beneficial insects and no health hazards. It also induces earliness. These points are briefly discussed below:

» Bt cotton does not contain yield enhancing gene. However, it leads to increased yield due to effective control of three types of bollworms viz. American, Spotted and Pink bollworms.

» The Bt gene is very much effective in controlling insects related to Lepidoptera order. Cotton bollworms belong to the order Lepidoptera and therefore are sensitive to Bt Cry I, Cry II and Cry V proteins which are specific to these insects.

» The cultivation of Bt cotton leads to significant reduction in the use of pesticides. The Bt insecticidal proteins are very effective in controlling bollworms. As a result, there is hardly need of spraying insecticide in Bt cotton.

> » In Bt cotton, there is drastic reduction in the number of insecticidal sprays to the tune of 75%~80%, which in turn leads to reduction in the cost of cultivation.

> » In Bt cotton, the insecticides are less used. This helps in reduction of environmental pollution.

> » Bt Cotton has inbuilt or genetic resistance which protects the crop from the attack of bollworms. This is a permanent type of resistance that is not affected by environmental factors such as rainfall, temperature, humidity *etc*. If there is a rainfall after the spray of insecticide, the effect of insecticide is washed away. This is not the case with Bt Cotton.

> » Bt cotton is ecofriendly. It does not have adverse effect on parasites, predators, beneficial insects like honey bee, silkworm, lac worm *etc*. and microorganisms such as earthworm, blue green algae, nitrogen fixing bacteria *etc*.

> » Bt cotton promotes multiplication of natural enemies of the insects, pests such as *Chrysoperla*, *Trichogramma*, vasps, *etc*. These insects help in controlling the bollworms by feeding on their larvae and eggs.

> » Use of insecticide has adverse effect on the human health particularly who is engaged in spraying of such insecticides. Bt cotton does not have any such adverse effect on the human health.

> » It has been observed that the Bt cottons are early in maturity as compared to their non-Bt counterparts. The earliness ranges from 20~30 days in different cotton hybrids tested in India.

7.6.7. Role of Bt cotton in environmental protection

The major emphasis was given to the control of boll worms in Bt hybrids as against their non-Bt counterparts by considering the number of times the Economic Threshold Level (ETL) crossed, total number of sprays given for the control of various insect pests under protected and unprotected conditions *etc*. In all these aspects, the released Bt cotton hybrids were found to be more efficient as compared to their non-Bt counterparts. As compared to insecticidal control of bollworms, Bt cotton technology will not harm non-target beneficial insects, reduction in production cost, increased profit, reduced farming risk and improved economic outlook for cotton. Use of this technology is also helpful in improving wild life population, reduced run off insecticides, reduced air pollution and improved safety to farm workers and neighborhood.

7.6.8. Biosafety tests and assessment of toxicity to non-target organisms

Feed-safety studies with Bt cottonseed meal were carried out with goats, buffalos, cows, rabbits, birds and fish. Biosafety tests indicated absolute safety to goats, cows, buffaloes, fish and poultry. The results revealed that the animals fed with Bt-cotton seed meal were comparable to the control animals in various tests and showed no ill-effects.

The Cry1Ac is mainly toxic to the bollworms (cotton bollworm, pink bollworm and spotted bollworm), semi-loopers and hairy caterpillars. Bt-cotton expressing Cry1Ac is absolutely non-toxic to all other non-target organisms such as beneficial insects, birds, fish, animals and human beings. There was some evidence of a reduction in numbers of predators and parasitoids which specialize on the Bt controlled bollworms, but also of increases in numbers and diversity of generalist predators such as spiders. Generally, the decrease in the parasitoid and predator populations was associated with decrease in the densities of the pest populations on account of Bt-cotton. Due to these changes in pest complex, farmers had to spray 3-5 times on Bollgard as compared to 6-8 times on non-Bt cottons.

7.7. EMERGING PESTS ON Bt COTTON IN INDIA

Due to large scale cultivation of Bt cottons since 2002 in India, changes in insect pest complex are evident. Mealy bugs and mirid bugs are emerging as potential threat.

7.7.1. Mealy bugs

Phenacoccus solenopsis was the dominant species of mealy bugs found across the country during its first year of incidence (2007-08) (Fig. 7.4). Subsequently *Paracoccus marginatus* emerged as a pest on cotton and other crops of Tamil Nadu in 2009 since its first report in 2007 (Table 7.9).

Table 7.9. Incidence of emerging pests

State	Locations	Emerging pest	Incidence*	Damage
Haryana	Odhan, Sirsa, Kaleriwal, Dhabwali and Baraguda	Mealy bug	5-44%	Grade 3: 1.66-19% Grade 4: 1.33-19%
Maharashtra	Nanded	Mealy bug	33.70**	NR
Gujarat	Surat	Mealy bug	---	Grade 4 in August. Grade 2 till Jan
Andhra Pradesh	Guntur	Mealy bug	5.76 - 35.35 %	1.71- 4.00
Karnataka	Haveri Belgaum	Mirid bugs	43.85 bugs/25 squares	---
Tamil Nadu	Coimbatore	Mealy bug	55-83.1%	1.00 -1.22
		Mirid bugs	16-85.1 bugs/100 squares	---

* Refers to number of plants harboring mealy bugs and causing more than Grade 1 damage.
**Refers to number of mealy bugs on 2.5 cm stem length.

Fig. 7.4. *Left* - Cluster of Solenopsis mealybug on cotton terminal. *Right* – Biological control of cotton mealy bugs with predatory lady bird beetle, *Cryptolaemus montrouzieri*.

7.7.1.1. Management: *Aenasius bambawalei* is the most effective parasitoid. The predatory beetles *Cryptolaemus montrouzieri* (Fig. 7.4), *Brumus suturalis* and *Scymnus* spp. are prominent in the ecosystems in India and Pakistan. The entomopathogenic fungi, *Metarrhizium anisopliae, Beauveria bassiana,*

Verticillium lecanii and *Fusarium pallidoroseum* are effective in infecting mealybugs. Botanical mixtures containing neem oil, citrus peel extracts and fish oil rosin were found to be effective in controlling the mealybugs. The insect growth regulator, Buprofezin is effective in control. Insecticides such as Malathion and Acephate, which are considered by the WHO as only slightly hazardous (WHO III category) can be used as soil application near the root zone.

Mealy Kill (botanical bioprotectant) is very effective as it dissolves the waxy coating on the mealy bug thus making it vulnerable to desiccation and bioagents. The technology has special relevance to citrus belt of Vidarbha as citrus peels are a rich source of the insecticidal component of Mealy Kill. A single spray of Mealy Kill costs about Rs. 200 per acre.

7.7.2. Mirid bugs

Three species of mirid bugs (*Creontiaedes biseratense, Hyalopeplus lineifer, Campyloma livida*) have been found to cause damage of varying intensities on cotton. Mirids feed on tender shoots, squares and cause excessive shedding of flowers, small squares and parrot beaking of bolls (Fig. 7.5). They occur in large numbers moving rapidly on the plant and often miss the eye. Mirid bugs are reported to cause maximum damage in Haveri and Belgaum districts of Karnataka with up to 2 mirids per square in the months of October and November. Mirids were reported to cause an avoidable loss of 290 kg/ha in Dharwad. Yield loss due to mirids in Nagpur of Central India ranged from 25-30%. Avoidance of broad spectrum insecticides seems to have assisted in their establishment as emerging pests of cotton.

Fig. 7.5. Green mirid adult (*Left*) and its damage on boll (characterized by shiny black spots) (*Right*).

7.7.2.1. Management: Fipronil 5 SC @ 1.0 ml/l has recorded significantly lowest mirid bugs population at 3 days after second spray both during 2008 and 2009 (0.46 and 0.20 bugs/plant, respectively). The next best treatments were acephate 75 WP @ 1 g/l and fipronil 5 SC + table salt @ 0.5 ml + 5.0 g/l which were statistically at par with each other and recorded 1.11 and 1.32 mirid bugs per plant during 2008 and 1.33 and 1.27 mirid bugs per plant during 2009, respectively. Seed cotton yield was significantly highest in fipronil 5 SC @ 1.0 ml/l both during 2008 (24.75 q/ha) and during 2009 (20.30 q/ha), followed by acephate 75 WP @ 1 g/l during 2008 (22.84 q/ha) and fipronil 5 SC + table salt @ 0.5 ml + 5.0 g/l during 2009 (19.19 q/ha) where the latter two were statistically at par with each other during both the years.

7.8. DEVELOPMENT OF RESISTANCE

One major concern with the large-scale production of Bt-cotton is that target pests will develop resistance against the Bt-toxin. This phenomenon is widespread for chemical pesticides in China

(Wu *et al.*, 1997) and has also been reported for the cotton bollworm against Bt-toxins in Australia. Resistance to Bt toxins linked with increased damage to Bt crops in the field has now been documented in at least four target Lepidopteron pests.

Pink bollworm (*Pectinophora gossypiella*) has developed resistance to its genetically modified (GM) cotton variety, Bollgard I, in Amreli, Bhavnagar, Junagarh and Rajkot districts in Gujarat. This was detected by the company during field monitoring in the 2009 cotton season. The Bt cotton variety in question was developed using a gene-Cry1AC-derived from *B. thuringiensis*. It was supposed to be resistant to pest attacks. But, of late, the pest has developed resistance to the gene.

A second generation variety, Bollgard II, introduced by Monsanto in 2006, contains two proteins, Cry1Ac and Cry2Ab. The company says no resistance has been observed in the variety anywhere in the country, including Gujarat.

As with the recently reported resistance of pink bollworm to Bollgard II cotton in India, laboratory diet bioassays provide the critical data documenting genetically based decreases in susceptibility of bollworm field populations to Bt toxins Cry1Ac and Cry2Ab.

The percentage of field populations of boll worm (*Helicoverpa zea*) resistant to Cry2Ab (Bollgard II) increased from 0% in 2002 to 50% in 2005. In bioassays from 2003 to 2006, 14 field-derived strains of bollworm had <50% mortality at 150 µg Cry2Ab per ml of diet, the highest toxin concentration tested.

7.9. RESISTANCE MANAGEMENT

The rationale behind developing strategies to slow down resistance development is to preserve effective pest control options even though there is general belief that new innovations will be available in the future. Bt-toxin is one such effective control option since it works extremely specifically, only affecting target pests, and has no reported negative human health impact. The main resistance management concept is the high dose/refuge strategy. Although development of resistance is not completely prevented, the build-up of pest resistance is considerably retarded. The pace of resistance build-up crucially depends on selection pressure and thus all resistance management strategies aim at a reduction of that pressure. Potential resistance management strategies listed are the use of refuge areas (the concept of using a refuge area is more difficult to implement and enforce under the conditions of a developing country and with small-scale farming in general), high doses of toxin, gene pyramiding/gene stacking (genes of two or more insecticidal proteins or different genes are introduced in one plant), regulation of gene expression (tissue or time specific expression) and all other control measures that reduce the pest pressure (for example destruction of overwintering pupae or larvae, control of alternate hosts, natural enemies). Most of the latter options are standard within an integrated pest management (IPM) system.

In Bt-cotto,n the concentration of the endotoxin is not high enough to kill most partially resistant insect individuals especially of the cotton bollworm, while it may be sufficiently high for control of tobacco budworm. Hence, in the USA, a refuge area strategy is followed to ensure cross-mating with susceptible insects. In China, no such refuge area scheme is implemented but use of unique multi-cropping system has been proposed as a natural refuge in Northern China (different crops planted in the same or neighboring fields). Because there is no Bt-corn grown in the country until today, this crop (also a host of the CBW) can act as refuge area by producing enough susceptible individuals that mate with resistant insects from Bt-cotton plots and hence dilute the build-up of resistance. In addition, late-season CBW larvae can be reduced with insecticides to further decrease the frequency of resistance alleles.

If proper pest management measures are followed to ensure that at least 90 per cent of the surviving larvae of the American bollworms in Bt-cotton fields, are killed with biopesticides, resistance can be delayed to 45 years even with 40 per cent area under Bt cotton. The strategies that would enable extending the usefulness of Bt technology would be:

» Use eco-friendly methods such as cultural control or hand-picking of surviving bollworms in Bt cotton fields. Deep-plowing of fields immediately after Bt-cotton harvest to destroy resistant pupae.

» Biopesticides that are neem based or HaNPV (virus) were found to be more effective on larvae surviving on Bt-cotton because of their slower growth as compared to those on the conventional non-Bt cotton. Hence these would be useful to manage younger larvae on 60-90 days old crop. Alternatively, eco-friendly insecticides such as spinosad, emamectin benzoate, novaluron or indoxacarb can be used on 90 and 120 days old crop to reduce populations of resistant insect genotypes.

» Use of attractive synchronous alternate host crops such as marigold, sunflower, chillies etc. for *H. armigera* which could be used as intercrop or trap crop refuges.

» Use alternate genes that do not share common resistance mechanisms as that of Cry1Ac, in transgenic plants either in rotation or alternation or mixtures such as the dual-gene or triple-gene based Bt-cotton varieties.

7.10. SUSTAINABLE COTTON PRODUCTION

The implementation of green agriculture can help with the following objectives, all at the same time.

» Improve the efficiency of cultivation techniques.

» Increase agricultural produce manifold.

» Maintain ecological balance through good agricultural practices.

Here are a few ways in which farmers can implement sustainable agriculture for cotton cultivation.

7.10.1. Minimize the potential harmful effects of pesticides

Generally, cotton cultivation does not require extensive pesticide application for protection against diseases. While only young cotton plants are somewhat vulnerable and will require the most insecticide application, it accounts for nearly 24% of the global insecticide use. In India, anywhere between 40 and 50% of all pesticides used in the country are in cotton production.

Farmers can optimize crop protection by practicing precision agriculture to reduce the unnecessary use of harmful chemicals. SmartRisk helps cotton-producing companies with the exact identification of areas with low crop performance via 360-degree monitoring. It helps prevent water and soil pollution via fertilizer and pesticide run-off and also increases crop productivity.

7.10.2. Using fresh water resources sustainably

Cotton is a *Kharif* crop, which makes it highly resistant to hot weather conditions. It is a drought-resistant crop and does not require significant irrigation. Light rainfall or proper irrigation can contribute to most of the water requirements of cotton plants. However, according to research by Water Footprint Network , producing 1 kilogram of cotton in India consumes 22,500 liters of water, on an average, compared to the global average water footprint of 10,000 liters. The difference in water consumption is due to inefficient water use and high rates of water contamination due to pesticide run-off. The excessive water consumption also translates to a proportional volume of the loss of virtual water through exports to other countries.

Efficient water conservation practices can help producers overcome these challenges. In addition, farmers can get accurate weather predictions via SmartFarm, plan for irrigation when needed, and take preventive measures in case of unfavorable weather conditions.

7.10.3. Take better care of soil health

Regenerative agriculture encourages soil-renewing farming practices. These include minimizing soil disturbance by reducing tillage, providing soil armor, and cultivating a wide range of crops throughout the year. A protective layer of green plant roots facilitates symbiosis, acting as a natural fertilizer. Similarly, practices like crop rotation, cover cropping, and companion cropping can help the soil retain a balance of different nutrients synthesized by various plants.

7.10.4. Mitigate biodiversity loss due to agricultural activities

The mentioned farming practices constituting sustainable agriculture, like reducing tillage and diversifying plant species, are intrinsic to maintaining biodiversity loss. Another technique that can cater to this cause is intercropping. In this process, cotton cultivators can grow along with other plants on the same farm to promote beneficial interactions. Additionally, they can plant cover crops and trees with deep roots to prevent soil erosion by water and wind.

7.10.5. Promote decent work for socio-economic development

Cotton farming employs about 7% of all labor in developing countries and is a source of livelihood for more than 250 million people worldwide. However, the production involves maximum risk and minimum reward for the farmers. In addition to considering environmental impact, sustainable cotton production needs to address its socio-economic impact on farmers, their families, and the extended community at large. It includes concerns such as the demanding working conditions of farmworkers, their health and safety, economic insecurity, the incidence of child labor or bonded/forced labor, and the treatment of women/girl workers. Several international textile and fashion brands are supporting projects that ensure decent and ethical working conditions. In this regard, implementing traceability to the source can effectively promote the welfare and rights of the farmworkers.

7.11. OPPORTUNITIES FOR FUTURE

Keeping in mind the development of resistance to Cry 1Ac protein in insects, notable progress has already been made to diversify the transgene and to pyramid genes which are having different mode of action so that development of resistance is delayed. The genes available for exploitation include Bollgard II of Monsanto (Cry 1Ac + Cry 2Ab), VIP COT of Syngenta (VIP 3A), and Wide Strike of Dow AgroSciences (Cry 1Ac + Cry 1F). The problem with regard to fiber quality of Bt hybrid as noted in MECH 12 and MECH 184 may be circumvented by involving a proper combination of parents to produce superior Bt cotton hybrid for both yield as well as fiber quality.

Biotech-enhanced insect resistant Bollgard III being developed by Monsanto is a three-gene Bt cotton technology (Vip3A + Cry1Ac + Cry2Ab) which will provide the cotton crop added protection from a broad spectrum of bollworm complex (bollworm, army worm and pink bollworm) and *Spodoptera*. It will help farmers get more productivity from the same acreage, get savings on pesticides sprays, use resources efficiently and thus, earn higher income.

7.12. CONCLUSION

The increase in cotton production in the country after the introduction of Bt cotton is not merely by GM technology alone but there are other factors like enhancement of area under cotton especially irrigated area, low pest activities, well distributed rainfall, and better market price. The benefits of Bt cotton in India are in line with those enjoyed by farmers worldwide who have cultivated Bt cotton. The area under Bt cotton cultivation is expected to increase leading to increased production and reduced costs in an environmentally favorable manner. This will positively affect the livelihood of millions of small farmers by improving their net incomes.

Rigorous scientific studies conducted in India and abroad demonstrate that Bt cotton and its products are safe for the environment, humans, animals, and agriculture. In fact, the use of Bt cotton is a positive step towards environmental protection because it makes possible the reduction of the insecticide load in the environment and reduces handling of such chemicals by farmers. This reduced use of insecticides will enhance the effectiveness of biological controls and implementation of Integrated Pest Management (IPM) programs. The higher farm income observed in the experiments has now been demonstrated by the large-scale use of Bt cotton by Indian farmers, and the incorporation of the gene is proving an effective and environmentally friendly plant protection tool resulting in greater cultivation of Bt cotton in the coming years. The cotton trade is looking forward to the productivity and quality benefits of Bt cotton seed. Efforts are being made to incorporate three genes (Vip3A + Cry1Ac + Cry2Ab) in Bollgard III to improve efficacy and postpone possible resistance problems. As newer products are approved in the regulatory system, it is likely that farmers will have greater choice to plant Bt hybrids according to market quality requirements.

Chapter - 8

Golden Fiber
(Jute Production) Revolution

8.1. INTRODUCTION

Jute is a natural fiber with a golden, soft, long, and silky shine. It is the cheapest fiber procured from the skin of the plant's stem. Because of its colors and high cash value, jute is known as a golden fiber. Hence, the Golden Fiber Revolution in India is related to jute production.

After cotton, jute is the most important vegetable fiber in consumption, production, usage, and availability. During the industrial revolution, jute started being used as a raw material in the fabric industry and until today, the processed jute is used for making strong threads and jute products.

Jute is an important natural fiber crop in India next to cotton. In trade and industry, jute and mesta crops together known as raw jute as their uses are almost same. Raw jute plays an important role in the country's economy. It was originally considered as a source of raw material for packaging industries only. But it has now emerged as raw material for diverse applications, such as textile, paper, building and automotive industries, use as soil saver, as decorative and furnishing materials, *etc*. Raw jute being bio-degradable and annually renewable source, it is considered as an eco-friendly crop and it helps in the maintenance of the environment and ecological balance. Further attraction of jute lies in its easy availability, inexhaustible quantity at a comparatively cheaper rate. Moreover, it can be easily blended with other natural and man-made fibers.

8.1.1. Background of jute industry

Though jute was available in entire Southeast Asia, the jute industry was one of the most prominent industries in undivided Bengal as the use of jute products was more widespread in undivided Bengal. Bengal being a riverine state is congenial for jute cultivation hence the rate of jute cultivation and production was high in Bengal.

Furthermore, the British East India Company also started jute cultivation commercially and used jute-woven bags to distribute food grains all over the world for the same reason. Jute being one of the prized industries of Bengal created a special socio-economic environment for several decades and it still prevails. The economy rests on the growth of the jute industry, which is also reflected in the culture of Bengal.

The history of jute manufacturing in India dates back to 1854, when the first mill was set up at Rishra near Kolkata, by an Englishman, George Aukland. By 1860, the Hoogly basin became the center of jute mill industry in India.

Before independence, India had a monopoly in the production of raw jute and jute manufacturing in the world. The partition of the country inflicted a severe blow to the jute industry, with the result that jute mills remained in India, whereas a major portion of jute producing area went over to Pakistan.

In 1947-48 production of jute was 16.5 lakh bales as against pre-partition output of 65.7 lakh bales. This situation created a crisis in the jute textile industry and it was further worsened, when the Pakistan Government stopped supply of raw jute to India.

India is the largest Jute producing country with annual production estimated to be around 1.986 million tons. The improvement in crop cultivation and the use of technology in jute farming has made India prominent in global production of Jute. West Bengal accounts for almost 50% of the country's total jute production. Other major jute producing regions in the country include Uttar Pradesh, Bihar, Assam, Meghalaya, and Orissa.

Thus, efforts were made by the Government of India for the extension of area under jute in the country after independence. From a paltry production of about 16.5 lakh bales in 1947-48 the production of raw jute rose to 60 lakh bales in 1961-62 and 120 lakh bales in 1985-86. There are 69 jute mills in the country with a total installed capacity of 44,376 looms. The industry has absorbed about 2-5 lakh industrial workers.

India, along with the major producers, is also the largest consumer of jute and jute products in the world. So much so that it had to import around 337,000 tons of Jute and jute products in the year 2011 to meet the domestic demands.

8.2. AREA, PRODUCTION AND YIELD

The major jute producing states in India are West Bengal, Assam, Bihar, Orissa and Andhra Pradesh, but the Indian jute industry is mainly dependent on West Bengal. Although jute acreage has stabilized around 8 lakh hectares in the country, jute productivity has more than doubled to 2.212 tons/ha in 2010-11, since partition of the country from 1.1 tons/ha and the total production of raw jute in the country is about 10.58 million bales of 180 kg each. This has happened due to development of short duration high yielding varieties as well as other production technologies developed by ICAR including fertilizer management, weed management, farm machineries and pest management. National Institute for Research on Jute and Allied Fiber Technology (NIRJAF) (ICAR) has developed around 15 high yielding varieties of jute and allied fibers. Production and distribution of certified seed of newly released varieties are the need of the hour for further improvement in productivity. ICAR has taken the initiatives for production of quality seed of newly released varieties of jute in its farms which may increase the yield by at least 20-30%. Important issues namely, improvement in productivity, quality improvement, retting of jute, farm machineries and quality seed production are addressed by ICAR through various on-going projects, which have contributed to the present level of production with adequate support from the Jute Technology Mission and Directorate of Jute Development of the Ministry of Agriculture.

Area, production and yield of jute in major States during 2020-21 is presented in Table 8.1.

Table 8.1. Area, production and yield of jute in major States during 2020-21

State	Area ('000 ha)	Production ('000 Bales)	Yield (kg/ha)
West Bengal	12.47	162.10	2340
Bihar	12.86	194.73	2726
Odisha	3.77	18.72	894
Assam	3.00	18.10	1086
Chhattisgarh	0.66	1.31	358
Others	4.40	29.41	1229
Total	38.16	434.81	2051

With its insignificant coverage of total cultivated area, it plays a predominant role in the country's economy by generating employment, earning foreign exchange, solving many of the socio-economic problems *etc*. In earlier years, jute was considered as a golden fiber, but there after it had to pass through different critical situations. The main problem came in the way with the introduction of synthetic fiber by the end of sixties/early seventies. After the development of diversified product of jute fiber and due to growing concern about the environment pollution, the importance of jute has again revived.

The most important feature of the jute mill industry in India is its concentration in the lower Hoogly basin, within a radius of about 64 kms from Kolkata. This region alone has 55 jute mills. Jute manufacturing in this region has responded to a variety of factors. These are:

» Raw jute is locally available.

» Coal is available from Raniganj and Asansol Coalfields, which are situated close by.

» Humid climate favors manufacturing of jute yarn.

» Availability of cheap labor from Bihar, Jharkhand, Assam, Orissa and Uttar Pradesh. Labor is also available in the region on account of dense population.

» Availability of port facilities for import of machinery and export of finished products.

» Development of means of transportation particularly water transportation.

» Developed banking and marketing facilities.

» Availability of capital for investment in jute textile industry from local business men.

» Hydro-electricity from Damodar Valley Corporation.

Thus, the above-mentioned factors have played a very significant role in developing jute mills in the Hoogly Basin of West Bengal. Jute Centers have sprung up on both sides of the river.

8.3. PROBLEMS FACED BY JUTE INDUSTRY IN INDIA

The jute industry of India is confronted with many problems and jute mills hardly operate to their full capacity.

8.3.1. Shortage of raw materials

During the partition of India, most of Jute mills remained in India while major Jute producing areas went to Bangladesh (then East Pakistan). Thus, created the problem of a shortage of raw material. Despite the substantial efforts of the Government to increase the area under Jute production, India is not self-sufficient in superior quality raw material and has to import the same from Bangladesh and other countries to meet this shortage. Efforts are being made to augment the domestic production by introducing jute cultivation in non- traditional areas.

8.3.2. Competition from substitutes

Jute industry is largely an export oriented industry. Prior to Independence, India had virtual monopoly in raw jute and manufactured products. Not only countries like Bangladesh, Philippines, Japan and Brazil *etc.* have come out as stiff competitors but use of paper, cloth and plastic *etc.* as substitute for jute goods which is badly affecting the export market. There is a tough competition from synthetic packing materials of advanced countries of Europe and N. America. Therefore, market of jute goods is shrinking. These products are durable and cheaper. To overcome these problems, there is a need to modernize the jute industry, bring down the cost of production and diversify production to find out new uses for jute goods.

8.3.3. Lack of modernization of plant and machinery

The jute industry is faced with the problem of obsolete and worn-out plant and machinery and need technological advancements. Due to frequent breakdowns, defective and inferior quality products are being made. No modernization has been made in machinery development and automation. New mills in Bangladesh are producing better quality goods which are diminishing the competitiveness of Indian mills. This hampers the quality and quantity of jute manufacture. The continued dependence on an antiquated labor intensive technology has now turned 70 per cent of the jute mills sick (Ganguli, R, Economic Times, July 7, 1994).

8.3.4. High prices

The jute industry is also plagued with high prices due to obsolete machinery, existence of inefficient and uneconomic units, high price of raw jute and unreliable supply of raw jute for the mills. A sizeable number of jute mills are sick and the profitability of the industry has been low.

8.3.5. Fluctuating production

The production of jute manufactures has been fluctuating from year to year. This is due to irregular and inadequate supplies of raw jute, shortage of power, slackness of export demand, and lack of incentive. Measures have been taken to improve the production and regular supply of raw jute.

The other problems faced by jute textile industry are as follows:

» Jute industry in India mainly suffers from inadequate supply of capital, raw material, scarcity of water and dearth of skilled labor etc.

» Jute mills in India generally follow traditional methods for producing jute products which involve high production cost.

» Competition from the other Asian countries like China, Bangladesh is increasing rapidly.

» Low cost jute products from other Asian countries are capturing international market.

» Lack of proper administration and governance system.

» Dearth of proper agricultural strategies and rapid urbanization are putting harmful effects on jute cultivation in India.

» Shortage of superior quality jute.

» Emergence of Thailand, Myanmar, Philippines and Brazil in the field of jute production and jute manufacturing.

» Overall demand of jute in international market is reducing.

» Within the jute goods market, Bangladesh mills are competing along with China with better quality goods.

» Bulk handling practices in US, Argentina, Canada, Japan and Europe have decreased demand for jute.

» Cost of Indian jute product is comparatively higher.

» Availability of local fibers like Mesta (especially in Andhra Pradesh) etc., have reduced jute demand.

» Infrastructural bottlenecks, power, transportation, and capital also pose several threats to the sustainability of Jute industry.

» In West Bengal, strong labor unions frequently lead to problem of strikes, lockout and other labor problems, this has further decreased the competitiveness.

Jute industry is a vital industry. The government has taken a number of measures to tackle these problems. The Government of India in 2006 launched a Jute Technology Mission with four mini Missions that included jute research, development of raw jute agriculture and its extension; processing and marketing of raw jute. But this mission was unable to achieve targets and used the allocated funds. The Jute Corporation of India at Kolkata in 1971 is engaged with stabilizing the prices of raw jute.

The UNDP caters to the needs of the jute sector from production to export stage. The National Centre for Jute Diversification (NCJD), established in late 1995, has played an important role in commercialization of technologies for the manufacture of jute diversified products and creating awareness about the uses of jute in non-conventional applications. With the launching of the National Jute Program (NJP), a few years ago, aided by the Government and the United National Development Program jute has received the fillip it deserved to surge ahead.

The International Jute Organization (IJO) held a conference recently for drafting specifications for hydro-carbon free jute bags to be used as packing material for food grains. There are 23 research institutions and over $23 million has been spent by the UNDP with matching grants by the Government of India under its National Jute Program (NJP).

8.4. SIGNIFICANCE OF THE GOLDEN FIBER REVOLUTION

» Jute increases the organic fertility of the soil for other crop plantations.

» Organized processing and cultivation of jute help farmers to earn and save substantial money from carbon credits.

» The valuation of the carbon credit of jute cultivation is pinned at farmers' savings made in purchasing inorganic fertilizer.

» Renewable resource application of jute has made it a key raw material in the paper industry.

» Burning one ton of plastic bags emits 63 gigajoule of heat and 1,340 tons of carbon dioxide whereas burning one ton of jute bags emits only 2-gigajoule heat and 150 kg carbon dioxide.

8.5. RESEARCH AND DEVELOPMENT

Research on jute technology is being carried out at the National Institute for Research on Jute and Allied Fiber Technology (formally known as Jute Technological Research Laboratory), Kolkata.

8.5.1. Grading of raw jute for quality evaluation

Grading of raw jute and its implementation for the benefit of both cultivators and industry is one of the significant contributions of the institute. The present BIS system (IS: 271-1975) of grading of jute is the outcome of detailed study on physical properties of fiber. Important physical characters of fiber were identified on the basis of which a systematic and scientific grading method had been developed. The system has been implemented by both farmers and traders.

8.5.2. Development of grading aids

Graded jute albums containing model samples of jute of different characters and graded jute sample box have been developed as ready reference and standard grading aids for field level graders.

8.5.3. Upgradation of low quality barky jute

The technology for upgrading the barky jute by a special fungal culture (*Aspergillus* sp.) has been developed. The fungus can be supplied in a solid carrier based form in small polyethylene bags. The work was awarded by the NRDC and WIPO (World Intellectual Property Organization, a constituent body of the United Nations). The technology has been transferred to M/s Assam Agro Industries Development Corporation, Guwahati, for commercial production of fungal culture.

8.5.4. Bacterial cultures for accelerating the retting of jute

A mixed bacterial culture has been developed which added to the retting system, accelerates the retting process without affecting the fiber properties. Wide demonstrations throughout jute growing states have established the acceptability of the method.

8.5.5. Development of jute Ribhoner machine for improved retting and extraction of fiber

The main constraint in jute retting is the lack of available retting water. A portable ribboner machine which can strip off the bark from the harvested jute plants before retting has been designed and developed so as to facilitate the retting of green ribbons in smaller volume of water conveniently and produced fiber of improved quality. The process has been field demonstrated and is awaiting implementation by jute growers.

8.5.6. Development of instruments

Various testing instruments, viz. Bundle strength tester, Air flow fineness tester, Bulk density meter, Color and luster meter, Automatic irregularity tester, Ultrasonic emulsifier, Yarn hairiness meter, Instrument for measuring thermal insulation value of fabric have been designed, developed and commercialized on small-scale by the institute.

8.5.7. Development of small-scale jute spinning plant

Small scale jute spinning machinery have been developed which can produce sutli and other high value blended products at the decentralized sectors. The field trials unit has been installed at Kamarpukur Pallimangal unit of R.K. Mission.

8.5.8. Utilization of different agricultural wastes

The technologies for processing agricultural waste fibers viz., pineapple leaf fiber, banana fiber, aeronaut fiber *etc.* to produce yarn either alone or in admixture with other natural and synthetic fibers have been developed. The technologies are awaiting adoption.

8.5.9. Development of different jute products

Various diversified products viz., cement bag, book binding cloth, decorative fabric, union fabric, high performance jute fabric *etc.* have been developed. The technologies need large-scale trials on pilot scale before release to the entrepreneurs.

8.5.10. Development of different types of carpets from texturized jute/polypropylene blended jute

Chenille, hand-knitted, tufted and woven carpets can be made from jute or its blends with polypropylene. These carpets are quite comparable with all-wool carpets in respect of tuff withdrawal force and recovery from compression. The technologies can open up new markets for jute.

8.5.11. Development of caplon blankets from chemically texturized and jute-polypropylene blended yarn

The blankets developed from texturized jute/PP blended jute are commercially known as "caplon" blankets. In respect of strength, moth-resistance and thermal properties, the caplon blankets are to some extent superior to all-wool blankets. The technology has been transferred to entrepreneurs both in corporate and decentralized sectors.

8.5.12. Development of blended yarn technology for jute-natural fiber/synthetic fiber for manufacturing non-conventional jute products

The Institute has worked exhaustively on blended yarn technology using jute. Some of the blended products are as follows:

» Jute-mesta blend for reducing the cost of conventional products such as Hessian and sacking.
» Jute-ramie blend for making upholstery, furnishing and industrial fabric.
» Jute-viscose blended yarns for development of decorative fabric.
» Jute-polypropylene blended yarn for upholstery, furnishing fabric *etc.*
» Core-spun jute-HDPE/ HDPP monofilament for development of heavy-duty fabric.
» Blending jute yarn and HDPE tape at weaving stage for development of an improved jute cloth in flat-bed loom and circular loom.

The methods of spinning like (a) wrap spun, (b) twin spun, and (c) core spun 'yarn' of jute were studied in detail to overcome the shortcomings in conventional spinning and to make value added products. The findings made are well in advance of time and can be transferred to industries.

8.5.13. Adhesive bonded non-wovens

Pioneering work has been done in the field of non-wovens with the major objectives (i) to utilize short jute fibers which are not spinnable; (ii) To prepare light-weight jute-based fabrics; and (iii) to develop fabrics which will be able to fetch much higher prices than the conventional products.

The adhesive bonded non-wovens have been utilized for the products like (i) decorative fabrics, e.g., wall covering, window screen *etc.*, (ii) filter media, and (iii) substrates for coating, laminates and for fiber reinforced plastic.

Needle-punched non-wovens from jute and its blends may be used for (i) plaiting cloth and wool packing, (ii) bitumen-impregnated roof covering materials, (iii) substrates for reinforcing synthetic resin matrices, and (iv) Floor covering, sound insulating medium, blankets, carpet underlays, *etc.*

Furthermore, non-conventional raw materials *viz.*, goat hair and date-palm leaf have been found to suitable for making needle-punched non-wovens.

The non-wovens are finding new applications for making fabric to control soil erosion which have been tried and demonstrated satisfactorily under field conditions.

8.5.14. Packaging from non-wovens

Apart from various industrial applications of non-wovens, the institute has developed varieties of jute non-woven consumer bags which range from 'ladies' fancy hand bags to big shopper carry bags.

8.5.15. Particle board from jute stick

Technologies have been developed for manufacturing particle boards from jute sticks by chemi-mechanical processes using natural/synthetic hinders under various curing conditions. As substitute of costly wood particle boards, these boards have been found suitable in application for false ceiling, partition walls, furniture, and packaging materials. A number of entrepreneurs have started the commercial manufacture with the know-how of the institute.

8.5.16. Pulp and paper from jute sticks

Whole mesta plant can be pulped by chemi-mechanical processes to make newsprint. Pilot plant trials have been conducted both at Central Pulp and Paper Research Institute, Dehram and at "HINDU" where the feasibility of the process was established.

8.5.17. Biogas from jute wastes

Biogas containing more than 55% methane by volume can be generated using jute caddis as substrate. The raw material is slow decomposing and as such the generation of gas continues for a longer period than when conventional cellulosic raw materials are used. Furthermore, the residual slurry left in biogas plants after biogas production can be used as manure for jute cultivation as well as for production of both summer and winter mushroom.

8.6. SCHEMES FOR JUTE DEVELOPMENT

8.6.1. Mini Mission-II of Jute Technology Mission

The Government of India has launched the Jute Technology Mission (JTM) for the overall development of the jute sector in the country during 2006-07. Mini Mission-II (MM-II) of JTM was implemented by DAC&FW with the objective to increase the productivity and to improve the quality of fiber in 10 States viz. Andhra Pradesh, Arunachal Pradesh, Assam, Bihar, Meghalaya, Nagaland, Odisha, Tripura, Uttar Pradesh and West Bengal from 2006-07 to 2013-14. The sharing of funding of the components was 90: 10 between Central and State Governments. However, in some of the components implemented by ICAR/ other central agencies, 100 per cent share was borne by the Central Government. Under the scheme, assistance was provided for production and supply of certified seeds, transfer of technology through field demonstrations and training of farmers, establishment of retting tanks, supply of fungal culture, farm implements and soil ameliorants, popularization of IPM practices, *etc*.

8.6.2. National Food Security Mission- Commercial Crops (Jute) Program

Jute Development Program is being implemented under National Food Security Mission - Commercial Crops (NFSM-CC) for enhancing production and productivity w.e.f 2014-15 in 9 States viz., Andhra Pradesh, Assam, Bihar, Meghalaya, Nagaland, Orissa, Tripura, Uttar Pradesh and West Bengal. Under this scheme thrust has been given on transfer of technology through frontline demonstrations and training in order to extend benefits to the farmers. From, 2015-16, in view of increased devolution to the States on account of implementation of recommendations of 14th Finance Commission, NFSM is being implemented on sharing basis between Government of India and States on 60: 40 basis for general category states and 90: 10 basis for North East and hilly states. However, the Central Agencies are funded 100% by GOI.

8.6.3. Jute development under Rashtriya Krishi Vikas Yojana (RKVY)

There is no earmarked allocation of funds for jute under RKVY. However, the States can undertake jute program under RKVY with the approval of State Level Sanctioning Committee (SLSC) under the chairmanship of Chief Secretary of the State.

8.6.4. National Jute Board

The NJB is statutorily mandated to undertake measures to:-

» Evolve an integrated approach for jute cultivation in the matters of formulation of schemes, extension work, implementation and evaluation of schemes aimed at increasing the yield of jute and improving the quality thereon.

» Promote production of better quality raw jute.

» Enhance productivity of raw jute.

» Promote or undertake arrangements for better marketing and stabilization of the prices of raw jute.

» Promote standardization of raw jute and jute products.

» Suggest norms of efficiency for jute industry with a view for eliminating waste, obtaining optimum production, improving quality and reducing costs.

» Propagate information useful to the growers of raw jute and manufacturers of jute products.

» Promote and undertake measures for quality control or raw jute and jute products.

» Assist and encourage studies and research for improvement of processing, quality, techniques of grading and packaging of raw jute.

» Promote or undertake surveys or studies aimed at collection and formulation of statistics regarding raw jute and jute products.

» Promote standardization of jute manufactures.

» Promote the development of production of jute manufactures by increasing the efficiency and productivity of the jute industry.

» Sponsor, assist, coordinate, encourage or undertake scientific, technological, economic and marketing research pertaining to the jute sector.

» Maintain and improve existing markets and to develop new markets within the country and outside for jute manufactures and to devise marketing strategies in consonance with the demand for such manufactures in the domestic and international markets.

» Sponsor, assist, coordinate or encourage scientific, technological and economic research in the matters related to materials, equipment, methods of production, product development including discovery and development of new materials, equipment and methods and improvements in those already in use in the jute industry.

» Provide and create necessary infrastructural facilities and conditions conducive to the development of diversified jute products by way of assisting the entrepreneurs, artisans, craftsman, designers, manufacturers, exporters, non-Governmental agencies in the following manner:

- Transfer of technology from research and development institutions and other organizations in India and abroad.
- Providing support services to the entrepreneurs for the implementation of their projects including technical guidance and training.
- Organizing entrepreneurial development programs.
- Planning and executing market promotion strategies including exhibitions, demonstrations, media campaigns in India and abroad.
- Providing financial assistance by way of subsidy or seed capital.
- Providing a forum to the people engaged or interested in diversified jute products for interacting with various national and international agencies, engaged in the jute and jute textile sector.

» Organize workshops, conferences, lectures, seminars, refresher courses and set up study groups and conduct training programs for the purpose of promotion and development of jute and jute products.

» Undertake research on jute seed to improve quality and to shorten the gestation period of jute crop.

» Incorporate measure for sustainable Human Resource Development of the jute sector and to provide necessary funds for the same.

» Modernization of jute sector and technology development.

» Take steps to protect the interests of jute growers and workers and to promote their welfare by improving their livelihood avenues.

» Secure better working conditions and provisions and improvement of amenities and incentives for workers engaged in the jute industry.

» Register jute growers and manufacturers on optional basis.

» Collect statistics with regard to jute and jute products for compilation and publication.

» Subscribe to the share capital of or enter into any arrangement (whether by way of partnership, joint venture or any other manner) with any other corporate body for the purpose of promoting the jute sector or for promotion and marketing of jute and jute products in India and abroad.

8.6.5. Jute Corporation of India (JCI) Ltd., Kolkata

JCI is a GoI enterprise set up in 1971. JCI is the official agency of the Ministry of Textiles (MoT) responsible for implementing the MSP policy for jute producers and serves as a stabilizing agency in the raw jute market. JCI also undertakes commercial operations, procuring jute at prices above the MSP on commercial consideration to generate profits. JCI's price support operations involve procuring raw jute from small and marginal farmers at MSP without any quantitative limit as and when the prevailing market price of jute falls below the MSP. These operations help create a notional buffer in the market by siphoning off excess supply, in order to arrest inter-seasonal and intra-seasonal fluctuations in raw jute prices. JCI's Departmental Purchase Centers (DPCs), which are situated in rural areas, purchases raw jute directly from the farmers.

8.6.5.1. Mission/Vision: To act as the Price Support Agency of the Govt. of India and to undertake Minimum Support Price (MSP) Operation to provide remunerative price of raw jute to the jute growers and gradually increase its market share in the domestic trade.

8.6.5.2. Main functions

» To take up Price Support Operation on behalf of the Government as and when the raw jute prices touch the level of the Minimum Support Price fixed by the Government of India without any quantitative limit.

» To undertake commercial operations for jute mills of NJMC as and when required.

» To undertake distribution of certified jute seeds under subsidy scheme of NJB and gradually increase the quantity with an objective to make available certified jute seeds to farmers.

» To conduct other extension activities like, demonstration of new retting techniques and display of daily market rate by allocating centers under JTM MM-III and NJB schemes for the benefit of the jute growers.

» To carry out the role of the implementing Agency for Mini Mission III and also provide necessary support on activities of Mini Mission IV and other Mini Missions of the Jute Technology Mission.

» To plan and implement schemes under Corporate Social Responsibility.

8.6.6. Indian Jute Industries Research Association (IJIRA), Kolkata

IJIRA is headquartered in Kolkata, with its regional centers being at Cherthala (Kerala), Vijayanagaram (Andhra Pradesh) and Guwahati (Assam). Guwahati Regional Centre also has the Power loom Service Centre co-located with it. There is a Testing Laboratory too at Shantipur (West Bengal). Since its inception, IJIRA has, over the years, grown into a world class research institute on jute.

8.6.6.1. Major areas of research

» Development of an eco-friendly substitute of JBO for jute processing.

» Light fast bleached and dyed jute product development.

» Jute bamboo composites.

» Development of technical textiles such as wider braided jute fabric.

» Jute-ramie blended finer yarns and fabric development.

» Energy efficient green sizing technology for jute yarns.

» Development of aroma based home textiles.

» Multifunctional ceramic based nano-finishing outdoor textiles by sol-gel-methods.

» Development of low cost jute bags for food grains and sugar.

» Quality assurance for food grade jute products.

In additional to rendering various technical services to Jute Sector, IJIRA is presently involved in Jute Technology Mission Projects DDS 7.1 and DDS 6.4 projects, promotion and techno-marketing of jute geo-textiles in association with National Jute Board, quality assurance of food grade jute products and consultancy.

8.6.7. National Research Institute for Jute Allied Fibers (NRIJAF)

The following are the mandate of NRIJAF:

» Improvement of jute (*Corchorus capsularis* and *C. olitorius*) and allied fiber crops like mesta

(*H. cannabinus* and *H. sabdariffa*), sun hemp, (*Crotalaria juncea*), rame (*Boehmeria nivea*), sisal (*Agave sisalana*) and flax (*Linum usitatissimum*) for yield and quality.

» Improvement of jute and allied fiber crops for biotic and abiotic stresses.

» Development of economically viable and sustainable production technology and cropping systems with jute and allied fiber crops.

» Transfer of technology and human resource development in relation to jute and allied fiber crops.

» Development of proper post-harvest technology for improving the quality of fiber.

8.6.8. UNDP - Govt. of India National Jute Development Program

The UNDP - Govt. of India National Jute Development Program was launched in 1992 with a UNDP assistance of US $ 23 million and a Govt. of India commitment of Rs 500 million. It was a program to revitalize the jute sector and to diversify its products to realize higher values for the jute fiber. Its aim was to search for and develop diversified, high value-added jute-based products through an ambitious R and D program, and to encourage entrepreneurs to commercialize it through an incentive credit facility. Machinery needed to modernize the jute muth was also to be indigenized. During the progress of the program, several changes were made and new activities were accommodated by the suggestions made by evaluation teams.

The terminal evaluation of the program was carried out in May 1998. The evaluation teams finds that the National Jute Development Program was highly successful to meet its principal objective to revitalize India's jute industry and its R and D. The Ninth Five-year Plan (1997-2002) document emphasis on economic growth, industrial production, productivity and exports, infrastructure, building and agricultural production.

It also stressed on employment generation particularly for weaker sections, promote environment friendly practices. The National Policy framework on Generation of Productive Employment (NPFGPE) aims to provide opportunities for employment, technological advancement and to support economic activity for villages, small scale and micro enterprise.

The Government and UNDP outlined the country cooperation framework for 1997-2002. Under this framework a fiber and handicrafts program and its jute sub-program was designed. The Fiber and Handicrafts Program (FHAP) consists of a number of sub-programs pertaining to jute, non-mulberry silk, angors wool, hand knotted carpets and cane and bamboo.

8.6.8.1. The Jute Sub-Program: The jute sub-program proposes to carry on the key activities of the National Jute Development Program and provide support to microenterprises, human resource development and jute agriculture. The sub-program also support as quality assurance for jute products, machinery development for the organized and the decentralized sectors and assistance to the entrepreneurs from the special jute development fund.

The jute sub-program is structured to provide focused intervention on priority areas to enable proper utilization of UNDP resources. The sub-program fully utilizes the manpower, technical knowledge and the network of National Jute Development Program.

The Ministry of Textile, Govt. of India is responsible for policy formation, regulation development and export promotion of textile sector. Ministry of textiles decentralized the sectors into Development Commissioner for Handlooms, Handicrafts; Textile Commissioner; and the Jute Commissioner. The ministry also coordinates eight research associates related to textiles.

The Jute Manufactures Development Council has been set up as a statutory body for the development and production of jute manufactures the National center for jute diversification gives a focused attention to diversification efforts in the jute sector. The jute sub-program aims to follow up the results of the National Jute Development Program:

» Support NGO, identity and promote new NGO to generate at least 3000 jobs.

» Implementation of strategic marketing plan for jute goods of high value.

» Mill workers to be trained in modern technologies and process.

» Promotion of fine jute fiber cultivation, identification and procurement of texture equipment *etc.*

8.6.9. Jute Integrated Development Scheme

8.6.9.1. Objectives

» To highlight / create more awareness for Jute products on the merit of environment related issues.

» To increase the number of skilled workforce for production of jute diversified products in the decentralized sector.

» To set up a greater number of production units for use of jute as raw material in order to expand the production base for the diversified jute products.

» To create employment for the rural masses in the newly established Jute Diversified Products (JDP) production units.

» To give thrust for design and development of new and innovative jute products to make the jute products and even the traditional products acceptable in the domestic as well as the foreign markets.

» To increase demand for consumption of raw jute for producing JDPs, thereby ensuring continuous market for the produce of jute farmers.

» To build up supply chains for certain and smooth supply of jute raw materials to the production units

» Establishment of trade channels for promotion of decentralized JDP units.

» To increase sale of JDPs through market promotion activities to ensure continuous markets for jute products

8.6.9.2. Expected outcome: This Scheme will be implemented by the NJB on annual basis for six years, beginning 2015-16. It is estimated that by the end of the 6 years period i.e. 2020-21, 100 clusters will have taken shape. This will directly and indirectly benefit at least 160 individuals per cluster. It is estimated that during this implementation period, the total involvement of individuals mostly women (WSHGs), who will be gainfully engaged in employment in the jute diversification, will be to the tune of at least 15,000.

The outcome of the scheme would also establish the competitive environmental advantages of jute globally as well as boost the sale of jute goods globally and increase the share of JDPs too.

Over a period of time, consumption of raw jute will be increased. The environment will remain safe for the mankind as jute is a natural and earth-friendly material.

8.6.9.3. Scheme details: To achieve the above objectives an integrated and also modular approach has been envisaged and conceptualized with emphasis on the following key elements.

>> Developmental schemes
>> Raw material and retail outlet scheme
>> Marketing scheme

8.6.9.4. Developmental schemes: This will involve setting up of JDP Cluster for the jute carry bag, jute handloom and handicraft sector to take care of a wide gamut of activities such as product development, design support, institutional support, training to weavers and artisans and marketing support *etc.* at micro level in an integrated, modular and coordinated manner for an overall development of the decentralized jute product sector. Further, it will provide escort support services in order to make them full-fledged jute diversified product entrepreneurs over a period of time.

i) Implementation strategy

>> Creating awareness and capacity building workshops in different parts of the country on the use of jute and encouraging small and tiny entrepreneurs to manufacture jute diversified products with a sustainable approach.

>> Promoting use of jute in more diversified application areas by conducting specific training programs, organizing awareness programs and through product/design development programs *etc.*

The collaborating agencies for operations under the scheme will be selected from Govt. / Semi-Govt. organizations, autonomous bodies, reputed public sector organizations, NGOs, SHG Federations, Institutes, and Entrepreneurs having adequate experience in carrying out development activities for promotion of entrepreneurship and industries. These collaborating agencies will act more as facilitators for rendering the backward and forward linkages to existing and potential entrepreneurs.

All activities are to be carried out with proper linkages with the State Governments/lead NGOs/ Co-operative Societies/Agencies of the Central and State Government and implemented with the help of the District Collector, PD, DRDA, DIC *etc.* Linkages will be established with handicrafts and handloom development programs of the Government for better results.

8.6.9.5. Raw material and retail outlet support scheme: Supply outlets for distribution of jute raw material are to be developed for making available jute fiber, fabric and yarn in small quantities as required by the micro-enterprises, crafts persons and artisans at their areas at mill gate prices.

i) Implementation strategy

>> Setting up of outlets in association with the collaborating partners for selling of jute yarn, fiber and other raw materials across the country at reasonable price, preferably mill gate price.

>> Providing information on applications of jute in various uses and awareness generation.

Supply outlets for raw material are to be opened where JDP activities have taken off and jute based raw materials are not available in the local market. Raw materials and services shall be provided to suit the requirement of small artisans, entrepreneurs, women self-help groups, weavers and handicraft artisans. Linkages will be established with handicrafts and handloom development programs of the government for better result.

8.6.9.6. Marketing schemes: Market development due to acceptance of jute products in various segments of the domestic and international markets constitute the background for the activities to be undertaken for promotion of JDPs in the decentralized sector. In consonance with the growth of the decentralized JDP sector and the emergence of a varied range of products to cater to different segments of the consumers, a concentrated campaign bundled with existing SOP (Student Outreach Program) may be launched and carried out on a long term basis. The primary beneficiaries will be small and tiny entrepreneurs of jute diversified products with its spread across the country.

i) Implementation strategy

»   Development of market information.

»   Creating awareness of the market and products through seminars and workshops.

»   Development of entrepreneurs by providing escort services and providing various marketing platforms *viz.* promotion by participation in trade shows, National level fairs, Organization of jute fairs at State/district and village levels.

»   Brand promotion of jute product (Made of jute, the Indian natural fiber).

»   Focused advertisement campaigning giving thrust on product specific campaigns and highlighting latest developments in jute diversification throughout the year.

»   Eco-friendliness and carbon positive attribute of jute will be highlighted.

»   Bundling with the existing Retail Outlet Scheme.

»   Business generation with sustainable approach.

8.7. IMPORTS AND EXPORTS

8.7.1. Imports

India imports jute mainly from Bangladesh and Nepal only.   It is reported that there is large scale import of cheap yarn, B-Twill fabric and sacking from Bangladesh and Nepal which is seriously distorting the Indian market. Bangladesh provides export incentives to its producer's which along with other factors such as low wages and power cost make their products significantly cheaper than Indian products in Indian market. Bangladesh Government offers a 7.5% cash subsidy on the export of jute bags, which coupled with cheaper labor, cheap power and low capital cost makes imported jute bags from Bangladesh cheaper than bags manufactured in the country.

Orders are issued from time to time under the JPM Act by the Government of India, Ministry of Textiles, stipulating that jute packaging material for reserved commodities should be manufactured in India from raw jute produced in India. Orders have been issued for compliance during 2002 under the JPM Act directing all manufacturers / importers / processors / traders to mark / print/ brand or stitch a jute cloth/eco-friendly cloth label printed with the country in which manufactured. Customs authority has also been requested from time to time to ensure that import of jute goods takes place only after the directions in the above Notifications are complied with.

8.7.2. Exports

The jute industry is traditionally export oriented. India ranks first in raw jute and jute goods production. It ranks second in export of jute goods in the world. Jute packing materials are facing tough competition from other low-cost synthetic substitutes.

About 35% of the manufactured jute items are exported. In 1982-83, the country earned Rs. 202.76 crores by way of exporting jute goods, whereas in 1985-86, India exported jute manufacturing's worth Rs. 270 crores.

The markets are Australia, New Zealand, U.S.A., Canada, Indonesia, Japan, Argentina, and Russia, European, African and Middle East countries (Table 8.2).

Table 8.2. Export to top 5 countries from India

Country	Quantity (Tons)	Value (million Rs.) (Apr'15 to March'16)
Nepal	24943	1126.495
USA	12	3.218
Ethiopia	40	1.728
Japan	67	1.718
Tunisia	29	1.681

The Exports trends of different jute items during the year 2014-15 and 2015-16 is presented in Table. 8.3.

Table 8.3. Exports trends of different jute items during the year 2014-15 and 2015-16

Apr-Mar	2014-15		2015-16*	
Item	Quantity ('000 tons)	Value (Rs. in Crores)	Quantity ('000 tons)	Value (Rs. in Crores)
Hessian	80.2	769.5	2.2	15.7
Sacking	46.9	296.6	1.1	6.9
Yarn	23.6	138.7	4.0	25.6
JDP	0	508.6	0	43.9
Others	7.7	100.4	0.4	4.5
Total	161.7	1813.8	7.7	96.6

Source: National Jute Board   (*) up to April 2015

8.8. FUTURE PROSPECTS

The future prospects of the Jute industry, however, is bright due to the following:

- » Diversification of jute products
- » Environmental awareness
- » Ban on polythene and plastic bags
- » Increasing use for oil conservation
- » Construction of bunds, river embankments, landslide protection
- » Along with cotton, jute is also being used for apparel manufacturing.

8.8.1. New initiatives

- » Possibilities are being explored for large scale use of jute in making paper and composites.
- » The use of Jute Geo-textiles is being promoted.
- » R&D is being promoted for the use of jute in different sectors like automobiles, roads, construction etc.
- » New retting technology have been developed and trials of the same is being done.

>   » Efforts are being done to modernize the jute industry.
>   » Efforts are being done to explore new uses of jute fiber on mass scale through National Institute of Design (NID).
>   » The Jute Diversified Products (JDPs) are being promoted for domestic consumption as well as for exports.

8.9. CONCLUSION

Indian jute industry is facing two big challenges in recent times which are high production cost and inadequate supply of capital. Therefore, new technologies should be introduced to produce standard jute products at low cost to capture the growing international market. Besides, supply of raw material should be brought under control, labor rate should be held in check, and proper policies are to be framed to maintain a sustainable growth. Experience suggests that the jute industry in India had flourished in the past because of its favorable environment, availability of labor and demand of its jute products from national and international markets *etc*. Therefore, chances are still there to make Indian jute industry a grand success and for this purpose some true initiatives as suggested above, need to be taken to replace this present inconsistent growth by a consistent one and hence help the industry grow further and sustain for a longer period.

Chapter - 9

Pink
(Onion Production) Revolution

9.1. INTRODUCTION

Onion (*Allium cepa*) is extremely important vegetable crop not only for internal consumption but also as highest foreign exchange earner among the vegetables. It occupies an area of 1.431 million ha, with production of 22.819 million tons (Indian Horticulture Database, 2019-20). The export of onion during 2020 -21 was 1,434.925 million tons with a value of Rs 2496.68 crores. It is used either fresh as a salad or in preserved form (Islam *et al*, 2007). Leading countries in onion production are China, India, Pakistan, Bangladesh, Indonesia and Turkey.

In India onion is grown under three crop seasons i.e. *kharif*, late *kharif* and *rabi*. More than 60 per cent production comes from *rabi* crop and rest from the other two seasons. Maharashtra, Madhya Pradesh, Karnataka, Gujarat, Rajasthan, Bihar, Haryana, Andhra Pradesh, Tamil Nadu, and West Bengal are main onion growing states in India (Table 9.1). Productivity in late *kharif* and *rabi* is 25 t/ha and in *kharif* is 8-10 t/ha. The area and production of onion has increased many fold since 1970, but productivity enhancement is slow.

Table 9.1. Top 10 states in onion production (2017-18)

State	Area ('000 ha)	Production ('000 tons)	Productivity (tons/ha)
Maharashtra	507.96	5355.39	10.543
Madhya Pradesh	150.87	3859.83	25.584
Karnataka	195.28	3197.40	16.373
Gujarat	22.49	1303.07	57.940
Rajasthan	64.76	1292.20	19.953
Bihar	53.77	1248.96	23.228
Haryana	29.93	905.80	30.264
Andhra Pradesh	42.00	695.12	16.550
Tamil Nadu	28.36	187.50	6.611
West Bengal	35.20	560.65	15.927

The pungency in it is due to the presence of allyl propyl disulphide, the bioflavonoid present in the yellow color of the outer skin of onion bulb is due to Quercetin. It contains anti-fungal property viz. catechol. The carbohydrate present in it is fructan and is the richest source of Vanadium. It is helpful in preventing arthritis, heat stroke, coronary heart disease, diabetes, cancer, asthma, *etc*. It induces tears but still liked by everyone. It becomes a part and parcel of Indian cuisine.

9.2. AREA, PRODUCTION AND YIELD

Onion is grown in an area of 1285 thousand hectares with production of 23,262 thousand tons (Anonymous, 2018). India ranks second in onion production after China (Anonymous, 2018). The area under onion cultivation in decades remain constant. But the production data shows variation in the decades. Highest area was found under Maharashtra (507.96 thousand hectare) followed by Karnataka (195.28 thousand hectare) and least under Tripura (0.16 thousand hectare). Maharashtra produces highest onion (8854.09 thousand tonnes) followed by Madhya Pradesh (3701.01 thousand tons) and least production was found at Kerala (0.31 thousand tons). Productivity was observed highest at Sikkim (56.45 t/ha). The major importer of onion are Bangladesh, Malaysia, Sri Lanka, United Arab Emirates, Nepal, Indonesia, Qatar, Vietnam, Social Republic, Kuwait, Oman, *etc*.

Although second in onion production after China at world level, we are far behind in productivity compared to many countries. The average productivity of onion in India now stand at only 16.11 t/ha, which is lower than world average of 18.67 t/ha. The highest productivity of onion has been reported to be 62.50 t/ha in Ireland (DES, 2013).

The main reasons for low productivity of onion in India are listed below:

» Inherent low yield potential of short-day onion varieties.

» Non -availability of suitable FI hybrids.

» Susceptibility of all cultivars to diseases, pests and abiotic stresses.

» Tropical climate is more congenial for diseases and pests.

» Non - availability of genuine seeds of released varieties.

» Sub - optimal standards of cultivation adopted by farmers.

» Shortage of irrigation at critical stages.

» Poor storage capacity of present-day varieties and poor storage facilities.

» Kharif crop always pulldown country's average productivity.

» Fluctuation of prices distracts the attitude of farmers towards use of inputs and modern technology.

The major ten onion producing states in India with respect to area, production and productivity are shown in Fig. 9.1, Fig. 9.2 and Fig. 9.3, respectively. Highest area was found under Maharashtra (507.96 thousand hectares) followed by Karnataka (195.28 thousand hectares) and Madhya Pradesh (150.87 thousand hectares) (Fig. 1). Maharashtra produces highest onion (8854.09 thousand tons) followed by Madhya Pradesh (3701.01 thousand tons) and Karnataka (2986.59 thousand tons) and least production was found at Kerala (0.31 thousand tons). Productivity was observed highest in Sikkim (56.45 tons/ha), which is really an astonishing fact followed by Madhya Pradesh (24.53) and Gujarat (24.29). Since, Sikkim has been declared as an organic state, they are utilizing only organic products.

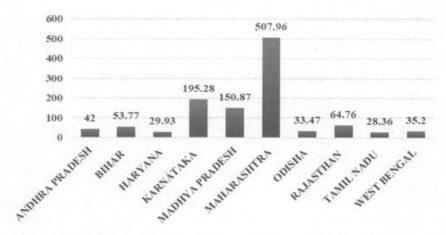

Fig. 9.1. Area-wise top 10 onion producing states ('000 ha).

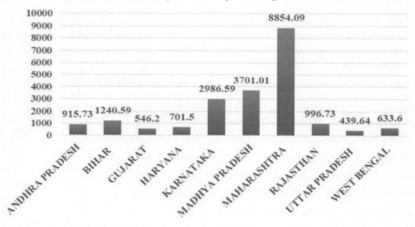

Fig. 9.2. Production-wise top 10 onion producing states ('000 tons).

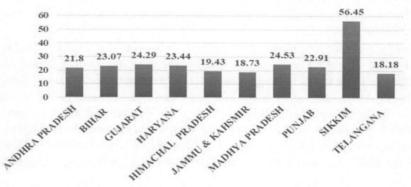

Fig. 9.3. Productivity-wise top 10 onion producing states (tons/ha).

The production has increased more than five times during the past three decades. The reason for increase in production is mainly due to expansion in area under onion cultivation from 0.33 million ha during 1991-92 to 1.20 million ha in 2013-14 with rise in production and productivity. India possesses many innate advantages over other onion producing countries, its large genetic base, varied soil and climatic conditions and skilled human power and numerous public organizations which worked on onion improvement and overall growth of onion industry in India. Still public and private organizations are involving in onion improvement to increase the productivity of onion in India.

9.3. IMPROVED VARIETIES/HYBRIDS

Three main cultivars of onion are found in India based on the outer color as Red color varieties (Arka Akshay, Arka Bheem, Arka Kalyan, Arka Kirthiman, Arka Lalima, Arka Vishwas, etc.); Yellow color varieties (Arka Pitambhar, Arka Sona etc.) and white color varieties (Arka Swadista, Pusa White Flat, etc.). Among these, red varieties have more demand for consumption purpose and white cultivars for processing purpose.

9.3.1. Improved varieties

Varieties developed by various organizations have been tested at different locations under the All India Coordinated Vegetable Improvement Project (AICVIP) and based on their performance; the following varieties have been recommended for cultivation (Table 9.2).

Table 9.2. Improved onion varieties suitable for different seasons

Variety	Organization	TSS (%)	Yield (tons/ha)
Punjab Selection	PAU, Ludhiana	14	*Rabi* - 20
Punjab Red Round	PAU, Ludhiana	---	28-30
Pusa Red	IARI, New Delhi	12-13	Late *Kharif* & *Rabi* - 30
Pusa Madhavi	IARI, New Delhi	---	*Rabi* - 35
Pusa Ratnar	IARI, New Delhi	11-12	*Rabi* - 30
Pusa White Flat	IARI, New Delhi	12-14	*Rabi* - 32.5
N-2-4-1	Agrl. Dept., Maharashtra State	12-13	30-35
Arka Kalyan	IIHR, Bangalore	11-12	33.5
Arka Niketan	IIHR, Bangalore	12-14	*Rabi* - 34
Arka Akshay	IIHR, Bangalore	---	45
Arka Sona	IIHR, Bangalore	10	*Rabi* - 45
Arka Bheem	IIHR, Bangalore	---	47
Kalyanpur Red Round	CSAUAT, Kanpur.	13-14	25-30
Agrifound Dark Red	NHRDF, Nashik	12 13	*Kharif* - 30 to 40
Agrifound Light Red	NHRDF, Nashik	13	*Rabi* - 30-32.5
Bhima Super	DOGR, Rajgurunagar	10.0 to 11.0	*Kharif* - 20 to 22, late *kharif* - 40 to 45

Bhima Raj	DOGR, Rajgurunagar	10.0 to 11.0	*Rabi* - 25 to 30
Bhima Red	DOGR, Rajgurunagar	10.0 to 11.0	*Kharif* – 19 to 21, late *Kharif* - 48 to 52, *Rabi* - 30 to 32
Bhima Shweta	DOGR, Rajgurunagar	---	*Kharif* - 18 to 20, *Rabi* – 26 to 30
Bhima Shubhra	DOGR, Rajgurunagar	---	*Kharif* - 18 to 20 t/ha, late *Kharif* – 36 to 42 t/ha
Bhima Kiran	DOGR, Rajgurunagar	12	*Rabi* - 30.3 t/ha
Bhima Shakti	DOGR, Rajgurunagar	11.8	*Kharif* - 45.9, *Rabi* - 42.7 in Maharashtra; *Rabi* national av. - 29.18
PKV White	PDKV, Akola	10	*Rabi* - 25 to 30

9.3.2. Processing varieties

Processed products of onion are in demand in many countries. Dehydration industries demand white onion varieties with globe shaped bulb and high TSS (>18%). Some of the varieties identified as suitable for dehydrated products were Punjab-48 (Bajaj *et al.*, 1979; Verma *et al.*, 1999), Roopali (Maini *et al.*, 1984), S74 (Kalra *et al.*, 1986), Texas Yellow (Raina *et al.*, 1988) and PWO-1 (Saimbhi and Bal, 1996). After assessing Indian varieties and landraces which do not offer TSS range more than 12%, Jain Food Park Industries, Jalgaon introduced White Creole, which was further subjected to selection for high TSS and developed V-12 with TSS range of 15-18% (Mahajan *et al.*, 2011).

9.3.3. Hybrids

Only two hybrids, Arka Kirtiman (47 t/ha) and Arka Lalima (50 t/ha), have been released by IIHR after development of CMS lines along with the maintainer, by Pathak *et al.* (1986). Aghora and Pathak (1991) reported heterosis in bulb-yield of up to 28.5% over the best commercial variety in short-day onion.

Some of the exotic hybrids perform well during late *kharif* under Indian conditions and yields are almost double that of the Indian varieties at Directorate of Onion and Garlic Research, Rajgurunagar, but these have very low TSS, storage life and are yellow in color, which has no consumer preference in India. These can be used to capture the European and Japanese markets where there is a great demand, but it is possible only through a cool chain. Of the 90 exotic varieties tested during 2000-2008, more than 20% higher yield was recorded in 10 varieties during late *kharif* over Bhima Super, and 16 during *rabi* season over N-2-4-1 under Maharashtra conditions (Table 9.3). Yield increase was recorded up to 60.87% in late *kharif* and 57.41% in *rabi* over the respective checks of best open-pollinated varieties (Table 9.4).

Table 9.3. Performance of exotic hybrids/varieties of onion in *kharif* season under the Indian plains (2000-2008)

Exotic onion variety	Late *kharif* yield (t/ha)	% Increase over Bhima Super
HN 9539	54.03	22.34
Hy 3404	57.36	29.89
DPS 2023	60.87	37.84
Early Supreme White	54.65	23.75
Cougar	56.50	27.95
DPS 1034	59.66	35.10
Linda Vista	50.58	14.53
Lexus	59.66	35.10
Kalahari	53.10	20.23
Rio-Tinto	54.37	23.11
Serengeti	55.87	26.52
Bhima Super (Check)	44.16	---

Table 9.4. Performance of exotic hybrids/varieties of onion in *rabi* season under the Indian plains (2000-2008)

Exotic onion variety	*Rabi* yield (t/ha)	% Increase over N-2-4-1
HN 9733	65.90	52.55
HN 9935	68.00	57.41
Cougar	67.84	57.04
DPS 1009	64.45	49.19
DPS 1024	66.05	52.89
DPS 1043	61.45	42.25
Mercedes	63.27	46.46
Lexus	63.83	47.75
Reforma	66.53	53.99
N-2-4-1 (Check)	43.20	---

9.4. GOOD AGRICULTURAL PRACTICES

9.4.1. Spacing and time of planting

Among the different spacing, 15 x 10 cm showed significant effect on marketable yield (32.67 t/ha), total yield (33.59 t/ha) and number of bulbs (644786) harvested per hectare as compared to remaining treatments. Onion transplanted on 15 November and 15 x 10 cm spacing significantly improved the number of bulbs harvested per hectare (650085), marketable yield (34.65 t/ha) and total yield (35.30 t/ha) (Devulkar *et al.*, 2015).

9.4.2. Micro-irrigation

Among the irrigation methods evaluated, drip irrigation at 100% Pan Evaporation (PE) significantly improved the marketable bulb yield in onion crop with 15-25%, higher per cent A grade bulbs, and water saving of about 35-40% and labor saving of 25-30% in comparison with surface irrigation (Sankar *et al.*, 2008, 2009).

9.4.3. Nutrient management

Onions and garlic have a relatively low nutrient uptake efficiency, due to their shallow root system characterized by thick roots with very few and short hairs. So to achieve optimum yields, it is important to apply nutrients as close as possible to the rooting zone frequently and in readily available forms.

9.4.3.1. Macronutrients: Major nutrient removal figures show that nitrogen and potassium are the nutrients needed in relatively greatest quantities in onion crop. They are fundamental for achieving high marketable yields (Fig. 9.4).

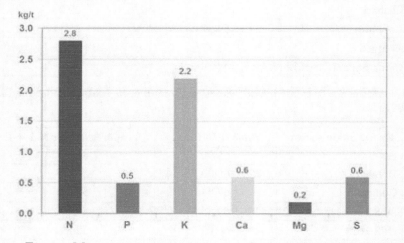

Fig. 9.4. Macronutrient removal by onions (kg/ton) (Bender, 1993)

i.   **Nitrogen:** High rates of nitrogen are needed to satisfy crop demand. However it is important not to over-apply nitrogen, particularly in bulb onions, as this can delay maturity, soften bulbs and lead to storage rots.

      A balanced N nutrition program can help to improve overall yield by increasing onion bulb weight (Fig. 9.5). Optimum nitrogen supply is important for onion bulb weight. Nitrogen supports production of leaves and hence, increases the number of bulb scales and the size and weight of the harvested onion bulbs.

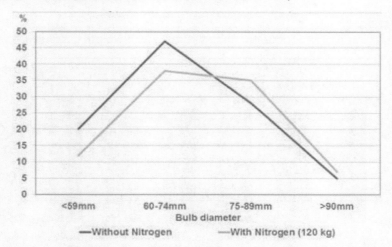

Fig. 9.5. Effect of Nitrogen rate on onion bulb size (Ruiz and Escaff, 1992).

ii. **Phosphorus:** Where crops are grown on soils with very low natural phosphorus levels, higher rates are required to boost yield, thereby increasing bulb weight and size (Fig. 9.6).

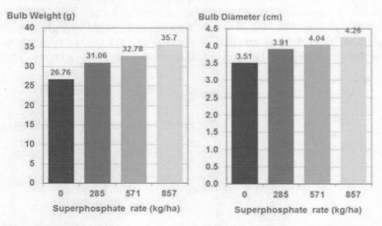

Fig. 9.6. Effect of superphosphate rate on onion bulb size and weight (Farghali and Zeid, 1995).

iii. **Potassium:** Potassium is particularly important where high rates of nitrogen have been applied, as potassium helps to maintain yield with reduced risk of storage problems that can be caused by excess nitrogen.

Onions are very prone to salt stress, so it is important to use sources of potassium that are less likely to cause damage. Toxic levels of chloride in the plant can lead to loss of yield and onion bulb weight in storage. Sulfate and nitrate forms of potassium usually result in lower bulb weight loss in storage (Fig. 9.7).

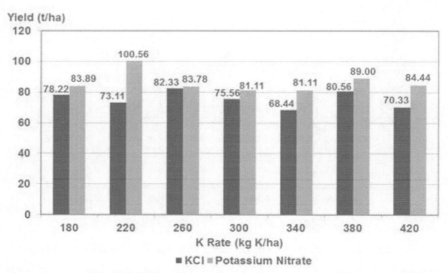

Fig. 9.7. Effect of potassium source on onion yield

iv. **Sulfur:** Sulfur is important for onion bulb weight and yield, and helps to improve the crop's utilization of nitrogen (Fig. 9.8). However, care has to be taken as high rates, particularly in S-rich soils, can have a detrimental effect on yield. It has also a marked effect on the pungency of the onion through increasing the pyruvic acid content of the bulb – a key quality characteristic.

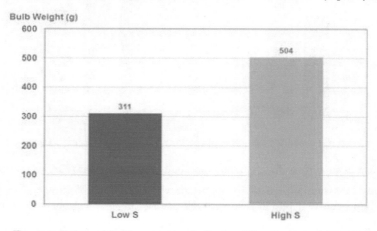

Fig. 9.8. Effect of Sulfur on onion bulb size (Hamilton *et al.*, 1997).

9.4.3.2. Micronutrients: Foliar application of urea + zinc + copper resulted in lowest decay and total loss in stored onions (Singh *et al.*, 2002). Improved plant growth and yield characters were observed at 0.03% boron and zinc at 0.025% (Sharangi *et al.*, 2003). Abd-El-Moneem *et al.* (2005) observed reduction in basal rot infection in garlic when cloves were treated with Zn and Cu before planting. Srivastava *et al.* (2005) reported that boric acid at 0.1% and zinc sulfate at 0.4% resulted in maximum bulb yield and total soluble solids.

Much lower levels of micronutrients are needed to satisfy yield and quality onion crop production and the correct balance of these trace elements is essential. The key micronutrients needed in greatest quantities are boron and iron. These have an influence on yield and quality. Zinc also pays a role in seed germination (Fig. 9.9).

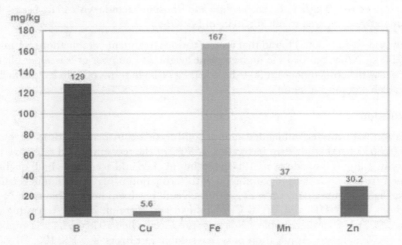

Fig. 9.9. Micronutrient removal by onions (mg/kg) by bulbs only

i) Boron: Trials confirm that boron has a positive effect on onion yield by increasing bulb weight. It is important to maintain the correct balance of calcium, nitrogen and boron in the soil as high Ca and high N levels can reduce boron uptake (Fig. 9.10).

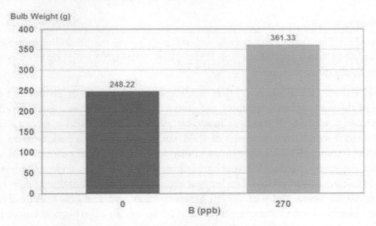

Fig. 9.10. Effect of Boron on onion bulb size.

9.4.3.3. Integrated nutrient management (INM): Proper application of organic manures, crop residues, green manure, suitable crop rotation, balanced application of fertilizers based on soil-testing is important. This can be achieved through integrated nutrient management practices.

According to Goto and Kimoto (1992), the highest commercial yields of onion bulbs were obtained by application of castor-bean cake along with P and K and, FYM combined with NPK.

Warade *et al.* (1996) obtained the highest bulb yield (22.7 t/ha) with 40 tons of FYM and biofertilizer inoculation along with NPK, thereby saving 25% on nitrogen alone. Bhonde *et al.* (1997) revealed that treatment of FYM at 15 t/ha + Azotobactor seedling dip and Nimbicidin application indicated a possibility of replacement of inorganic fertilizers under organic farming. Thilakavathy and Ramaswamy (1998) also opined that 2 kg/ha of Azospirillum and Phosphobacteria with 45 kg N and 45 kg P was more remunerative compared to 60: 30: 30 kg of NPK/ha.

Reddy and Reddy (2005) found that among various treatment combinations, vermicompost at 30 t/ha + 200 kg N/ha recorded the highest plant height and number of leaves per plant in onion, but was at par with vermicompost at 30 t/ ha + 150 kg N/ha in terms of bulb length, bulb weight in an onion-radish cropping system.

9.4.4. Fertigation

Drip irrigation with the recommended rate of solid fertilizer in 2 applications gave the highest bulb yield (49.635 t/ha) while drip fertigation at 50% of the recommended rate gave the highest bulb quality in onion (Chopade *et al.*, 1998). Optimum yield and acceptable bulb quality of onion was obtained from drip irrigation, combined with fertigation using NPK liquid fertilizer @150: 125: 200 kg/ha (Balasubramanyam *et al.*, 2000). Overall results indicated that with N fertigation improved bulb yield, NUE, and WUE (Mohammad and Zuraiqi, 2003). According to Rumpel and Dysko (2003), higher marketable yields were produced when 50 kg N/ha was applied through fertigation (41% increase). Application of water-soluble fertilizers @ NPK 100: 50: 80 kg /ha as basal + 50 kg N in onion through drip irrigation was the best treatment in terms of yield and cost: benefit ratio (Sankar *et al.*, 2005b).

Drip irrigation system not only helps in water saving but also reduces nitrogen losses by leaching into ground water as in fertigation fertilizer nutrients are applied in root zone only.

9.4.5. Cropping systems

Due to increased fertilizer prices and consideration for ecological sustainability, interest is focused on intensive cropping system, especially legume crops, in a sustainable crop sequence as an alternative or supplement to chemical fertilizers.

Arya and Bakshi (1999) observed that onion cultivation was more profitable when okra and radish, as one of the component vegetables, are grown in the vegetable sequence. Studies conducted at DOGR revealed that in *kharif* season followed by onion in *rabi* are ideal cropping sequences under western Maharashtra conditions in terms of yield, soil health and cost benefit ratio (Sankar *et al.*, 2005). There was a tremendous improvement in physical and chemical properties of soil in legume based cropping sequences particularly soybean followed by rabi onion and groundnut followed by late kharif onion (Sankar *et al.*, 2014). The highest intercrop yield was obtained when sugar beet was sown in ridges, 60 cm apart, and with 25 cm between sugar beet and onion. A gradual decrease in onion yield was observed with increasing inter- and intra-spacing. The highest land equivalent ratio (LER) was obtained from sugar beet-onion intercropping.

9.4.6. Intercropping

Onion is very much suited to grow as an intercrop in sugarcane under paired row planting system during winter season (November - December planting) (NRCOG, 2004). Since this crop is shallow rooted bulb forming vegetable having tow canopy, it does not compete with deep-rooted long duration crop like sugarcane. Sugarcane-onion intercropping is a common practice in some pockets of Haryana,

Maharashtra and Tamil Nadu. Singh (1996) reported that cane equivalent yield and net returns were high when sugarcane planted in autumn was intercropped with onions.

In Karnataka, onion is grown as an intercrop with chilli or cotton. Chilli intercropped with one row of multiplier onion cv. Co 2 recorded the highest yield of chilli pods and more net income per unit area per unit time compared to monoculture (Elangovan *et al.*, 1985; Dodamani *et al.*, 1993).

Khurana and Bhatia (1991) reported higher net returns in potato cv. Kufri Badshah intercropped with onion cv. Hisar-2 than fennel crop. Kothari *et al.* (2000) reported that mint (cv. Hy-77) intercropped with one, two and three rows of onion (cv. Nasik 58) increased the net return, land utilization efficiency, improved soil moisture (0-15 cm) and utilization of solar radiation than sole cropping. Ibrahim *et al.* (2005) reported the highest intercrop yield when sugar beet plants were arranged in ridges at 60 cm apart, and with distance of 25 cm between sugar beet and onion. Mollah *et al.* (2007) reported the highest groundnut equivalent yield and benefit: cost ratio from groundnut was intercropped with two rows of onion.

9.5. PEST MANAGEMENT

9.5.1. Weed management

According Abdel and Haroun (1990), Goal (Oxyfluorfen 23.6%) at 0.75 liters/ha applied 3 weeks after transplanting or Stomp Pendimethalin 50% at 2.0 liters applied after transplanting and before irrigation gave the best control, resulting in highest bulb yield. Tamil Selvan *et al.* (1990) reported that post-sowing application (3 days after sowing) of Oxyfluorfen at 0.1-0.6 kg/ ha gave more effective control of weeds in onion. Application of 0.2% Oxyflurofen 23.5 % EC before planting + one hand weeding at 40-60 days after transplanting is recommended for marketable bulb yield and weed control efficiency (65-80%).

9.5.1.1. Integrated weed management: Vinay Singh (1997) suggested that mulching at 30 DAT gave maximum bulb yield (26.33 t/ha), followed by 3 hand weeding's at 30, 60 and 90 DAT. Pendimethalin at 1.0 kg a.i./ha + 1 manual weeding at 60 DAT proved to be the most economical with a cost: benefit ratio of 2: 3.1. Well prepared and pre-irrigated seedbed plots covered with 50 μm-thick, transparent, polyethylene mulch for 6 weeks prior to onion planting gave the lowest number and weight of weeds/ m^2 and higher seedling emergence (Abdallah, 1998).

Combined application of 110: 40: 60: 20 kg NPKS along with organic manures equivalent to 15 tons FYM and *Azospirillum* and Phosphorus solubilizing bacteria at 5 kg each/ha gave higher yield and cost benefit ratio, and increased soil available nutrient status after harvest. By adopting this recommendation, the use of costly inorganic fertilizers can be reduced by 25 %.

9.5.2. Insect pest management

9.5.2.1. Thrips, *Thrips tabaci*: Onion crop planted in the months of September and October (late kharif) had less severe attack by thrips and required little crop protection. Reflective mulch with aluminum paint (Scott *et al.*, 1989) repelled 33-68% of the thrips and was found to be more effective, particularly, at the seedling stage rather than at plant maturity (Lu, 1990). Adequate irrigation throughout the growing season is critical in minimizing damage (Fournier et al, 1995). Field trials at DOGR suggested that sprinkler irrigation reduced thrips population considerably, compared to drip and surface-irrigation.

Thrips are weak fliers and can be carried by wind. Therefore, planting live-barriers like maize can effectively block adult thrips from reaching onion plants. Two rows of maize or an inner row of wheat and outer row of maize surrounding onion plots (250 sq. m) blocks adult thrips up to 80% (Srinivas and Lawande, 2006). This practice brings down insecticide application by half. Highest benefit: cost ratio can be obtained with maize + wheat barrier around onion. Recent trials at DOGR suggested that mineral oil sprays @ 2% could bring down thrips population by 48%.

Seedlings root-dip with Carbosulfan (0.025%) or Imidacloprid (0.04%) for 2h before planting protects young plants up to 30 days (Srinivas and Lawande, 2007a). Among different insecticides evaluated, occurrence of re-infestation was very low with Fipronil and Profenofos. Among the relatively new insecticides, Carbosulfan, Methomyl, Lambda cyhalothrin, Profenofos, Spinosad and Fipronil were found effective in suppressing thrips population.

The commercial onion variety Pusa Ratnar was found resistant to *T. tabaci* in Punjab and Maharashtra (Darshan Singh *et al.*, 1986; Brar *et al.*, 1993)

9.5.3. Disease management

Rahman *et al.* (2000) reported that leaf blight diseases caused by *Alternaria porri*, *Colletotrichum* sp., *Stemphylium* sp. and *Cercospora* sp., singly or combined, could be controlled by four sprays of Mancozeb at 0.3%, starting from 45 days after transplanting. Among the newer fungicides, two sprays of Hexaconazole (0.1%) were found most cost-effective (Barnwal *et al.*, 2006).

9.5.3.1. Fusarium basal rot, *Fusarium oxysporum* f. sp. *cepae*, *F. oxysposum* f. sp. *allii*: Crop rotation of 4-5 years with non-host crop has been found effective in eliminating the disease. Mixed cropping with tobacco and sorghum is also effective in reducing disease severity (Srivastava and Pandey, 1995). Green-manuring increases antagonistic microbial population in the soil. Good drainage, deep plowing in hot summer and avoiding injury during cultural practices reduces disease incidence.

Pre-harvest spray of Carbendazim (0.1%) resulted in least decay of stored onion after 5 months from storage (Srivastava *et al.*, 1996).

9.5.3.2. Purple blotch, *Alternaria porri*: Seed-treatment with Thiram (0.25%), crop rotation and summer ploughing are recommended for control of the disease (Gupta and Pathak, 1987). Application of Mancozeb (0.25%) and Captafol at 0.2% (Gupta *et al.*, 1986a), Iprodione @ 0.25% (Gupta *et al.*, 1996), Metalaxyl and Dinocap (Upadhyay and Tripathi, 1995; Srivastava et al., 1996) were found effective in controlling the disease.

Onion varieties, Agrifound Light Red (Sharma, 1997), 53-3 (Pandotra, 1965), Agrifound Dark Red, Red Globe (Sugha *et al.*, 1992), VL Piyaz 3 (Mani *et al.*, 1999) and RO 59 (Mathur *et al.*, 2006) were reported to be moderately resistant.

9.5.3.3. Stemphylium blight, *Stemphylium vericariuno*: Cultural control methods include long rotations with non-host crops, good field drainage and reduced plant density to contain the diseases. Since the pathogen survives on dead plant tissues, sanitation of the field and collecting and burning the crop refuse reduces disease incidence. Barnwal and Prasad (2005) observed lowest disease intensity in a crop sown in the last week of November as compared to that sown in October. Irrigation at 10-day intervals and high doses of nitrogen resulted in reduced disease incidence (Srivastav *et al.*, 2005).

3 to 4 sprays of 0.25% Mancozeb offer best control, with higher benefit: cost ratio (Gupta *et al.*, 1996b). For onion seed crop, fortnightly sprays of 0.25% Mancozeb or 0.25% Iprodione are recommended (Srivastava *et al.*, 1995).

9.5.3.4. Anthracnose, *Colletotrichum gloeosporioides*: Since the pathogen survives on crop refuse, sanitation and destruction of infected plant-debris helps reduce the disease. Application of Benomyl at 0.2% as soil treatment is recommended (Remiro and Kirmati, 1975). Spraying Mancozeb at 0.25% also gives good control.

9.5.3.5. Downy mildew, *Peronospora destructor*: Bulbs used for seed production should be selected from healthy fields for management of the disease. Crop rotation for 3-4 years with non-host crop should be practiced. Late planting, poor drainage, higher dose of fertilizer and frequent irrigation should be avoided, as these practices encourage high disease-incidence (Ahmad and Karimullah, 1998).

Spraying Mancozeb at 0.25% and Ziram at 0.1% at 10 to 12-day intervals is recommended (Marikhur *et al.*, 1977). Bulb and seedling-dip in Ridomil MZ at 0.25% for 12 hrs followed by 2 foliar sprays gave effective disease control. Metalaxyl and Cyomaxanil proved most effective in reducing disease severity up to 88% (Palti, 1989).

9.5.3.6. Black mold bulb rot, *Aspergillus niger*: Pre-harvest spray (0.2%) of Carbendazim + Mancozeb and Iprodion, 20 days before harvesting, proved effective (Ahir and Maharishi, 2008).

9.6. FARM MECHANIZATION

The shortage of labor at the crucial time and increasing labour cost make onion mechanization inevitable. This intervention is mainly solicited in labour intensive work viz., sowing, transplanting, harvesting etc. low seed rate, easy sowing, saving in time and early maturity of onion were observed in sowing with pneumatic seed drill. However, transplanting method of onion production recorded the highest marketable yield, which was significantly higher over the direct sowing with pneumatic seed drill.

A six-row tractor operated onion transplanter for flatbed has been designed and fabricated (IIHR, 2010). The six roller wheels press the root of the seedlings in soil and shovels cover the roots with soil. The row spacing in the present prototype is 15 cm and seedling spacing is 10 cm. The expected working speed is 1 km/hour and field capacity is 0.8 ha/day.

Manual onion harvesting is also full of drudgery and the mechanization is essentially needed. Prototype of onion digger with length 1.2 m, speed ratio 1.25:1 and slope of the elevator 15 degrees, was found to have i digging efficiency 97.7%, separation index 79.1%, bulb damage 3.5%, fuel consumption 4.1 1/ha (12.81 1/ha) and draft 10.78 kN (Khura *et al.*, 2011).

Onion detopper was designed and developed at Haryana Agricultural University, Hissar to facilitate the digging and top removal (Rani and Srivastava, 2012).

For mechanical extraction of onion seeds, spike tooth extraction mechanism developed at IARI, New Delhi (LARI, 2010) gave an extraction efficiency of 99% and cleaning efficiency of 97%. The seed loss ranged between 2.2% and 3.1% at cylinder speeds of 3-5 m/s. The costs of seed extraction by mechanical onion seed extractor and manual/conventional method were Rs. 1,800 and Rs. 9,000 per tonne of onion umbels, respectively.

To reduce the cost of grading and increase the precision, two onion graders viz. manually operated and motorized graders were designed and evaluated by DOGR (Tripathi and Lawande, 2009). These have increased efficiency of 5 and 20 times, respectively, over hand grading. The precision of grading achieved by graders is 98% against 50% in hand grading. The capacity of manual grader is 5 quintals per person per hour with 90% accuracy. The capacity of motorized grader is two tons per hour with 90% accuracy.

Onion peeling machine would enhance efficiency of the processing. Central Institute of Agricultural Engineering, Bhopal has developed a batch type multiplier onion peeler (Naik *et al.*, 2007). The multiplier onion needs to have the ends cut with a sharp knife and soaked in clean water for a period of 10 minutes to assist the loosening of peel followed by air drying for 1-2 minutes to remove the surface water. With 92% peeling, and unpeeled and damaged percentage being 6% and 2%, respectively, the capacity of the peeler was found 50-60 kg/hr.

9.7. POST-HARVEST MANAGEMENT

9.7.1. Post-harvest losses

The post-harvest losses can range from 45-50% if proper care of the harvested produce is not taken. These losses mainly consist of physiological weight loss (20-25%), sprouting (8-10%) and decay (10-12%) (Gopal, 2014). The estimation of seasonal variation in storage losses revealed that the kharif onions were more prone to losses than late kharif and rabi seasons produce. The total losses which include physiological loss of weight, rotting and sprouting reached almost 70% in kharif after three months storage (DOGR, 2013).

A significant reduction in storage losses was observed when the last irrigation was applied five days before harvesting over irrigation applied just before harvesting (Sharma *et al.*, 2007). The crop grown with drip irrigation was reported to have significantly lower losses than the crop grown with surface irrigation (Tripathi *et al.*, 2010).

9.7.2. Curing

The windrow method of field curing for 3-5 days followed by shade curing for 7 to 10 days has been recommended. The curing of bulbs under poly-tunnel in kharif season and pits in rabi season was found effective in reduction of losses. Artificial curing of bulbs in curing chamber with full load at 35°C and airflow velocity of 3.2 m/s cured the bulb efficiently.

9.7.3. Prevention of sprouting

Pre-harvest application of isopropyl-N (3-chlorophenyI) carbamate (CIPC) (2%) at 75 days after planting has been found to reduce sprouting significantly in kharif onion varieties viz. Bhima Raj and Bhima Red after three months of storage (DOGR, 2012, 2013). The gamma-irradiation of some varieties revealed that it could effectively check the sprouting and rotting in all onion varieties (Tripathi *et al.*, 2011).

9.7.4. Storage

Traditional storage practices result in substantial losses in stored onions. Nearly 80% of the farmers are dependent upon primitive and old storage. As a result high percentage of wastages occurs during storage (nearly 40 to 50 percent).

The mud plastered top and bottom ventilated storage structure was superior in reduction of losses i.e. weight loss, rotting and sprouting over other structures irrespective of packing materials. Among single row structures, low cost bottom ventilated structure was found to be the best in reduction of losses (reduced by 20 -30 percent) and increase in net profit. Low cost storage model of 5 to 10 t capacity and high cost model of 25 to 50 t capacity with bottom and side ventilation recommended by DOGR have become popular among the farmers (Murkute and Gopal, 2013). Cold storage is the most efficient way to restrict physiological weight loss. However, conducive atmosphere for sprouting in cold storages restricts its use.

9.7.5. Processing

Processing is an efficient way to increase shelf life without compromising the freshness and quality. Dehydrated products such as flakes, rings, granules, powder etc. and processed onions like onion in vinegar and brine are the important products being prepared and marketed worldwide. Onion can also be processed into oil, vinegar and wine etc. However, dehydration of onions is the oldest method of producing concentrated product which has longer shelf life when packaged properly, and can be simply reconstituted without any substantial loss of flavour, taste, colour and aroma. Onions are generally dried from an initial moisture content of about 86% (wet basis) to 7% or less for efficient storage and processing (Sarsavadia *et al.*, 1999). Specific characteristics recommended for drying are white flesh, 15-20% total solid content, high pungency, high insoluble solids and low reducing to non-reducing sugars ratio (Mitra *et al.*, 2012). Based on the recovery and quality of red and white onion flakes, cabinet drying method has been recommended (DOGR, 2011).

9.8. EXPORT

The major importer of onion is Bangladesh, Malaysia, Sri Lanka, United Arab Emirates, Nepal, Indonesia, Qatar, Vietnam, Social Republic, Kuwait, Oman, *etc.*. Among them, Bangladesh has imported 333165 tons worth Rs. 59951.83 Lakhs during 2017-18. They specially have more demand for the small cultivars of onion. Next importer is Malaysia followed by Sri Lanka which imported 276162.25 tons and 227965.35 tons, respectively. We can observe from Table 9.5 that total 1588985.71 tons worth Rs. 308882.23 Lakhs was exported to different countries of the world from India.

Table 9.5. Country-wise export of onion (fresh/chilled) from India

Country	Quantity ('000 tons)	Value (Rs. Million)
Bangladesh	333.165	5995.183
Malaysia	276.162	5885.895
Sri Lanka	227.965	5603.927
United Arab Emirates	226.248	4155.482
Nepal	100.151	1436.334
Indonesia	65478	1113.043
Qatar	53.942	1035.099
Vietnam Social Republic	4.405	744.699
Kuwait	52.082	744.199
Oman	43.338	636.743
Philippines	12.378	636.737
Singapore	30.196	633.565
Saudi Arabia	44.825	565.365
Others	78.996	1702.051
Total	1588.985	30888.223

Source: Anonymous, 2018. Horticultural Statistics at a Glance, 2018.

9.9. FUTURE STRATEGIES

There is need is to explore the innovative measures to improve productivity and stabilize production of onion in India. The following interventions may help to improve the productivity and prospects of onion cultivation in India.

- » Basic research in breeding for resistance, processing qualities and export worthy varieties are lacking. Thrust in these areas can help to improve onion productivity and export.

- » Biennial nature, high cross-pollination and sharp inbreeding depression in onion are still challenges for breeders using conventional approaches. There is thus an opportunity to use biotechnology, particularly molecular approaches and functional genomics to overcome these problems.

- » Due to poor maintenance of breeders' stock, many varieties are out of production chain or could not even make entry into the chain. Fanners find easy and economical to produce their own onion seed but due to ignorance of out-crossing they are not able to maintain purity. Due to supply of spurious seed by many seed merchants, the spread of good varieties has been hampered. Thus, there exist opportunity to produce and distribute good quality seed of true-to-type varieties and capture the market of onion seed. Seed multiplying agencies working in public sector need to be sensitized in this regard.

- » Thrust is required to increase the national storage capacity. Intrastructure facilities need to be created that about 30-40% produce is stored in the cold storages to significantly reduce the post-harvest losses.

Focus should also be to evolve a robust supply chain based on domestic demand, export and a quantum for processing to avoid price fluctuations by harnessing available resources, modem infrastructures, improved technologies and innovative endeavours. Policy makers will have to work hard to provide amicable solutions on pricing which should lead to higher profits to fanners but not at the cost of consumers.

9.10. CONCLUSION

Onion being an essential element of all types of dishes either vegan and non-vegan food. The demand for onion remains same throughout the year irrespective of seasons. The supply depends upon production of onion in *rabi* as well as *kharif* season. India is second largest populous country having huge demand for onion. The retail price is decided by the demand for and supply of a commodity. The gap between supply of and demands for it always fluctuate. Availability of onion for domestic consumption depends upon total production and net export, correspondingly the price of onion is highly volatile. The highest producer of onion is Maharashtra among all states. The largest importer of Indian onion is Bangladesh. In the national as well as international market, there is high demand for onion especially in the gulf countries. Therefore, government has to give more emphasis to increase the productivity of onion in India. There is need to develop technologies to increase the onion productivity under short day onion since Indian onions are short-day type and need to develop improved varieties or hybrids for abiotic stresses namely drought, heat, salinity and water logging tolerance and biotic stresses namely anthracnose, blight, purple blotch and thrips tolerance or resistant varietal development are the prime importance to sustain the onion productivity in India.

Chapter - 10

Round (Potato Production)

10.1. INTRODUCTION

Potato is a predominant vegetable in India, at present most of the domestic supply of potatoes is consumed as fresh (68%) followed by processing (7.5%) and as seed (8.5%). The rest 16% potatoes are wasted due to post-harvest losses. Potato plays a very important role in Indian agriculture as potato alone contributes about 21 percent of the total vegetable area and 26 percent of total vegetable production of India (DAC&FW, 2017b). It is a nutrient-rich crop which provides more calories, vitamins and nutrients per unit area than any other staple crops, thus, it is considered as one of the main staple crops for ensuring food and nutrition security (Knapp, 2008), especially for developing countries. FAO declared potato as the crop to address future global food security and poverty alleviation during 2008. As per FAOSTAT and NHRDF data for the year 2018-19, India with production of 52.60 million tons of potato from 2.16 million ha is ranked second in potato production in the world, only behind China with 91.88 million tons from 4.87 million ha.

India reported that at present level of farm management practices, India actually able to harvest only 42.45% of the achievable yield, which could be improved to 80 % by efficient and effective dissemination and implementation of improved technologies. Moreover, the information and insights of adoption studies are also vital for setting up research priorities, improving efficiency of agricultural research, extension services, and investment in new technologies. Many times, even if technologies are available, farmers are well aware about them and ready to adopt them; but still are not in a position to fully adopt them due to several constraints.

Current share of potato to agricultural GDP is 2.86% out of 1.32% cultivable area. The contribution of potato in agricultural GDP from unit area of cultivable land is about 3.7 times higher than rice and 5.4 times higher than wheat.

Potato is a highly nutritious, easily digestible, wholesome food containing carbohydrates, proteins, minerals, vitamins and high-quality dietary fiber. A potato tuber contains 80 per cent water and 20 per cent dry matter consisting of 14 per cent starch, 2 per cent sugar, 2 per cent protein, 1 per cent minerals, 0.6 per cent fiber, 0.1 per cent fat, and vitamins B and C in adequate amount. Thus, potato provides more nutrition than cereals and vegetables. Keeping in view the shrinking cultivable land and burgeoning population in India, potato is a better alternative to deal with the situation.

10.2. AREA, PRODUCTION AND YIELD

In Global and Indian scenario, Potato is the third most important food crop in the world after rice and wheat in terms of human consumption. Global annual potato production during the triennium ending (TE) 2013 was 370 million tons resulting in per capita availability of over 50 kg.

The area, production and yield in 1949-50 (the year of establishment of ICAR-CPRI) was 0.23 million ha, 1.54 million tons and 6.59 t/ha, respectively. As per NHRDF, India produced 52.60 million tons of potato from 2.16 million ha area with an average yield of 24.07 t/ha during 2018-19 (Table 10.2). There has been a phenomenal increase in potato area (8.5 times), production (29.4 times) and productivity (3.5 times) over six decades. As a consequence, India emerged as the second largest potato producer in the world after China (Table 10.1). However, the productivity in India is still low when compared with most of the potato growing developed countries. India ranks fifth in productivity (22.56 t/ha) behind USA (49.76 t/ha), France (39.37 t/ha), Netherlands (36.61 t/ha) and Germany (35.37 t/ha).

Table 10.1. Country-wise area, production and yield of onion (2018-19)

	Area ('000 ha)	Production ('000 tons)	Yield (tons/ha)
China	4,897.70	91,881	18.76
India	2,224.73	50,190	22.56
Russia	1,295.42	22,074	17.04
Ukraine	1,189.49	20,269	17.04
USA	385.47	19,181	49.76
Germany	299.74	10,602	35.37
Bangladesh	473.05	9,655	20.41
France	233.81	8,560	39.37
The Netherlands	190.13	6,961	36.61

Source: FAO Stat

Table 10.2. State-wise area, production and yield of onion (20218-19)

State	Area ('000 ha)	Production ('000 tons)	Yield (tons/ha)
Uttar Pradesh	610.50	15,813	25.10
West Bengal	436.04	12,782	25.23
Bihar	321.88	7,441	25.33
Gujarat	29.73	3,510	29.73
Madhya Pradesh	22.86	3,400	22.85
Punjab	26.43	2,870	26.43
Assam	7.49	1,140	7.49
Chhattisgarh	15.97	711	15.97
Jharkhand	14.18	709	14.18

Haryana	25.84	479	25.84
Uttarakhand	13.75	366	13.75
Karnataka	28.00	345	12.32
Others	111.22	2,104	18.91
Total	2161	52,599	24.07

Source: NHRDF

10.3. EMERGING PROBLEMS AND CHALLENGES

10.3.1. Lower potential yield

More than 90% potato in India is produced during October and March months that fall under the short days in northern hemisphere. Due to shortening of crop life cycle under short day conditions and lesser availability of sunshine hours, the physiological potential of potato productivity in tropics and sub-tropics falls drastically compared to temperate countries. To enhance attainable yield under the lower potential yield is an important challenge before the agencies involved in potato research and development.

10.3.2. Early harvesting

Small and marginal farmers tend to sandwich potato crop between paddy and wheat crops. As both wheat and paddy crops face less production risk and no market risk due to their coverage under support price scheme in India, the small and marginal farmers opt for early harvesting of potato after 60-70 days after planting with a compromise in yield of 30 to 40% of the normal crop. As considerably large proportion of potato crop is harvested early, it reflects poorly on the overall national potato productivity. To increase national potato productivity to 34.5 tonne/ ha by 2050 is another important challenge under this scenario. Gene(s) responsible for tuberization need to be identified in order to facilitate early maturity of potato for developing varieties to serve this sector of the potato.

10.3.3. Uneconomic land holdings

Very large number of potato growers in India produce potatoes on very small sized and uneconomical holdings. For most of these growers, farming is a way of life rather than a business and for livelihood they have to rely upon the remittance from one or more family members who work in cities. Most commonly the persons responsible for growing potato are old or less capable who do not follow latest package of practices for crop husbandry. To disseminate technical knowledge to this less responsive farming community and have adequate adoption of latest scientific potato technologies is really a tough challenge.

10.3.4. Capital intensive crop

Potato being an input and capital-intensive crop, requires higher amount of capital that is many times not available with the small and poor farmers with the result, crop is cultivated with suboptimal doses of various inputs. In a number of cases, potato is cultivated on less productive and even the problem soils. Further, the very fast escalation in the prices of farm inputs *viz.* fertilizers and plant protection chemicals in the recent past has further deteriorated the situation. In addition to inadequate inputs, very large proportion of potato farmers use inferior quality inputs specially the seed potato as they cannot afford to purchase the good quality one. This is more of a policy issue which needs to be addressed sooner than later in order to improve profitability of potato farmers.

10.3.5. Shortage of farm labor

Rural development schemes of the government especially the Mahatma Gandhi National Rural Employment Guarantee Scheme has diverted farm labor to other developmental activities. With the result, the labor-intensive potato cultivation is one of the most adversely affected crops in India during recent past. Development of suitable potato machines for small and marginal farmers should be very high on the future potato research agenda. More than half of farm labor on mechanised farms is currently put for picking potato tubers. For large farmers, there is a need develop combine potato harvester performing digging and tubers lifting activities in a single go.

10.3.6. Wrong doses of fertilizers

Very large proportion of farmers, including the potato farmers, apply chemical fertilizers as per their personal judgment and experience. Use of complex fertilizers is on the rise these days and in the absence of proper soil test reports farmers are using complex fertilizers quite commonly. With the result, some of the nutrients are over supplied while the others are under supplied disturbing nutrient balance of soils in addition to environmental degradation and reduced farm profitability. Creation of reliable and adequate soil testing mechanism needs to be established at the earliest possible by the developmental agencies related to agriculture in all parts of the country.

10.3.7. Inadequate and inefficient transport infrastructure

Lack of state-of-the-art transport infrastructure adversely affects potato producers and consumers across the country due to its bulky nature and season as well as region dominated production. This deprivation is not only responsible for wider price differences in this agri-commodity across the regions but also high post-harvest losses. Quick resolution of this constraint by the policy makers and development agencies will ensure cheaper and better-quality potato availability in all parts of the country.

10.3.8. Enhanced pesticide resistance

Demonstration of enhanced resistance of pests to pesticides (both diseases and insects) during previous years is another important issue to be addressed with a very high priority by potato research and developmental agencies using conventional as well as modern molecular techniques. Application of molecular techniques for better diagnostics and understanding the genetic makeup of potato pests is likely to bring desirable results in the future.

10.3.9. Global warming

Rising average night temperature in several potato growing areas in the country is lowering already constrained potential yield of Indian potato due to limited sunshine hours and shorter duration of crop. Development of potato varieties tolerant to elevated temperature will help in addressing this emerging problem. After successful release of heat tolerant potato variety, Kufri Surya, CPRI is likely to release more such varieties to tackle the problem of heat stress.

10.3.10. Climate change impact

The impact of climate change on potato in different regions was assessed using INFOCROP Potato model. The yield is estimated to be severely reduced in southern and peninsular India (9-47%), moderately reduced in Indo-Gangetic plains (3-13 %) and slightly increased (3-7%) in the north western Indo-Gangetic plains due to milder winters. The potato production in India may decline by 2.61 and 15.32 % in the year 2020 and 2050, respectively (Fig. 10.1). The least vulnerable region will be North-western plains (Punjab, Haryana and areas of western UP and northern Rajasthan) with possible increase of 3.46 to 7.11% and the most vulnerable region will be West Bengal, plateau regions and other areas in south India with a possible decrease of 9 to 55% in productivity. There is a need to adopt adaptation and mitigation measures, discussed under other heads, for ensuring higher potato productivity in order to counter the effect of climate change on future potato productivity.

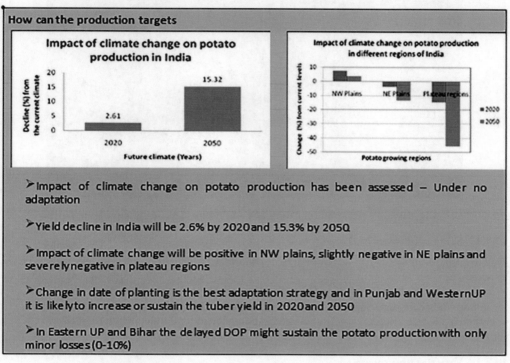

Fig. 10.1. Impact of climate change on potato production

10.3.11. Inefficient cold storage facilities

In many parts of the country the cold storage facilities are unavailable due to lack of reliable electricity supply which negatively affects potato growth in those areas. In other parts of the country, the cold stores are using outdated technology and are responsible for higher post-harvest losses of potatoes. National Horticulture Board has been helping cold storage up-gradation with big schemes; however, the targeted results have still not been achieved. Ensuring regular electricity supply in all parts of the country will help creation of cold storage in other areas where it is required.

10.3.12. Emergence of new pests

Climate change and import of potato under Open General Licence (OGL) in nineties have brought newer pathogen load in Indian potato scenario. Several new viruses appeared after the import of potatoes under OGL. It is one of the most important and tough challenges before Indian potato research scientists to make potato crop free from these pests. Improvement of seed potato quality has to be achieved through state-of-the-art diagnostics using transmission electron microscope, molecular techniques and dip-stick. Subsequently rapid *in vitro* multiplication of planting material has to be done using areoponics (Fig. 10.2) and other biotechnological approaches.

Fig. 10.2. *In vitro* multiplication of planting material using Areoponics

10.3.13. Scarcity of water

Due to climate change, erratic precipitation and shortage of irrigation water are expected to be the important developments in the future which are supposed to put tough challenges for potato cultivation in India. Research efforts, with higher precision, on development of micro-irrigation technologies for potato and their better dissemination in the regions with scarcity of irrigation water are to be undertaken. Biotechnological approaches need to be exploited for developing drought tolerant potato varieties for tackling this impending problem.

10.3.14. Varieties for processing

CPRI has produced 6 potato varieties for processing with the result the current potato processing in India has reached 3.2 million tons. However, by the year 2050 about 465 million people will be added to our existing 375 million urban population (in year 2010) which will generate a huge demand for processed potato products. The industry has to be supported by the required technologies including the improved processing varieties in order to enhance potato processing up to 25 million tons by the year 2050. There is a need to have greater number of processed potato varieties with improved processing attributes for tackling this rapid enhancement of processed potato products in fast developing India.

10.3.15. Food security and production target

Ensuring food and nutritional security to 1619 million people in the country during 2050 is assessed to be a tough challenge in the future where potato will have to shoulder much heavier responsibility to address the issue. We need to produce 125 million tons of potatoes by 2050 as per capita consumption of potato is increasing very fast in addition to the rising population. As the future potato growth has to be led by productivity enhancement, increase in production potential and bridging yield gaps will be another important challenge in the future. Enhancement of potato production potential in various parts of the country in the light of heat and moisture stresses under climate change regime and national food security will be addressed through the exploitation of biotechnological and molecular breeding and plant protection technologies under the highly advanced expertise and facilities at the CPRI. In order to achieve higher harvest index from crop plant, dwarf potato genotypes should also be developed using molecular techniques for cultivation under long photoperiod and or high temperature. Raising potato production efficiency will be achieved with the help of modeling research, precision farming, nanotechnology, manipulating crop geometry, developing ideotypes and improved mechanization (*e.g.* automatic potato harvester, grader and bag filler). Advanced information technology-based (*e.g.* on mobile phones and through decision support/ expert systems) dissemination of refined scientific potato technologies will be utilized. Efficient input (fertilizers, pesticides and irrigation *etc.*) delivery system and more accurate diagnosis of diseases using nano-science widely across the country will be one of the solutions for tackling this challenge. CPRI has already initiated concerted efforts in this direction. Development of drought and heat tolerant potato varieties will play crucial role in the future of Indian potato industry.

10.4. IMPROVED PRODUCTION TECHNOLOGIES

It was only because of indigenously developed technologies that potato in India has shown spectacular growth in area, production and productivity during the last five decades. The major achievements of potato research in India are as under:

10.4.1. Improved potato varieties and certified seeds

Potato being a vegetatively propagated crop is subjected to large number of seed-borne diseases responsible for degeneration of seed stocks over the years. It is therefore imperative to use good quality seed for economic production. As the potato seed tubers get degenerated quickly, it is generally recommended that they should be replaced in every 3-4 years with quality/certified seeds to avoid progressive reduction in yield of the crop. The quality seeds of improved potato varieties (IPVs) are the basic and crucial determinants of productivity and account for 30-50 percent of the total potato production cost.

The process of micro-propagation has become much more important in the case of potato for the purpose of production of disease-free plants from infected one. There is a tremendous scope to increase healthy seed production vertically by adopting aeroponic technology as well as apical rooted cutting technology.

Advantages of this system include:

» Tropical states which do not have isolated and virus-free potato growing areas can also produce quality seed.

» Early supply of nucleus seed to commercial growers by reducing the field exposure time.

» Improved tuber quality and reducing the load of degenerative diseases.

» Utilize the resources and trained manpower round the year.

» Vertical growth and reduction in pressure on land.

So far 47 potato varieties have been bred for different agro-climatic regions of the country with 28 varieties alone for north Indian plains. Varieties have also been developed for north Indian hills and other special problem areas viz. Sikkim, north Bengal hills and south Indian hills. Of the 47 varieties developed, 19 possess multiple resistance to different biotic and abiotic stresses. Besides, nine varieties are suitable for processing purposes. These are Kufri Chipsona-1, Kufri Chipsona-2,

Kufri Chipsona-3, Kufri Himsona, Kufri Frysona, Kufri Jyoti, Kufri Chandramukhi, Kufri Lauvkar and Kufri Surya. All these varieties fall in three maturity groups, i.e. early (70-80 days), medium (90-100 days) and late (110-120 days).

The potato varieties developed by CPRI are grown not only in India but also in several neighbouring countries. The variety Kufri Chandramukhi is grown in Afghanistan, Kufri Jyoti in Nepal and Bhutan, and Kufri Sindhuri in Bangladesh and Nepal. Besides, five Indian hybrids are also commercially grown in Sri Lanka, Madagascar, Mexico and Philippines.

10.4.2. Seed plot technique

This technique was developed in 1970s to enable healthy seed potato production in the sub-tropical Indian plains under low aphid period. This technique aided by biotechnological approaches for virus elimination, micro-propagation and effective viral diagnostics has sustained the National Potato Seed Production Program by producing about 2600 tons of breeder's seed annually. This breeder's seed is further multiplied to about 4,32,000 tons of certified seed by the State Departments of Agriculture/ Horticulture. Thus, the country saves about 484 million US dollars because most Asian countries like Pakistan, Bangladesh and even China continue to import seed potatoes from Europe.

The decentralization of potato breeding from hills to plains in India through the seed plot technique enabled the development of varieties suited to different agro-climatic regions of the country. The area under seed potato production also increased by 12 times and enabled the availability of seed potato throughout the country in proper physiological state.

10.4.3. Tissue culture for rapid multiplication

Efforts are being made to improve seed health standards and reduce the time required for production of breeder's seed by employing *in vitro* techniques of meristem culture and micro-propagation. Presently, about 5 per cent of Breeder's seed production program is fed annually by micro tubers produced through tissue culture. It is proposed to produce 100 per cent of breeder's seed through tissue culture propagated material in the years to come.

10.4.4. Agro-techniques

The development of package of practices for potato production in different agroclimatic zones has helped in improving potato productivity in these zones. The potato crop is input intensive and requires optimum cultural practices for achieving higher productivity. Optimum cultural practices depend on delineated phenological phases of crop growth and development *viz.* pre-emergence, emergence to tuber initiation, tuber initiation to tuber bulking and tuber bulking to termination of bulking.

The cultural practices are adjusted in the Indian plains in a way so that tuber initiation and development coincide with the period when night temperature is less than 20°C and day temperature is below 30°C. The phenological phase of tuber initiation to tuber bulking is mainly conditioned by nutrition and moisture. For this purpose, fertilizer and irrigation requirement in different agro-climatic zones have been worked out through multi-locational trials under AICRP (Potato). Termination of tuber bulking coincides with onset of foliage senescence. By manipulating the nutrition and moisture, the foliage senescence is delayed for ensuring continuation of linear tuber bulking phase resulting in higher yield.

10.4.5. Cropping systems

Several profitable potato-based inter-cropping and crop rotations have also been identified for different regions of the country. Potato can be profitably intercropped with wheat, mustard and sugarcane. These cropping systems have helped in the maintenance of soil fertility and have improved the fertilizer economy, crop yield and gross returns. Besides, potato cultivation has also been mechanized in selected regions through the fabrication and development of cost-effective tools and implements.

The potato based cropping sequence recommended for different regions of the country include:

Bihar: Paddy-Potato-Mung bean; Groundnut-Potato-Mung bean

Punjab: Maize-Potato-Wheat ; Paddy-Potato-Wheat; Potato-Wheat-Green manure crop

Assam: Potato-Mung bean-Paddy (Transplanted)

Gujarat: Groundnut-Potato-Bajra

Madhya Pradesh: Soybean-Potato-Okra

Central Plains Zone: Rice–Potato–Japanese mint; Rice-Potato-Onion

Uttar Pradesh: Rice-Potato-Cowpea; Rice-Potato-Okra; Rice-Potato-Green Gram; Maize + Black Gram-Potato-Onion

10.4.6. Micro-irrigation

Drip and sprinkler methods of irrigation can economize irrigation water by 30-50%. These modern methods could be advocated to the progressive potato growers because the initial cost of installation is high.

10.4.7. Integrated nutrient management

It is highly recommended in recent situation for recycling of organic residues, lowering the cost of cultivation and sustainable tuber yields. Use of organic sources (FYM, compost, crop residue incorporation, green manures and bio-fertilisers *etc.*) would not only complete requirements of crop for secondary and micro-nutrients, but also reduce demand for macro-nutrients considerably. This would further reduce cost of cultivation and import bills particularly for phosphate fertilizers which are costliest. Soil of application FYM (15 t ha^{-1}) takes care of half of P and K requirement of potato crop. Green manuring (sesbania, cowpea *etc.*) supplement nutrition by way of bringing nutrients from deeper soil layers and symbiotic N fixation. Further, these crops improve soil physical condition and usually reduce N requirement of potato crop by 15-20%. Crop residue incorporation of rice, wheat and leguminous crops also improves soil organic carbon and ultimately nutrient supply to the crops in long run. Microbial inoculants like Azotobacter and phosphate solubilizing bacteria have also been tried in potato crop. Their use can also result in saving on inorganic N and P up to 20% depending upon agro-ecologies.

10.4.8. Precision farming

The efficiency of using autopilot and parallel driving systems significantly enhances potato productivity by an average of 30% to a tune of 26.2 t/ha under Precision Cultivation Technology system in Russia. Hence, in conducting field trial, it is needed to focus on the implementation of new and progressive technologies, which undoubtedly include precision (coordinate) agriculture. Precision farming has become a revolutionary technology, which transforms farming related data into useful information for cropping decision-making. Overall, research indicates that precision farming can ameliorate input usage efficiency, and enhance net income in potato cultivation.

10.4.9. Climate change adaptation strategies

The various adaptation strategies to combat the impact of climate change on potato productivity may include breeding short duration and heat tolerant cultivars, and developing potato cultivars that form tubers at higher night temperatures. Mining for biodiversity to heat tolerance should be given priority. Breeding drought and salinity tolerant cultivars would be effective to face the future challenges of climate change. Use of wind breaks around fields and crop residue mulches for some period after planting, using drip and sprinkler irrigation in place of furrow and basin methods and altering cultural management in potato based cropping systems are few examples of agronomic management practices to reduce the impact of climate change. Besides, conservation tillage and on farm crop residue management are required to increase input use efficiency. Advance planning for possible relocation and identifying new areas for potato cultivation is needed. Improvement and augmentation of cold storage facilities and air- conditioned transportation from producing to consumption centers will be required for storage and transportation of this semi-perishable commodity. Strengthening education, research and development in warm climate production technology for ware and seed potato crop is also required to meet the production targets in future climates.

10.4.10. Pest management

Effective management practices have been devised for the major potato diseases and insect-pests in India. Late blight is the most notorious disease of potato which occurs almost every year in the hills and plains. Besides chemical control measures, several late blight resistant varieties have been developed. Potato varieties have also been bred which possess resistance to wart and cyst nematodes. Cultural and biological control measures have also been developed to control the diseases and insect-pests. The development of late blight forecasting systems for hills and plains has enabled the early warning mechanism for the appearance of late blight disease.

10.4.11. Contract farming

The PepsiCo's Frito Lay contract farming for potatoes is a good example of how international quality requirements are met by small farmers in India. A very strong extension network by PepsiCo company helps to monitor and maintain quality at every level. Evidently the farmers working as contract grower's benefit on several fronts: there is extensive training and education of farmers for proper timing and method of sowing, harvesting and other field operations; farmers' overall management capabilities are enhanced by meetings and visits by agricultural experts from time to time. Gross margins for contract farmers are higher.

Furthermore, because the company announces prices ahead of the production season, they are sure of covering at least their production costs and can invest in agrochemicals and other inputs, which in turn leads to enhanced productivity. Other risks from insect pest infestation and weather changes are also minimized as the company's extension agents are constantly working with the farmers to give timely input on these issues. Finally, weather insurance is also available for the company contract farmers, which further minimizes risks. The obvious advantage for the company is getting an assured quantity and quality for chips making to enable utilization of the processing plant at optimal capacity. Direct involvement with farmers enables good communication that ensures availability of produce which meets the specific quality requirements for processing and indicators for the company's HACCP and ISO certification.

10.4.12. Storage

In India, 85 per cent of potato is produced in winter and stored during long hot summer. This requires storage of potatoes in cold stores at 2-4⁰C, which involves substantial cost. It also leads to accumulation of reducing sugar in the potato tubers resulting in sweetening of potatoes.

However, there are a number of traditional low-cost and non-refrigerated storage structures (essentially based on evaporative or passive evaporative cooling) in use in India with varying degrees of success. These traditional structures have been studied, validated and popularized for particular regions. In non-refrigerated storages, use of sprout suppressants have also been popularized to prevent excessive weight loss and shrinkage due to sprouting. The CIPC (isopropyl-N-chlorophenyl carbamate) is the most effective sprout inhibitor when applied @ 25 mg a.i. per kg tubers.

10.4.13. Processing and value addition

In addition to raw consumption, potatoes can be processed into several products like chips, French fries, cubes, granules and canned products. The primary determinants for potato processing include high dry matter and low reducing sugar content. A dry matter content of more than 20 per cent is desirable for chips, French fries and dehydrated products. Similarly, a reducing sugar content in tubers up to 100 mg/100g fresh weight is considered acceptable for processing. Nine varieties viz. Kufri Chipsona-1, Kufri Chipsona-2, Kufri Chipsona-3, Kufri Jyoti, Kufri Chandramukhi, Kufri Lauvkar, Kufri Surya and Kufri Himsona, Kufri Frysona have been developed for processing purposes. In India, potato processing in organized sector started about a decade ago, and the recent proliferation of this sector mainly results from the development of three indigenous potato processing varieties, viz. Kufri Chipsona-1 and Kufri Chipsona-3 by CPRI. These two varieties are now being used by the industries for processing into chips and French fries.

Novel potato products developed by ICAR-Central Potato Research Institute, Shimla include:,

»   Potato cookies
»   Low fat potato halwa premix
»   Lactose-free potato burfi
»   Potato porridge/Daliya and semolina
»   Dehydrated cubes and shreds

10.4.14. Computer applications

Simulation modeling is now widely used in various disciplines to work out tactical decisions. CPRI has developed INFOCROP-POTATO model to simulate the potato growth and development, to determine the best growing period, to optimize management practices under different agro-ecological regions, and to forecast the accurate yield estimates. An expert system (Potato Pest Manager) has also been developed for decision support with respect to identification and management of diseases and insect-pests.

10.4.15. Transfer of technology

Research achievements alone are not adequate to gauge the success of an agricultural system. The research information needs to be assessed and refined under various bio-physical and socio-economic situations through adaptive research before it is labeled as a technology. In this regard, the multi-locational trials under AICRP (Potato) and the TOT projects undertaken by CPRI such as Operational Research Project (ORP), Lab-to-Land Program (LLP), Tribal Area Development (TAD) program, and Institution-Village Linkage Program (IVLP) proved landmark in getting feedback from the field and development of appropriate technologies.

Transfer of technology to the end users is a complex task which consists of a number of components and dimensions. One of the important components is proper linkage between technology generating system and the client system. In this regard, innovative approaches like need assessment, participatory planning and implementation, and direct scientist-farmer interface facilitated faster dissemination of technologies and consequent adoption by the farmers/clients. The CPRI has built up linkages with farmers through demonstrations, trainings, Kisan Melas, potato school on All India Radio, supply of literatures and other extension activities. Besides, studies have been conducted to measure the socio-economic impact and constraints in transfer of potato technology.

10.4.16. Export

Although India contributes 7.55% to the total world potato production, its 0.7% share in world's potato export is quite insignificant. Indian potatoes are truly free from the prohibited disease like wart, black scruf, and pests like tuber moth and nematodes, which are the barometer for phytosanitary standards. India has also the natural advantage of exporting fresh table potatoes during January to June when supply from European countries dwindles. It can also supply fresh potatoes round the year because India has diverse agro-climates and potato is grown throughout the year in one or the other part of the country.

Potato has a good future in India under the changed scenario of global economy. Globalisation has resulted in many developing countries becoming much more integrated into the international potato trade. With the phasing out of quantitative restrictions on agricultural commodities, the imports and exports of potato would be based on the differences in price and production cost between the importing and exporting countries involved. Due to low production cost in the country as a result of availability of cheap labor, India will have competitive advantage in the international potato trade.

10.5. POTATO IN THE NEW MILLENNIUM

With the improvement in the living standard of people in India, the dietary habits will shift from cereals to vegetables. Under such a situation, it is estimated that India will have to produce 49 million tons of potato by 2020. This target could be achieved only by improving the productivity level. The productivity of potato in India is quite low (22.56 t/ha) as compared to that of USA (49.76 t/ha), France (39.37 t/ha), Netherlands (36.61 t/ha), and Germany (35.37 t/ha). This is due to shorter crop duration in India. There is a wide-ranging variations in the agro-ecological setting of different parts of the country, which results in wide variations in the productivity levels of different states (Table 1). Therefore, all our efforts may be put in to develop location-specific and problem-specific varieties and technologies.

Most of the people in India have either no knowledge or wrong notions about the nutritive value of potato. With low fat (0.1 per cent) and calorie contents, it does not cause obesity. Due to misconception the potato consumption, the per capita consumption of potato in India is only about 16 kg/year. On the other hand, the per capita consumption in Europe is 121 kg/year and as high as 136 kg/year in Poland. Hence, there is ample scope for improving the consumption of potatoes in India. For this purpose, a publicity campaign like eggs and milk needs to be launched through mass media such as television, radio and newspapers highlighting its nutritional value. Moreover, the possibility of using surplus potatoes as animal feed also needs to be explored.

The surplus potatoes in a season are stored in cold stores at 2-4°C in the country. This makes stored potatoes just unfit for processing and loses preference for table purposes due to accumulation of sugar content. To avoid sweetening of potato, they are required to be stored at 10-12°C. Only seed potatoes should be cold stored at 2-4°C. This would release at least 60 per cent of cold storage space that can be converted to store potatoes for processing and table purposes at 10-12°C with CIPC treatment leading to considerable savings on energy and storage costs.

Processing is a fast-growing sector in the potato world economy. Due to increased urbanization, rise in per capita income and expanding tourism, the demand for processed potato products in India and international market has risen at a fast pace. However, in India, processing of potatoes constitutes less than 2 per cent of the total annual production as compared to 60 per cent in USA, 47 per cent in the Netherlands and 22 per cent in China. Hence, there is great scope to expand the potato processing industries in India and also to diversify the processing to produce flour, cubes, granules, flakes and starch.

Under the changed global scenario, the potato production and utilization pattern is changing very fast. These changes harbour many opportunities which could be tapped through effective extension system. The use of modern information and communication technologies (ICT) to create awareness is highly pertinent in the contemporary times. This would enable us to reach directly to the end users by eliminating the intermediate channels which create distortion of information. Efforts are also needed to devise market-based extension strategies in order to promote entrepreneurship among potato growers with regard to potato production and marketing.

Some of the major contributions of CPRI, Shimla in this regard are as follows:

» In the last decade, 4 improved potato varieties namely Kufri Kanchan, Kufri Arun, Kufri Pushkar, Kufri Sailja, Kufri Surya, Kufri Chipsona-3, Kufri Himalini, Kufri Girdhari, Kufri Himsona, Kufri Khyati, Kufri Sadabahar, Kufri Frysona, Kufri Gaurav and Kufri Garima have been released. A True Potato Seed population 92-PT-27 has also been released.

» The Institute has developed 'Seed Plot Technique' in 1970s to carry out disease-free seed production in the sub-tropical Indian plains under low aphid period. This technique, aided by biotechnological approaches for virus elimination, has sustained the National Potato Seed Production program.

» An alternate technology for crop production through botanical seed called 'True Potato Seed' (TPS) has been developed by the CPRI, which is suitable for regions where quality seed tubers cannot be produced. The TPS technology offers low cost on seed, case in storage and transportation, and lower incidence of diseases and insect-pests. Three TPS hybrids namely, TPS C-3, HPS I/13 and 92 PT-27 have been recommended for commercial cultivation.

» Decision support systems/tools have been developed for recommending location specific best management practices for diverse agro-ecological conditions under which potato is grown in India as well as to extend support to improve efficiency of crop improvement programs. These include:

  • Potato Pest Manager: brings expert knowledge on plant protection within the reach of farmers.

  • Decision Support System for management of late blight: for cost-effective and eco-friendly management of late blight. This system has already been validated for western Uttar Pradesh.

  • Potato Potential Yield Estimation tool: to estimate the yield gap under different agro-climatic conditions.

- • Potato E-book: brings potato information to a wide range of clientele interested in potato.
- » Late blight forecasting model JHULSACAST has been developed by the Institute for hills and plains. Based on this model, Decision Support System has also been developed which has three components *i.e.* forecasting, need based application of fungicides, and yield loss assessment. Besides, aphid forecasting model for plains has also been developed.
- » Portable dipstick kits have been developed for detection of five major potato viruses, *viz.* PVY, PVX, PVA, PVS, and PVM at field level.
- » A biofertilizer-cum-bio-pesticide formulation (Bio-B5) has been identified and patented for eco-friendly management of soil- and tuber-borne diseases as well as for yield enhancement.
- » Integrated package of practices for management of late blight, bacterial wilt, viruses, soil and tuber-borne diseases, tuber moth and cyst nematodes are developed. These packages include host resistance, sanitation, crop rotation, use of safer chemicals, predators, bio-control agents and agronomic practices.
- » Organomercurial tuber treatment is replaced with boric acid (3%), a safe alternative, to check tuber-borne diseases particularly black scurf and common scab.
- » As far as breeding resistant varieties is concerned, out of 44 potato varieties, 24 are resistant to late blight, 9 to early blight, 4 to wart, 1 to cyst nematode and 6 are tolerant to viruses.
- » The Institute has developed the elevated temperature storage technology of storing table and processing potatoes at 10-12 °C, which is widely being practiced in the country.
- » The techniques for processing of potato into products like chips, French Fries, cubes, granules and canned products have been standardized.
- » Tissue culture based hi-tech seed production system has been standardized that now accounts for 40% of the breeder seed produced by CPRI.
- » The aeroponic seed production system has been developed at the Institute which increases the mini-tuber multiplication rate from 5:1 to 50:1.

10.6. CHALLENGES

First and foremost challenge is to enable the country to achieve potato production of 125 million tons at an ACGR of 3.2% with the help of 34.51 t/ha productivity and 3.62 million ha area under the crop during 2050. Availability of additional agricultural area, fragmentation of land holdings, abiotic and biotic stresses are expected to pose serious constraints for the future growth of potato in India. Modeling research shows that problem of bacterial wilt, late blight, potato tuber moth and white flies will aggravate under the regime of climate change. Development of short duration and biotic stress resistant potato varieties will be employed to mitigate these constraints. The challenges in the way of achieving these targets and technological interventions in order to meet these challenges need to be addressed separately for different time horizons.

10.6.1. Short-term challenges

Rising input prices and severe shortage of farm labor are posing very serious short-term challenges for potato cultivation in India. Steady increment in the prices of potato's competing crops over last several years, such as wheat, sugarcane and other vegetables; have gradually lowered comparative profitability of this crop. Further, the input prices have been increasing steadily while the output prices are relatively stable affecting farm profitability adversely. Enhanced resistance of pests (both disease pathogens and insect pests) to the chemicals in the past is another important issue to be addressed in the short run.

Rising average night temperature and fog in several potato growing areas in the country is lowering already constrained productivity of potato due to limited sunshine hours and shorter crop duration. Lack of state-of-the-art transport infrastructure adversely affects potato producers and consumers across the country due to its bulky nature and seasonal as well as regional concentration of production scenario. This deprivation is not only responsible for wider price differences in this agri-commodity across the regions but also high post-harvest losses. Shortage of cold storage capacity and also their functionality in some parts of the country has also negatively affected potato growth.

Comparative farm profitability studies should be conducted to explain these issues to policy makers. Enhancement of crop productivity through more intensive and concerted efforts on dissemination of potato technologies to the end users will be used as a strategy to bridge yield gaps and to augment potato profitability of farmers. Improvement of seed potato quality will be achieved through state-of-the-art diagnostics using transmission electron microscope, molecular techniques and dipstick. Subsequently rapid *in-vitro* multiplication of planting material will be done using areoponics and other biotechnological techniques. Development of suitable potato machinery for small and marginal farmers will be very high on the agenda. More than half of farm labor on mechanized farms is currently used for picking potato tubers. For large farmers, developing combine potato harvester performing digging and tubers lifting activities should be expedited. Application of molecular techniques for better diagnostics and understanding the genetic makeup of potato pests in the recent past is likely to bring desirable results in the short term. After successful release of heat tolerant potato variety, Kufri Surya, more such varieties should be developed to tackle the problem of heat stress in the short-term. Gene(s) responsible for tuberization will be identified in order to facilitate development of early maturity of potato.

10.6.2. Medium-term challenges

Climate change, erratic precipitation and shortage of irrigation water would be important medium-term challenges for potato cultivation in India. Creating supportive conditions for taking potato processing levels from the current 2.8 million tons (including the processing at household, cottage industry and unorganized sector) to 6 million tons (near 10% of anticipated potato production in the country) in the medium term is another challenge. Food security is likely to gain much higher importance in India in the medium-term and potato will have to shoulder much heavier responsibility to address this impending challenge. As the future potato growth has to be led by productivity enhancement, increase in production potential and bridging yield gaps in spite of the impact of climate change on potato productivity *per se* and also due to its effect of higher biotic stresses.

Research efforts on development of micro-irrigation technologies for potato and their better dissemination in the regions with scarcity of irrigation water should be undertaken. Development of improved processing varieties of potatoes under the ongoing breeding program at the Research Institutes will help the nation raise potato processing levels. Enhancement of potato production potential in various parts of the country in light of heat and moisture stresses under climate change regime and national food security will be addressed through the exploitation of biotechnological and molecular breeding and crop protection technologies. In order to achieve higher harvest index from crop plant, dwarf potato genotypes should also be developed using molecular techniques for cultivation under long photoperiod and or high temperature. Raising potato production efficiency should be achieved with the help of modeling research, precision farming, nanotechnology and improved mechanization. Augmentation of seed potato quality and dissemination of technical knowhow to various stakeholders should be targeted through appropriate means in order to raise the crop productivity.

10.6.2.1. Precision farming: Precision farming as a tool of yield enhancement is ecologically better option and increased attention should be paid to such technologies that offers enhanced resource use efficiency. The efficiency of production resources is enhanced by providing precise quantities of the inputs to the crops in right compositions. In the initial phase, focus should be on precise application of inputs based on weather, soil and crop requirements to maximize productivity, quality and profitability in a more sustainable way.

10.6.3. Long-term challenges

Most of the medium-term challenges are expected to extend in long-term too. In fact, some of the challenges like climate change and its adverse impacts are expected to aggravate further. Over next 40 years, 465 million people will be added to our existing 375 million urban population (NCAP estimates) which will generate a huge demand for processed potato products. The industry has to be supported by the required technologies including the need for improved processing varieties in order to enhance potato processing up to 25 million tons by the year 2050. Food security of 1619 million people in the country will really be a tough challenge after 30 years. Increasing production potential of potato under the era of enhanced biotic and abiotic stresses to potato crop under global warming and climate change scenario is another complicated challenge before us in the long-term. Dissemination of refined potato technologies to a very large number of small and marginal farmers in India is a challenge of its own type in India.

Efficient input (fertilizers, pesticides and irrigation *etc.*) delivery system and more accurate diagnosis of diseases using nano-science widely across the country will be one of the solutions for tackling the challenges of food security and enhancement of production potential in next 40 years. Concerted efforts have already been initiated in this direction. Use of biotechnological and molecular techniques to develop potato varieties with higher yield potential and resistance to disease pathogens and insect pests will be another solution for this problem. Identifying yield enhancing traits through modeling research, manipulating crop geometry and developing ideotypes should also be employed. Development of drought and heat tolerant potato varieties will play crucial role in the future of Indian potato industry. Precision farming and improved mechanization (e.g. automatic potato harvester, grader and bag filler) should be used to enhance potato farm profitability. Advanced information technology-based dissemination (e.g. on mobile phones and through decision support/expert systems) of refined scientific potato technologies should be utilized.

10.7. WAY FORWARD

Potato production still faces challenges in terms of production, health management, marketing and utilization due to unrelenting changes in agroclimates. The economic growth of the country has also accelerated potato consumption, which is expected to continue for years to come. This has created demand for processed, ready-to-eat convenience food, which has led to diversification in potato consumption and utilization. To meet the growing demand, India needs to produce 70 million tons of potato from an area of 2.6 million ha by 2030 with a productivity of 27 t/ha. To bridge the gap between demand and production, there is a need to intensify research efforts in the following areas:

10.7.1. Management and enhancement of genetic resources

Germplasm constitutes the backbone of any crop improvement program. For sustaining variety development program, a germplasm collection with wide gene pool is required. Although about 3,900 potato accessions are presently being conserved, it is necessary to augment this collection to feed the basic requirement of different breeding programs.

10.7.2. Varietal improvement

Development of high yielding varieties during second half of 20th century has contributed largely for fast growth of potato cultivation. However, there is hardly any new variety with a quantum jump in yield potential. Besides, absence of varieties with specific desirable attributes like shorter crop duration, early bulking, heat and drought tolerance, high input use efficiency, cold-chipping, French fry preparation, and better nutritional quality is now posing a hurdle for extending potato in new areas. These issues need to be addressed on priority.

10.7.3. Exploitation of Biotechnology for potato improvement

There is a need to undertake focused activities on development of transgenic potato with improved resistance to diseases and quality improvement, structural and functional genomics of potato with a view to improve productivity, developing robust molecular markers/QTLs for resistance, yield and quality attributes for initiating marker-assisted selection and cloning of desirable genes and promoters and their use for developing transgenic varieties with desired attributes.

10.7.4. Enhancing availability of quality planting materials

In potatoes, a part of each year's crop - ranging from 5 to 15 per cent, is set aside for re-use in the next planting season. Seed production in potato is beset with problems of low seed multiplication rate, low-proportion of seed-sized tubers in the produce, high production and high rate of degeneration. The seed related problems are further aggravated due to limited availability of suitable seed producing areas. The most deficient states in this respect are NEH region, Maharashtra, Karnataka, West Bengal, Orissa and Gujarat. There is a need to identify areas suitable for producing quality tuber seed. TPS should be encouraged a viable alternative for producing quality planting material in low productivity states. For that, effort should be made to identify early bulking TPS populations with higher transplant survival. Moreover, basic research should be undertaken to develop technologies for producing true-to-the type botanical seeds of commercial potato cultivars through apomixis. Producing healthy mother stock of seed potato using high-tech methods such as biotechnology and aeroponics should receive more emphasis.

10.7.5. Adaptation and preparedness for climate change

Climate change is causing rise in temperature, mainly due to increase in CO_2 concentration in the atmosphere. There is need to have detailed knowledge about how climate is going to change the productivity of potato in future, what will be the impact of changing climate on important potato diseases and on population dynamics of insects/vectors. Besides, possible effect of enhanced CO_2 levels on potato crop growth and production under sub-tropical growing condition is to be critically investigated.

10.7.6. Eco-friendly management of natural resources

Ruthless exploitation of natural resources like land, water and soil has resulted in declining potato productivity and profitability. To avoid wastage of precious national resource and to minimize the environmental damage there is a need to develop and demonstrate balanced use of nutrients. It should consider the crop removal and available nutrient status of soil, economics of fertilization, agro-techniques, soil moisture regime, weed control, plant protection, seed rate, sowing time, soil's physical, chemical and microbial properties, cropping sequence, *etc.* so as to increase crop yield, quality and farm income, correct inherent soil nutrient deficiencies, maintain/improve soil fertility and avoid damage to the environment.

10.7.7. Resource based planning for different agro-ecological zones

The usual cropping season of many crops are undergoing gradual changes due to global climate change which has resulted in widening gap between technology generation and adoption. It is necessary to develop eco-region-specific technologies based on maximum productivity of available natural resources like climatic conditions, soil fertility and water. IT based enabling mechanism should be used for technology transfer and socio-economic impact analysis.

10.7.8. Eco-friendly management of pest and diseases

The problem of late blight and viral degeneration of seed stocks are aggravated by incidence of wide range of other biotic stresses and emergence of new pathogens and pests due to changes in cropping systems and the global climate change. The use of large quantities of pesticides has resulted in problems related to environmental degradation, residual toxicity in foods, and resistance development in pests. Thus, there is a need to develop eco-friendly disease/pest management schedules for different regions of the country for minimizing the use of pesticides and alternatively use bio agents/plant synthesized bioprotectants as pesticides. To promote export of Indian potatoes, there is an urgent need to develop pest risk analysis for both export and import purposes.

10.7.9. Diversified utilization and value addition to potato

In India, existing cold storage capacity is sufficient to store 70% of total potato produced in the country. Although technology for storing of potato has been developed to store potato at 10–12 °C and a number of traditional low-cost and non-refrigerated storage technologies are available, there is a need for intensive research to check the sugar accumulation and reduce the losses in storage. To avoid glut and market fluctuation in India, it is necessary to ensure year-round utilization of both fresh and processed potatoes. Major emphasis should be given for development of novel potato products, which are nationally/internationally acceptable, economically viable, healthy and environment friendly. The potato processing sector is emerging as a fast-growing industry in the country with more entrepreneurs joining and existing ones increasing the capacity of their processing units. It is necessary to step up our effort to cater to this fast-emerging sector; lest suitable varieties and technologies for the processing sector may be licensed and borrowed from overseas.

10.7.10. Effective exploitation of genetic resources for varietal improvement

- » Molecular characterization and development of core collection of the germplasm.
- » Development of mapping population and pre-breeding including somatic hybrids for exploiting wider gene pool.
- » Heterosis and hybrid vigor leading to enhancement in production potential of potato.
- » Development of potato varieties and populations for short duration, processing, starch making, heat and drought tolerance, biotic stress tolerance, nutrient use efficiency, *kharif* season, exports, early bulking, and TPS populations.

10.7.11. Safe application of biotechnology for potato improvement

- » Structural genomics and bioinformatics for developing robust molecular markers for qualitative and quantitative traits.
- » Functional genomics for gene discovery for targeted traits like late blight durable resistance, heat tolerance, high temperature tuberization, better water and nutrient use efficiency.
- » Proteomics and metabolomics for basic studies on tuberization, photosynthesis, partitioning

of photo-assimilates, starch metabolism, carotenoid and flavonoid synthesis, storage protein quality, and processing quality.

» Technology development for marker-free and site-specific integration of transgenes.

» Development of transgenic potato with improved resistance/tolerance to biotic/abiotic stresses and to improve nutritional and processing qualities.

10.7.12. Encouraging production of quality planting material

» Development and standardization of low cost and efficient mass propagation methods – aeroponics, bio-reactor technology.

» Vector dynamics and its implications on seed quality.

» Development of homozygous TPS populations using apomixes and mono-haploidy.

10.7.13. Resource based planning and crop management

» Development of IT based Decision Support Systems/tools for crop scheduling and management of weeds, nutrients, water, diseases and pests under climate change scenario.

» Standardization of technologies leading to improved carbon sequestration and soil health.

» Development of technologies for enhancing inputs use efficiency through precision farming and micro-irrigation.

10.7.14. Eco-friendly crop protection

» Cataloguing genome variability and dynamics of new pathogen/ pest populations (Pathogenomics).

» Development of diagnostics for detection of pathogens both at laboratory and field level using micro-array and nano-technologies.

» Ecology and management of beneficial microorganisms for enhancing crop productivity and disease management.

10.7.15. Encouraging energy efficient storage and diversified utilization

» Technology refinement for elevated temperature storage for both on- and off-farm situation.

» Development of new processes, products and utilization technologies for diversified use of potatoes including waste utilization.

» Food fortification to enhance nutritional quality of processed foods.

» Technologies for lowering glycemic index.

10.7.16. Strengthening institute-farmer interface for technology dissemination

» Comparative farm profitability studies vis-a-vis ability to contribute to GDP by various crops, for providing efficient policy input.

» Proficient technical dissemination through an optimal mix of traditional and modern extension tools.

10.8. CUTTING EDGE RESEARCH THEMES

There is a need for using the cutting-edge research themes for future R&D agenda.

» Cent per cent breeders' seed production through tissue culture nucleus stock.

» Extension of potato cultivation to newly created irrigated areas, as *kharif* crop in plateau region, as component of inter-cropping, relay-cropping and multiple cropping systems.

» Development of early and medium maturing potato varieties; varieties suitable for French fry, flakes and flour production; varieties with durable resistance to multiple diseases; heat, drought and salt tolerant varieties; and varieties with efficient nutrient and water use efficiency.

» Potato genome sequencing and functional genomics for tuberization and late blight resistance.

» Development transgenic potato resistant to late blight, bacterial wilt and reduced cold induced sweetening; potato genome sequencing and functional genomics for realizing yield potential, marker assisted selection, diagnostics, conservation of genetic resources and micropropagation. Biosafety and food safety of transgenic crop are to be determined prior to their release.

» Soil health improvement to overcome the widespread macro- and micro-nutrient deficiencies – the "hidden hunger." Balanced and efficient fertilizer use has to be supplemented with increasing soil organic matter content by incorporating crop residues, green manuring, application of FYM, compost, vermi-compost, biofertilizers and other bio-digested products. Biosafety and biosecurity aspects of introduced bioagent or living modified organism (LMO) are to be studied critically.

» Eco-region-specific technology generation based on maximum productivity of available natural resources like climatic conditions, soil fertility and water. Information technology (IT) tools like geographic information system (GIS), crop modeling, precision farming are to be used for sustainable utilization of natural resources of the specific agro-ecological zones. The technology packages to be developed for potato would be an integral component of multi-functional agriculture of the specific zone. IT-based decision support systems would be used for technology transfer.

» Integrated management of emerging diseases and pests. Emphasis would be on identification of new and effective bio-molecules for management of biotic stresses and deployment of resistances sources. Gene pyramiding and multiple disease resistance would be encouraged for eco-friendly and sustainable management of diseases and pests.

» Diversification of potato utilization, renewed emphasis on storage and post-harvest processing, encouraging export.

» Strengthening quality seed production in seed deficit potato growing regions. Tissue culture techniques would be effectively integrated for disease-free seed production.

» IT based enabling mechanism for technology transfer and socio-economic impact analysis of technologies developed.

» Development of transgenic potatoes to address high risk areas viz. biotic and abiotic stresses, quality enhancement and wider adaptation.

» Processing sector: development of cold chipping varieties.

» Seed sector: production of seed potatoes in non-traditional areas.

» Health sector: development of potatoes with low glycaemic index and high antioxidant contents.

» Identification of new genes and markers for important traits.

» Fully automatic potato harvester to economize on labor.

» Studies on potato proteomics and phenomics with reference to tuberization.

» The next generation molecular marker, SNP, with reference to disease resistance and quality traits, should be developed by allele mining and resequencing.

» Bio-risk intelligent system (surveillance of racial pattern of different pathogens and pests and early warning systems) should be developed for taking informed decision at the local, regional and national levels.

» ICT, GIS and remote sensing options will be used to understand and mitigate ill effects of climate change and global warming, identify new potato growing areas and to develop decision support systems to meet impending complex challenges.

10.9. CONCLUSION

India is second largest producer of potatoes in the world. During recent past the indices for potato production and harvested area have shown tremendous growth. However, there is relatively milder growth in potato productivity statistics.

In both rural- and industrial-based systems, innovations resulting from potato research should be incremental through a step-by-step improvement of an existing structure promoting technologies adapted to different agri-food systems. This is particularly true for smallholder family agriculture in developing countries where there is a great need to increase potato production in a sustainable way for both men and women who may face different productive challenges and opportunities.

To reach the strongest impact through potato-based agri-food systems, potato research and development efforts need to move towards cooperation between disciplines (and filling up the gaps that exist between them) to allow integration of knowledge rather than focus explicitly on single technology/solution development. Multidisciplinary approaches are required to recognize and solve practical problems at the levels of the cropping system as well as the whole potato value chain to enhance potato's contribution to sustainable agri-food systems with a better understanding of the evolving food systems that are changing rapidly with changing consumer needs and value chain requirements.

Information remains a serious constraint to farmers' modernization. The upcoming of new communication technologies such as smartphones, expansion of mobile broadband, and access to local online platforms integrating large amounts of local data and links to digital tools offer interesting new options, but there is a need for experimentation to fully exploit the potential of information technologies particularly with young people, without exacerbating inequality gap especially in developing countries.

Supporting general rural infrastructure directly or indirectly influences growth of agriculture including the potato cultivation. In order to make potato cultivation and general agriculture more efficient and profitable, rural development initiatives of the current union government in India needs to be proficiently implemented in all states. Future of Indian agriculture or potato cultivation depends, to a very large extent, on general development programs like irrigation infrastructure, assured supply of quality electricity, quality and magnitude of rail and road transport *etc.* Concerted efforts of agencies involved in potato research and development in India are required to develop and deliver more potent potato technologies.

The technologies should be developed specific to the farmers' needs and preferences. Such technologies need to be efficiently delivered as large proportion of small and medium potato farmers in the country generally do not adopt improved technologies at their own.

Chapter - 11

Gene Revolution

11.1. INTRODUCTION

Food insecurity is a worldwide problem which cause many deaths in developing countries and is among the most serious concerns for human health. To be healthy, our daily diet must include ample high-quality foods with all the essential nutrients, in addition to foods that provide health benefits beyond basic nutrition. Around the world, one billion people are living in food insecurity (Giger *et al.*, 2009). Food insecurity or inability to access food is the most fundamental form of human deprivation. Millennium Development Goals stated that half of the world's hungry should be reduced by 2015. Despite all efforts taken to reduce global food insecurity, the published statistics of 2012 revealed that there are 852 million people in food insecurity around the world. The world's agriculture should respond creatively to the demand of increasing population with increased agricultural production.

Two waves of agricultural technology development have been seen in the past 50 years. The Green Revolution was the first wave which using technology development, targeted poor people in the poor countries (Pingali and Raney, 2005). Green Revolution by using high-yielding crop varieties developed through conventional plant breeding practices and agrochemicals was the cause of the increased crop productivity in some countries. However, for ever-rising global food demand, conventional plant breeding alone, can no longer response (Datta, 2013).

In the 1940s, Norman Borlaug, Father of the Green Revolution, began experimenting with wheat. With research fields in Mexico, he saw the difficulty of the non-capitalist world had in growing enough to eat. He made it his life's mission to use science to make crops grow more plentifully and cheaply.

His Nobel Prize-winning invention, a variety of dwarf wheat with a short stalk to support its enormous head of grain, did just that. Harvests soared worldwide, wheat yields tripled, and countries like India went from famine to surplus as Borlaug's wheat was planted there (Borlaug, 1972; Stuertz, 2002). It is estimated that "about half the world's population goes to bed every night after consuming grain descended from one of the high-yield varieties developed by Dr. Borlaug and his colleagues of the Green Revolution (Gillis, 2009)" i.e. around three billion people. Borlaug's work was only the beginning. He opened the door to a whole new field of research into making food more nutritious, hardier, safer, and easier to grow. Today's biological inventors, using the technology of genetic engineering, are building on Borlaug's Green Revolution with a "Gene Revolution."

In agriculture, scientists are combining their understanding of plant genetics with laboratory techniques of modern molecular biology to "unlock" the DNA of crop plants. By inserting genes from other plants or even from common microorganisms in just the right place, they are able to give plants with desirable traits, solving problems that farmers have faced for millennia.

Biotechnology offers vast potential for improving the efficiency of crop production, thereby lowering the cost and increasing the quality of food. The tools of biotechnology can provide scientists with new approaches to develop higher yielding and more nutritious crop varieties, to improve resistance to pests including weeds and adverse conditions, or to reduce the need for fertilizers and other expensive agricultural chemicals.

2015 marked the 20th anniversary (1996-2015) of the commercialization of genetically modified (GM) or transgenic crops, now more often called "biotech crops". An unprecedented cumulative area of 2 billion hectares of biotech crops were successfully cultivated globally in the 20-year period 1996 to 2015. Farmer benefits for the period 1996 to 2015 were estimated at over US$150 billion. The 1.815 billion accumulated hectares comprise 1.0 billion hectares of biotech soybean, 0.6 billion hectares of biotech maize, 0.3 billion hectares of biotech cotton, and 0.1 billion hectares of biotech canola.

In the first 20 years (1996 to 2015), biotech crops were planted by up to 18 million farmers (up to 90% were small/poor farmers) in 28 countries annually. With an increase of 100-fold from 1.7 million hectares in 1996 to 179.7 million hectares in 2015, this makes biotech crops the fastest adopted crop technology in recent times – the reason – biotech crops have the trust of millions of farmers because they deliver significant and multiple benefits. Accordingly, the number of biotech countries has more than quadrupled from 6 in 1996, to 16 in 2002 and 28 in 2015.

Biotech crops have delivered substantial agronomic, environmental, economic, health, and social benefits to farmers and, increasingly, to society at large. The rapid adoption of biotech crops, during the initial 20 years of commercialization (1996 to 2015) reflects the substantial multiple benefits realized by both large and small farmers in industrial and developing countries, which have grown biotech crops commercially.

11.2. BIOTECHNOLOGICAL APPLICATIONS IN AGRICULTURE

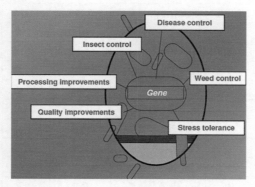

Fig. 11.1. Contribution of transgenic technology.

Although biotechnology efforts have focused primarily on herbicide tolerance and insect resistance, there are other potential applications to crop production (Fig. 11.1). The possibilities are highly varied, and new applications are developing so quickly that it is difficult to keep up with all of them.

Here are a few examples of current biotechnology research on plants.

»   Herbicide tolerance
»   Insect resistance
»   Disease resistance
»   Nematode resistance
»   Improved nutritional quality
»   Salt tolerance
»   Drought tolerance
»   Biofortification

11.3. AREA UNDER BIOTECH CROPS

In 2016, the 21st year of commercialization of biotech crops, 185.1 million hectares of biotech crops were planted by ~18 million farmers in 26 countries (ISAAA, 2016). Of the 26 countries that planted biotech crops in 2016, 18 countries were considered as biotech mega-countries, which grew at least 50,000 hectares. USA remained as top producer of biotech crops globally, which planted 72.9 million hectares in 2016, covering 39% of the global biotech crop plantings. Brazil landed on the second spot, with 49.1 million hectares or 27% of the global output. Brazil also had the highest biotech crop growth from 2015 to 2016 with a 4.9% increase. From the initial planting of 1.7 million hectares in 1996 when the first biotech crop was commercialized, the 185.1 million hectares planted in 2016 indicates ~110-fold increase (Table 11.1). Developing countries grew 54% of the global biotech hectares compared to 46% for industrial countries (Fig. 11.2). Thus, biotech crops are considered as the fastest adopted crop technology in the history of modern agriculture.

Table 11.1. Global area of biotech crops in 2015 and 2016: by Country (million hectares)

Rank	Country	2015	2016
1	USA*	70.9	72.9
2	Brazil*	44.2	49.1
3	Argentina*	24.5	23.8
4	Canada*	11.0	11.6
5	India*	11.6	10.8
6	Paraguay*	3.6	3.6
7	Pakistan*	2.9	2.9
8	China*	3.7	2.8
9	South Africa*	2.3	2.7
10	Uruguay*	1.4	1.3
11	Bolivia*	1.1	1.2
12	Australia*	0.7	0.9

13	Philippines*	0.7	0.8
14	Myanmar*	0.3	0.3
15	Spain*	0.1	0.1
16	Sudan*	0.1	0.1
17	Mexico*	0.1	0.1
18	Colombia*	0.1	0.1
19	Vietnam	<0.1	<0.1
20	Honduras	<0.1	<0.1
21	Chile	<0.1	<0.1
22	Portugal	<0.1	<0.1
23	Bangladesh	<0.1	<0.1
24	Costa Rica	<0.1	<0.1
25	Slovakia	<0.1	<0.1
26	Czech Republic	<0.1	<0.1
	Total	181.5	179.7

*Biotech mega-countries which grew more than 50,000 hectares, or more.

Source: ISAAA, 2016.

Fig. 11.2. Distribution of biotech crops in developing and industrial countries in 2016 (Source: ISAAA, 2016)

11.4. BIOTECH CROPS

Perhaps the most direct way to use biotechnology to improve crop agriculture is to genetically engineer plants—that is, alter their basic genetic structure—so they have new characteristics that improve the efficiency of crop production. The term genetic engineering is used to describe the process by which the genetic makeup of an organism can be altered using "recombinant DNA technology." This involves the use of laboratory tools to insert, alter, or cut out pieces of DNA that contain one or more genes of interest. The traditional goal of crop production remains unchanged: to produce more and better crops at lower cost. However, the tools of biotechnology can speed up the process by helping researchers screen generations of plants for a specific trait or work more quickly and precisely to transfer a trait. These tools give breeders and genetic engineers access to a wider universe of traits from which to select.

Over the last 50 years, the field of genetic engineering has developed rapidly due to the greater understanding of deoxyribonucleic acid (DNA) as the chemical double helix code from which genes are made. Although powerful, the process is not simple. Typically, researchers must be able to isolate the gene of interest, insert it into a plant cell, induce the transformed cell to grow into an entire plant, and then make sure the gene is appropriately expressed. For example, if scientists were introducing a gene coding for a plant storage protein containing a better balance of essential amino acids for human or animal nutrition, it would need to be expressed in the seeds of corn or soybeans, in the tubers of potatoes, and in the leaves and stems of alfalfa. In other words, the expression of such a gene would need to be directed to different organs in different crops.

Genetically modified or "GM" crops are being considered as having the potential to bring about the second green revolution. India is currently working on 111 transgenic crop varieties of various vegetables, fruits, spices, cereals, bamboo, *etc.* Transgenic crops such as eggplant (brinjal) and mustard are already in release stage; while cabbage, castor, cauliflower, corn, groundnut, okra, potato, rice, and tomato are under field trials stage. But it is cotton which was the first GM crop to be extensively worked upon and commercially released in India; the reason being that India has the largest area under cotton cultivation in the world.

11.4.1. Development

Developing plant varieties expressing good agronomic characteristics is the ultimate goal of plant breeders. With conventional plant breeding, however, there is little or no guarantee of obtaining any particular gene combination from the millions of crosses generated. Undesirable genes can be transferred along with desirable genes; or, while one desirable gene is gained, another is lost because the genes of both parents are mixed together and re-assorted more or less randomly in the offspring. These problems limit the improvements that plant breeders can achieve.

In contrast, genetic engineering allows the direct transfer of one or just a few genes of interest, between either closely or distantly related organisms to obtain the desired agronomic trait (Fig. 11.3 and Table 11.2). Not all genetic engineering techniques involve inserting DNA from other organisms. Plants may also be modified by removing or switching off their own particular genes.

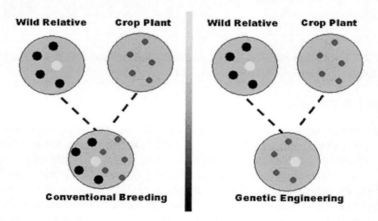

Fig. 11.3. Comparison of conventional breeding with genetic engineering (*Source:* ISAAA, 2014).

Table 11.2. Comparison of conventional breeding and genetic engineering

Conventional breeding	Genetic engineering
» Limited to exchanges between the same or very closely related species. » Little or no guarantee of any particular gene combination from the millions of crosses generated. » Undesirable genes can be transferred along with desirable genes. » Takes a long time to achieve desired results.	» Allows the direct transfer of one or just a few genes, between either closely or distantly related organisms. » Crop improvement can be achieved in a shorter time compared to conventional breeding. » Allows plants to be modified by removing or switching off particular genes.

Although there are many diverse and complex techniques involved in genetic engineering, its basic principles are reasonably simple. There are five major steps in the development of a genetically engineered crop. But for every step, it is very important to know the biochemical and physiological mechanisms of action, regulation of gene expression, and safety of the gene and the gene product to be utilized. Even before a genetically engineered crop is made available for commercial use, it has to pass through rigorous safety and risk assessment procedures.

The first step is the extraction of DNA from the organism known to have the trait of interest. The second step is gene cloning, which will isolate the gene of interest from the entire extracted DNA, followed by mass-production of the cloned gene in a host cell. Once it is cloned, the gene of interest is designed and packaged so that it can be controlled and properly expressed once inside the host plant. The modified gene will then be mass-produced in a host cell in order to make thousands of copies. When the gene package is ready, it can then be introduced into the cells of the plant being modified through a process called transformation. The most common methods used to introduce the gene package into plant cells include biolistic transformation (using a gene gun) or *Agrobacterium*-mediated transformation. Once the inserted gene is stable, inherited, and expressed in subsequent generations, then the plant is considered a transgenic. Backcross breeding is the final step in the genetic engineering process, where the transgenic crop is crossed with a variety that possesses important agronomic traits, and selected in order to obtain high quality plants that express the inserted gene in a desired manner.

The length of time in developing transgenic plant depends upon the gene, crop species, available resources, and regulatory approval. It may take 6-15 years before a new transgenic hybrid is ready for commercial release.

11.4.2. Commercial availability

Transgenic crops have been planted in different countries for twenty years, starting from 1996 to 2015. About 179.7 million hectares was planted in 2015 to transgenic crops with high market value, such as herbicide tolerant soybean, maize, cotton, and canola; insect resistant maize, cotton, potato, and rice; and virus resistant squash and papaya. With genetic engineering, more than one trait can be incorporated or stacked into a plant. Transgenic crops with combined traits are also available commercially. These include herbicide tolerant and insect resistant maize, soybean, and cotton.

11.4.3. Global adoption

The most planted biotech crops in 2016 were soybean, maize, cotton, and canola. Although there was only 1% increase in the planting of biotech soybean, it maintained its high adoption rate of 50% of the global biotech crops or 91.4 million hectares. This area is 78% of the total soybean production worldwide (Fig. 11.4).

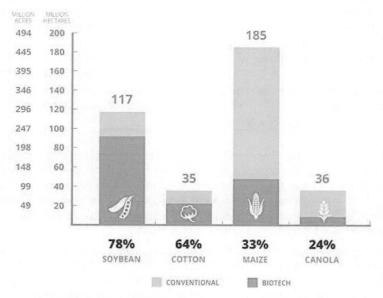

Fig. 11.4. Global adoption of 4 major biotech crops in 2016.

A significant increase of 13% was recorded for the global adoption rate of biotech maize from 2015. Biotech maize occupied 60.6 million hectares globally, which was 64% of the global maize production in 2016.

Biotech cotton was planted to 22.3 million hectares in 2016, which indicates a decrease by 7% from 2015. This reduction is attributed to the low global cotton prices, which also affected the global planting of non-biotech cotton.

Biotech canola increased by 1% from 8.5 million hectares in 2015 to 8.6 million hectares in 2016. This raise is attributed to the marginal increases in biotech canola plantings in the USA, Canada, and Australia, addressing the demand for edible oil.

In 2016, farmers in the USA and Canada planted biotech alfalfa. Approximately 1.2 million hectares of herbicide tolerant alfalfa and 21,000 hectares of low lignin alfalfa were planted in the US, while Canada planted ~1,000 hectares of low lignin alfalfa. Low lignin alfalfa was first commercialized in 2016, and offers 15 to 20% increase in yield.

Aside from soybean, maize, cotton, canola, and alfalfa; the biotech crops such as sugar beet, squash, papaya, eggplant, and potato were also planted in different countries.

11.4.4. Global value

According to Cropnosis, the global market value of biotech crops in 2016 was US$ 15.8 billion. This value indicates that there was a 3% increase in the global market value of biotech crops from 2015, which was US$ 15.3 billion. This value represents 22% of the US$ 73.5 billion global crop protection market in 2016, and 35% of the US$ 45 billion global commercial seed market. The estimated global farm gate revenues of the harvested commercial "end product" (the biotech grain and other harvested products) are more than ten times greater than the value of the biotech seed alone.

11.5. GENETIC ENGINEERING

Genetic engineering consists of introducing a foreign gene into a plant genome to create a new function. On the contrary, it could be used to reduce or suppress an existing gene function, as in the antisense strategy.

11.5.1. New agronomic traits

Two different techniques are commonly used to introduce genetic information (DNA) inside cells protected by pectocellulosic walls. A particle bombardment process, called ballistic, is used with success for monocots (gymnosperm, squash, and peas). Today, this technique allows the bombardment of meristems and avoids the difficulties of regeneration. But the mediation of *Agrobacterium* remains till now the most routine system to transform dicotyledonous plants.

The main agronomic traits introduced in crop plants and already commercialized are *Bt* toxin and herbicide resistance. Other studies concern virus resistance and male sterility.

Transgenic vegetable varieties suitable for processing have been developed. Bruise-resistant delayed ripening with higher sucrose and reduced starch content tomatoes are now available. High starch content in potato, dwarf lettuce for individual serving size and broccoli with slow ripening and longer staying (green) capacity, transgenics using genes for pest and disease resistance, nutritional quality of international standard, and for development of slow ripening cultivars for post-harvest management in tomato, brinjal, cabbage and cauliflower using anti-sense RNA have been evolved.

11.5.2. Antisense strategy

Calgene was created in 1994 [the first commercial transgenic tomato variety (Flavr Savr) with a long shelf life] by the suppression of polygalacturonase activity due to an antisense gene (Smith *et al.*, 1988). However, this Flavr Savr tomato variety was removed from the trade 3 years later, because of its disease susceptibility and its lack of productivity.

Later, other tomato varieties with long storage qualities were obtained by the utilization of an antisense RNA inhibition of ACC synthase or ACC oxidase, two ethylene precursors.

The antisense technique was also used to reduce the lignification of woody plants, by blocking the enzymes involved in the precursor of lignin biosynthesis. Another interesting application was the induction of white flowers in petunia and in different other ornamental plants by the suppression of chalcone synthase activity.

11.5.3. Transmission of the new traits

These traits induced by the transferred DNA are transmitted to the progeny as a dominant Mendelian character. Nevertheless, in some cases, transgenic plants with a strong gene expression can generate progenies with only a faint no gene expression. This problem is not clearly elucidated.

11.5.4. Marker, reporter, promoter and expression genes

To select the transformed cells, some marker genes, very often resistant to antibiotics or to herbicides, are attached to the coding sequence, as a promoter or an expression gene. These allow the new gene to express in the whole plant or in a specific plant tissue.

Some reporter genes are used to follow the evolution of the transformed cells. The most frequently used reporter genes include the gene 'gus' (glucuronidase) or the gene "lux" (luciferase).

11.6. STATUS OF GM CROPS

11.6.1. World scenario

The Flavr Savr tomato was the first genetically engineered whole food approved for commercial sale in 1994. Four other transgenic tomatoes with a delayed ripening were approved later (1995 and 1996).

In 1995, potatoes with *Bt* genes and a squash cultivar resistant to two viruses were released. Two years later, another squash cultivar resistant to 3 viruses and a papaya line also resistant to viruses were approved.

Two other horticultural crops are waiting the authority's approval: red hearted chicory (Radicchio) with male sterility and resistance to herbicide, and a tomato with Bt toxin.

Scientists claim the controversial potato, known as the "pomato", contains at least a third more protein than normal tubers, and that it holds "high-quality nutrients". It has been created by the addition of a gene called *AmA1* taken from the amaranth plant, which is native to South America and sold widely in western health food stores. The extra gene is said to give ordinary potatoes 30-50% more protein, as well as substantial amounts of the amino acids (lysine and methionine). The transgenic plant, developed by a team of scientists led by Asis Datta, who also leads the review committee on genetic modification in the Indian government's Biotechnology Department, is now in its third year of field trials and could be approved within six months.

The pomato would be given free to millions of poor children in order to reduce malnutrition in the country. It is planned to incorporate the vegetable into the government's free midday meal program for schools. There has been a serious concern that malnutrition is one of the reasons for the blindness (vitamin A and protein deficiency). So, it is really a very important global concern, particularly in the developing world.

Nowadays, it is also true that many transgenic plants exist in research laboratories, awaiting authorization for field testing, as they are very interesting model plants for learning physiology: lettuce with less nitrate by increasing nitrate reductase gene activity, a yeast ribonuclease gene to reduce viroid infection on potato, modification of lignin synthesis in plant to produce trees adapted to paper industry, or timber, or biomass production, regulation genes to control tree architecture.

The objectives of Agricultural Biotechnology for Sustainable Productivity Project (ABSP project) which is managed by Michigan University are the reduction of losses due to pathogens and pests by using transgenic plants (sweet potatoes, potatoes, cucurbits, and tomatoes) and by cloning commercial value-plants (bananas, pineapple, and coffee).

Biotech crops have the trust of millions of farmers because they deliver significant and multiple benefits. Accordingly, the number of biotech countries has more than quadrupled from 6 in 1996, to 16 in 2002 and 28 in 2015. Importantly, adoption rates for biotech crops during the period 1996 to 2015 were unprecedented– the majority is over 90% for major products in principal markets in both developing and industrial countries. High adoption rates reflect farmer satisfaction with the products that offer substantial benefits ranging from more convenient and flexible crop management, lower cost of production, higher productivity and/or net returns per hectare, health and social benefits, and a cleaner environment through decreased use of conventional pesticides, which collectively contribute to a more sustainable agriculture. There is a growing body of consistent evidence across years, countries, crops and traits generated by the public sector institutions that clearly demonstrate the benefits from biotech crops. These benefits include improved weed and insect pest control with biotech herbicide tolerant and insect resistant *Bt* crops, that also benefit from lower input and production costs; biotech crops also offer substantial economic advantages and conveniences to farmers compared with corresponding conventional crops.

Global population was approximately 7.3 billion in 2015 and is expected to reach up to ~9.7 billion by 2050. In 2015, 795 million people in the developing countries suffered from hunger, malnutrition and poverty (FAO, 2015). Biotech crops represent promising technologies that can make a vital contribution, but are not a panacea, to global food, feed and fiber security. Biotech crops have also made a critically important contribution to the alleviation of poverty under the aegis of the Millennium Development Goals (MDG).

A list of commercially available genetically modified crops is presented in Table 11.3.

Table 11.3. Commercially available genetically modified crops

Crop	Trait phenotype	Target trait gene(s)	Trait designation	Originating company	Year of first commercial sale	Trade name
Cotton	Resistance[1] to lepidopteron pests	cry1Ac	MoN531	Monsanto	1996	Bollgard Ingard
		cry1Ac, cry2Ab2	MoN15985	Monsanto	2003	Bollgard II
		cry1Fa, cry1Ac, pat	281-24-236 x 3006-210-23	Dow AgroSciences	2005	WideStrike
	Resistance to glyphosate herbicides	CP4 epsps	MoB1445/1698	Monsanto	1996	Roundup Ready
	Resistance to phosphinothricn herbicides	bar	LLCotton25	Bayer Crop Science	2005	LibertyLink
Maize	Resistance[1] to European corn borer and other lepidopteron insects	cry1Ab, pat	Bt11	Northrup King (now Syngenta)	1996	YieldGard Attribute
		cry1Ab	MoN810	Monsanto	1997	YieldGard (Corn borer)
		cry1F, pat	TC1507	Dow AgroSciences; Pioneer Hi-Bred Intl.	2003	Herculex I
	Resistance[1] to corn root worm	cry3Bb1	MoN863	Monsanto	2003	YieldGard (Root worm)
		cry1Ab, cry3Bb1	MoN863 x MoN810	Monsanto	2005	YieldGard Plus

	Resistance to glyphosate	Maize epsps	GA21	DeKalb (now Monsanto)	1998	Roundup Ready
		Two CP4 epsps expression cassettes	NK603	Monsanto	2001	Roundup Ready Corn 2
	Resistance to Phosphinothricin herbicides	pat	T14, T25	Aventis (now Bayer CropScience)	1996	LibertyLink
Soybean	Resistance to glyphosate herbicides	CP 4 epsps	GTS-40-3-2	Monsanto	1996	Roundup Ready
Canola	Resistance to glyphosate herbicides	CP 4 epsps, gox v247	GT73	Monsanto	1996	Roundup Ready
	Resistance to phosphinothricin herbicides	pat	Topas 19/2	Agrevo (now Bayer CropScience)	1995	LibertyLink
Alfalfa	Resistance to glyphosate herbicides	CP 4 epsps	J101, J163	Monsanto	2005	Roundup Ready
Squash	Resistance[2] to CMV, WMV2 and ZYMV	Coat protein genes of CMV, WMV2 and ZYMV	CZW3	Asgrow; Semini Vegetable Seeds (now Monsanto)	1998	Destiny III Conquerer III Liberator III
	Resistance[2] to WMV2 and ZYMV	Coat protein genes of WMV2 and ZYMV	ZW-20	Asgrow; Semini Vegetable Seeds (now Monsanto)	1995	Preclude II Patriot II Declaration II Independence II
Papaya	Resistance[2] to PRSV	Coat protein gene of PRSV	55-1 63-1	Cornell university; University of Hawaii, USDA	1998	SunUp, Rainbow

Source: Castle *et al.* (2006).

[1] Many insect resistance and herbicide tolerance traits are also available in combinations as stack traits.

[2] CMV - Cucumber mosaic virus; PRSV - Papaya ringspot virus; WMV2 - Watermelon mosaic virus; ZYMV - Zucchini yellow mosaic virus.

11.6.2. Indian scenario

India approved the commercial cultivation of *Bt* cotton in 2002. It was a breakthrough step to revive the ailing cotton sector in the country. The cotton industry at that time was characterized by stagnation in cotton production, decelerating trend in cotton yield and over reliance on cotton import for over many decades. Coincidental with the steep increase in adoption of *Bt* cotton between 2002 and 2015, the average yield of cotton in India, which used to have one of the lowest yields in the world, increased from 308 kg per hectare in 2001-02 to 570 kg per hectare in 2013-14. Cotton production increased from 13.6 million bales in 2002-03 to 39.8 million bales in 2013-14 and 39 million bales in 2014-15, which was a record cotton crop for India.

India continues to debate the relevance and need of biotech crops since it imposed the moratorium on *Bt* brinjal on 9th February 2010. The following three important biotech crops, which are ready to be commercialized, would trigger a new phase of growth and momentum in the crop biotech sector in the country:

» First, the approval of country's first vegetable crop *Bt* brinjal (eggplant) by revisiting its 5-year old moratorium in the context of the large-scale commercial planting of *Bt* brinjal in the neighboring country of Bangladesh; it is noteworthy that a noticeable increase in pesticide residues is occurring in important vegetables and fruits; biotech crops could help reduce the use of pesticides on food crops.

» Second, the approval of the country's first stacked trait – the insect resistant and herbicide tolerant cotton, Bollgard II Roundup Ready Flex cotton (BG-II RRF).

» Third, the approval of country's first biotech mustard (*Brassica juncea*) with enhanced heterosis (hybridization) in mustard, the most important edible oil crop in India.

11.6.2.1. *Bt* cotton: India got its first approval of genetically modified crop *Bt* cotton hybrid "Bollgard" in 2002. The total cultivated area under GM crops has increased many folds since then with India (11.6 million hectares in 2015) standing at 1st position in *Bt* cotton production. Monsanto first developed *Bt* cotton and the three genetically modified cotton hybrids (*Bt* Mech 12, *Bt* Mech 162 and *Bt* Mech 184) in collaboration with its Indian partner Mahyco, were released for commercial cultivation in central and southern India.

It is mostly the private companies (presently 22, in pipeline 8) which are releasing cotton varieties (Table 11.4). At the moment all the *Bt* cotton hybrids developed by private companies contain *Bt* cry 1 Ac gene developed by Monsanto. The first indigenous cotton variety released by public sector in 2005, developed by the Central Institute of Cotton Research (CICR) is "Bikaneri Narma (BN)" expressing *Bt* Cry 1Ac protein in the north, central and south cotton growing zones in India. The number of varieties released is increasing at a tremendous rate. Till July 2008, 142 cotton varieties had been released.

Table 11.4. Bt cotton genes used in India

Institutions/Companies	Bt cotton genes
Mahyco	Cry 1Ac
Monsanto	Cry1Ac+2Ab
Nath seeds	Cry1Ac modified (China)
JK seeds	Cry1Ac modified (IIT, Kharagpur)
Syngenta	Vip3A+Cry1Ab
Metahelix	Cry1 Ac

Dow Agri. Science	Cry1Ac + Cry1F
ICAR	Cry1Ac, Cry1Aa3, Cry1F, Cry1Ia5, Cry1Ab (Japan)
NBRI	Cry1Ac
ICGEB	Cry1Ac (Enc)

Source: ISAAA (2008)

The cotton growing region in India is restricted to northern, central and southern part. According to the latest issue of cotton statistics and news released by Cotton Association of India (CAI), maximum acreage of *Bt* cotton in the country is reported in the central zone, followed by the south zone and north zone. The total coverage of *Bt* cotton comes to 95% of the total area.

The major states growing *Bt* cotton in 2014-15, listed in order of hectarage, are Maharashtra (4.190 million hectares) followed by Gujarat (2.773 million hectares), Karnataka (0.875 million hectares), Andhra Pradesh (0.821 million hectares), Haryana (0.648 million hectares), Madhya Pradesh (0.547 million hectares), and Punjab (0.420 million hectares).

The high percentage adoption of *Bt* cotton by farmers across the different states reflects the priority of controlling the menace of the American bollworm complex, a group of deadly borer insects that caused heavy damage to cotton crop in the past. Evidently, the country achieved a near phasing out of single gene Bollgard-1 cotton hybrids, which has been almost replaced with dual gene Bollgard-II (BG-II) cotton hybrids introduced in 2006. The double gene *Bt* cotton hybrids provide additional protection to *Spodoptera* (a leaf eating tobacco caterpillar) while protecting cotton crop from American bollworm, pink bollworm, and spotted bollworm. It is reported that double gene *Bt* cotton farmers earn higher profit through cost savings associated with fewer sprays for *Spodoptera* control as well as increasing yield by 8-10% over single gene *Bt* cotton hybrids.

11.6.2.2. *Bt* Brinjal: In the entire South Asian region, the shoot and fruit borer (*Leucinodes orbonalis*) is identified as the most important limiting factor in brinjal production. Often the extent of damage due to this pest reaches 70-80%. Damage is severe during rainy months and early winter. Caterpillars feed inside the tender shoots initially, resulting in wilting of the affected shoots. When fruits are present, caterpillars bore into the fruits, contaminating the fruit with excreta and such fruits are unfit for human consumption.

The management of *L. orbonalis* is extremely difficult. Often per cent borer damage due to *L. orbonalis* in spite of weekly pesticide sprays exceeds 25%. Further, the pattern of pesticide use has resulted in resurgence of secondary pests such as whiteflies, mites, aphids and thrips.

Because of the ease with which pesticide crystal protein genes from *Bacillus thuringiensis* can be transferred into crops, this avenue of research was exploited to develop sustainable control measures to combat *L. orbonalis*. This will help to reduce pest population build up prior to fruiting.

Bt Brinjal is the first genetically modified food crop in India that has reached the approval stage for commercialization. It was developed by inserting a gene cry1Ac from a soil bacterium called *Bacillus thuringiensis* through an *Agrobacterium*-mediated gene transfer. It is a genetically modified brinjal developed by the Maharashtra Hybrid Seed Company Ltd. (Mahyco), a leading Indian seed company (Fig. 11.5). *Bt* Brinjal event EE1 has been developed in a Public Private Partnership mode under the aegis of the Agriculture Biotechnology Support Project (ABSP) from Cornell University where the *Bt* technology available with M/s Mahyco has been transferred (free of cost) to the Tamil Nadu Agricultural University (TNAU), Coimbatore; the University of Agricultural Sciences (UAS),

Dharwad; and the Indian Institute of Vegetable Research (IIVR), Varanasi. The Event EE 1 was introgressed by plant breeding into various local varieties (Table 11.5).

Fig. 11.5. Damage caused by brinjal fruit and shoot borer (left) and *Bt* brinjal resistant to borer developed by MAHYCO (right)

Table 11.5. Development of *Bt* brinjal hybrids and open pollinated varieties in selected institutions

Mahyco's 8 *Bt* brinjal hybrids	Public Sector's 16 *Bt* brinjal open pollinated varieties (OPVs)		
	UAS, Dharwad (6)	TNAU, Coimbatore (6)	IIVR, Varanasi (4)
MHB-4*Bt*	Malapur local (S)*Bt*	CO2-*Bt*	Pant Rituraj
MHB-9*Bt*	Manjarigota *Bt*	MDU1-*Bt*	Uttara
MHB-10*Bt*	Rabkavi Local *Bt*	KKM1-*Bt*	Punjab Barsati
MHB-11*Bt*	Kudachi Local *Bt*	PLR1-*Bt*	VR-14
MHB-39*Bt*	Udupigulla *Bt*	IVBL-9	
MHB-80*Bt*	GO112 *Bt*	VR-5	
MHB-99*Bt*			
MHB-112*Bt*			

The Genetic Engineering Approval Committee (GEAC) cleared *Bt* brinjal for commercialization on 14 October 2009. Following concerns raised by some scientists, farmers and anti-GM activists; the Government of India officially announced on 9 February 2010 that it needed more time before releasing *Bt* brinjal, with Indian Environment Minister Jairam Ramesh saying that there is no overriding urgency to introduce *Bt* brinjal in India. On 17 February 2010, Jairam Ramesh reiterated that the Centre had only imposed a moratorium on the release of transgenic brinjal hybrid, and not a permanent ban, saying that "until we arrive at a political, scientific and societal consensus, this moratorium will remain".

The imposed moratorium has been criticized by some scientists as not being based on any compelling scientific evidence and potentially setting Indian biotechnology decades back (Shantharam, 2010).

a. ***Bt* brinjal in Bangladesh:** The *Bt brinjal* project in Bangladesh may lay claim to be the first crop biotechnology transfer project to deliver a product to farmers. Bt brinjal was developed as an international public private partnership, between an Indian seed company Mahyco generously donating technology to the Bangladesh Agricultural Research Institute (BARI), facilitated by Cornell University led project ABSP-II and funded by USAID. Bangladesh approved Bt brinjal for commercial cultivation on 30th October 2013 and in record time – less than 100 days – on 22 January 2014, a group of small farmers (120) planted the first four commercial Bt brinjal varieties (Uttara, Kajla, Nayantara, and Iswardi/ISD 006) on 12 hectares. In 2015, a total of 250 hectares of Bt brinjal were planted by 250 farmers and the area is expected to increase substantially in 2016. Five additional Bt brinjal varieties (Dohazari, Shingnath, Khatkati, Chaga, and Islampuri). – three of them received NCB approval in 2015, will be available in 2016 and remaining two in the near future.

Bt brinjal increases marketable yield by at least 30% and reduces the number of insecticide applications by 70-90%, reduced cost of production, increases marketable yield and improves fruit quality; with a net economic benefit of US$ 1,868 per hectare, equivalent to a gain of up to US$ 200 million per annum. Agronomic performance data released by BARI from 19 *Bt* brinjal demonstration plots, established in 108 farmers' fields in 19 districts showed close to zero pest infestation, increased yield up to 100% compared to non *Bt* variety. Farmers have successfully sold Bt brinjal fruits in the open market labeled as "BARI Bt Begun #, no pesticide used".

11.6.2.3. GM mustard: India's first public sector edible oil biotech crop mustard (*Brassica juncea*) hybrid DMH-11 indigenously developed by University of Delhi South Campus from 1996 to 2015 with the funding from the Department of Biotechnology of the Ministry of Science and Technology (MOST) and the National Dairy Development Board (NDDB).

Delhi University team has used a "Barnese" gene isolated from a soil bacterium called *Bacillus amyloliquefaciens*. It codes for a protein that impairs pollen production and renders the plant into which it has been introduced male-sterility. This male-sterile plant is crossed with a fertile parental line, containing, in turn, another gene, "Barstar", from the same bacterium that blocks the action of the "Barnase" gene. The resultant progeny, having both the foreign genes, is a hybrid mustard plant that is not only high-yielding, but also fertile and capable of producing seed/grain (thanks to the "Barstar" gene in the second fertile line). The GM mustard hybrid gives 28% more yield than the best variety 'Varuna' and 38 % over the control varieties.

India's central biotech regulator, Genetic Engineering Appraisal Committee (GEAC), cleared the genetically modified (GM) herbicide (glufosinate) tolerant Dhara Mustard Hybrid-11(DMH-11) for commercial cultivation and recommended its approval to the environment ministry. The GEAC sub-committee had given its safety clearance while noting that the GM Mustard is safe for human consumption and environment.

India's dependence on edible oil imports made it necessary to harness GM-mustard. Annually, India spends about $12 billion on imported edible oil. The edible oil deficit will continue to widen with the increase in population and per capita income. To address this challenge, India needs to increase productivity of oilseed crops. DMH-11 is one of the promising technologies to improve mustard yield in India, which is almost stagnant since the last two decades.

Edible oils derived from soybeans are imported every year from USA and Argentina to contain ever increasing demand of India. It is observed that the oil imported from USA contains oil produced from GM soybean and GM rape seeds. It is very difficult to differentiate GM free oil from oil from

GM seeds. This shows that illegal GM oil can easily enter Indian borders in the name of non-GM oil.

The fate of this transgenic variety of oil-seed will now be in the hands of the MoS for environment Anil Madhav Dave who may accept it, reject it or sit over the file till the Supreme Court takes its call on a pending application on the matter.

11.7. GENETICALLY ENGINEERED INNOVATIONS

11.7.1. Herbicide-tolerant crops

Weeds compete with crops for light, nutrients, and water. They can also harbor destructive disease pathogens and disease carrying insects. When harvested along with the crop, they degrade it and ensure that next year's seed also will be contaminated with weeds.

By transferring a gene from a common soil bacterium into a soybean, the Monsanto Company created the first genetically engineered herbicide-tolerant plant (Anon, 1994). The "Roundup Ready" soybeans would survive when sprayed with glyphosate, a common herbicide and the active ingredient in Roundup weed killer. Since then, the herbicide-tolerant trait has become "the world's most widely adopted biotech trait," planted on billions of acres since 1996 (Anon, 2015a).

The plants have many advantages. Herbicide can be used throughout the growing cycle and just a coffee-cup full of a relatively inexpensive, easy-to-manufacture herbicide is all it takes to control weeds on an area of land the size of a football field (Walton, 2015). Glyphosate was chosen because it is relatively benign—it's about as toxic as baking soda, yet is extremely effective in killing weeds (Ryan, 2015; Lim, 2014; Charles, 2015). The trait does away with labor-intensive hand weeding, the standard in much of the world.

The Roundup Ready soybean is set to go off patent in 2015, but Monsanto has already engineered a second-generation variety that will be on patent for many years to come. In 2014, Dow AgroSciences' Enlist corn and soybeans were approved for planting in the United States (Dow AgroSciences, 2014). Paired with another common and relatively benign herbicide (2,4-D, commonly purchased as Scotts Turf Builder), the corn and soybeans will give farmers the option of rotation between herbicide-tolerant traits to slow weed resistance.

"In 2013 alone, herbicide-tolerant crops occupied 99.4 million hectares or 57% of the 175.2 million hectares of biotech crops planted globally." Of the major food crops in the United States, over 80% of soybeans, cotton, beets, alfalfa, canola, and corn planted were genetically engineered to be herbicide tolerant (ISAAA, 2015; USDA, 2015; Carpenter, 2014).

11.7.2. Insect-resistant trait

The stalk borer is a worldwide pest—it appears across large swaths of Europe, the United States, and Canada. In America, farmers have been dealing with the inconspicuous looking moth for almost a hundred years. In years where it is particularly active, it can take down up to 30% of a corn harvest (VIB News, 2015; Monsanto, 2015).

The larvae of the moth burrow inside the corn stalk where they feast, dropping ears to the ground prematurely or simply killing the plant entirely (University of Illinois, 2011). There are pesticides that can kill the bug, but spraying on the outside of the plant cannot reach inside of the stalk, where the insect is busy mining away.

Scientists first fortified plants against burrowing insects in 1987, with the invention of hornworm-resistant tobacco. By 2010, multiple varieties of corn that were toxic to burrowing insects, like the stalk borer, were marketed with a similar trait, including Monsanto's YieldGard, Syngenta's Agrisure, and Dow's Herculex (Nielsen, 2015).

Like hornworm-resistant tobacco, these corn seeds borrowed a gene from the *Bacillus thuringiensis* bacteria, instructing the plant to produce its own *Bt* insecticide. The toxin works by attaching itself to a receptor inside the insect's stomach, causing the insect's death, and was chosen because it is considered safe to people and non-target species (Cranshaw, 2015). Varieties differ by the specifics of the version of the *Bt* toxin they produce or the position the added gene occupies in the chromosome of the corn.

In the United States, tens of millions of acres of insect repellant corn have been planted, increasing yields and saving precious corn plants from burrowing insects. In India, cotton with the same trait single-handedly raised farmers' bottom lines by 3.2 billion dollars in 2011. The crop is so successful against the cotton bollworm that 10 million additional acres of cotton were planted in the 12 years since its introduction, creating a thriving export market. In Bangladesh, Bt eggplant was allowed for the first time in 2013. The eggplant is poised to be so successful that anti-biotechnology activists are out to stop the technology before it is widely adopted. Masked activists showed up at one eggplant field, threatening the farmer to falsely declare his crops a failure to a local newspaper in order to scare others out of planting it (Anon, 2014; Maxham, 2014).

11.7.3. "Vaccinated" papaya

The papaya ringspot virus gets its name from the symptoms it causes on the papaya tree. The fruit of an infected tree develop bumps and prominent ring-shaped spots, the leaves appear distorted and curled, and the trees can no longer produce papayas (Anon, 2014; Maxham, 2014).

The virus spreads by hitching a ride on aphids as they fly from tree to tree. The aphids pick up and spread the disease within seconds, making insecticides nearly useless—the bugs simply cannot be killed quickly enough to prevent the disease from spreading.

In Hawaii, where the papaya industry thrives, the disease has a long history. The virus was discovered on the Hawaiian island of Oahu in the 1940s. By the 1950s, it had caused severe damage to the papaya industry there. In the 1970s, the virus made the hop to the Big Island. Dennis Gonsalves, a plant pathogen researcher who hails from Hawaii, became concerned that the virus could wipe out the Hawaiian papaya for good (Voosen, 2011).

Gonsalves began investigating a genetic solution to the disease after news spread of the invention of a virus-resistant transgenic tobacco plant. He wondered if it was possible to "vaccinate" papaya trees against the virus using genetic engineering (Gonsalves, 2015; YouTube, 2015). The breakthrough came when Gonsalves and his colleagues found a way to include a small amount of the papaya ringspot pathogen in the chromosomes of the tree, thereby fortifying it against infection. They started their work none too soon.

In 1992, the virus engulfed the Big Island. But by 1998, the federal government approved the planting of two varieties of ringspot-resistant papayas that Gonsalves and his team invented. Known by their trademark names, seeds of "Rainbow" and "SunUp" varieties were pushed into virus-laden soil by Hawaiian farmers (Gonsalves, 2015; YouTube, 2015).

These papayas grew, saving Hawaii's papaya business, and are exported to grocery stores throughout the world to this day.

11.7.4. Saving bananas

Bananas in Africa have been devastated by a bacterial disease known as the "Xanthomonas wilt." This bacterium causes fruit to rot and ooze while still on the stalk and the plant wither and die (Discovery, 2013).

Seeking to alleviate the effects of the disease and give bananas a defense against the wilt, genetic engineers transferred two genes from peppers that confer resistance. The trees are 100% immune to the disease. The fortressed banana holds so much promise to alleviate crop losses in Uganda that the country temporarily lifted its ban on genetically modified crops. The banana wilt caused $500 million in damage annually to banana growers there (Namukwaya, *et al.* 2012; Tripathi *et al.*, 2010; Rice, 2011).

11.7.5. The Arctic Apple

Early this year, Okanagan Specialty Fruits announced that their patented technology, the Arctic Apple, had been approved for sale in the United States and Canada (Anon, 2015b; Anon, 2015c). Although they may look like any other Granny Smith or Red Delicious apple, these apples are different. Scientists at Okanagan invented the apples to solve a problem: browning.

When the cells of a typical apple are broken, two chemicals within the apple mix and react with oxygen to cause the flesh to brown (Anon, 2011). Sliced apples are used in fast food meals and on top of prepared salads, but those apples must be sprayed with an expensive coating that changes the flavor in order to suppress browning. By inserting genes, which control the production of one of these browning chemicals, into a common apple, scientists at Okanagan were able to suppress that chemical's production. Arctic Apples, named for the ever-white color of their flesh, don't turn brown when sliced, bitten, or bruised. It's a simple, yet spectacular, invention that could save the food service industry millions.

The apples will also be one of the first direct-to-consumer biotech-enhanced products, rather than a product geared toward farmers.

11.7.6. Non-browning potatoes

J.R. Simplot, a company that produces over 3 billion pounds of potatoes each year, recently added a genetically engineered variety to their roster. Simplot's new "White Russet" potato is similar to conventional spuds, but borrows DNA from five varieties of potatoes, giving it some desirable traits (Anon, 2013).

Among its selling points, Simplot scientists say they have fortified White Russet potatoes to be less susceptible to black spot—a type of bruising that occurs when potatoes are knocked about during harvest. As these spotted potatoes must then be picked out of the harvest, the bruising can account for wasting up to 5% of a potato crop (Anon. 2015d).

Similar to the Arctic Apple, White Russet potatoes are engineered to produce less of a chemical that would otherwise cause them to brown when sliced. For a company like McDonald's that sells 9 million pounds of French fries a day, these potatoes could mean an easy way to ensure all of their fries are that classic golden color and save them money in the process (Patton, 2015).

11.7.7. Triple-stacked rice

Sometimes patient coaxing and careful tending are simply not enough to see a plentiful harvest spring forth from your field—the unpredictability of the weather and poor soil conditions can make that impossible. Getting plants to more efficiently use resources, such as water, sunlight, and nitrogen in the soil, can ensure that food is plentiful, come rain or come shine.

Researchers at Arcadia Biosciences in Davis, California, recently unveiled a strain of rice that they hope will be less vulnerable to some common growing problems. If successful, rice farmers will be able to plant this same strain of rice, year after year, "regardless of the conditions." Researchers have triple-fortified the rice to combat three problems that make rice hard to grow: drought, salty soil, and lack of fertilizer. The rice borrows genes from barley, cress, and Agrobacterium to give it the triple-stacked properties (Anon, 2014).

11.7.8. Purple tomatoes

Scientists from the John Innes Centre in the U.K. have invented a tomato that boasts a royal purple color produced by the natural pigment anthocyanin—the same chemical that gives blueberries their color. A gene from a snapdragon was used to achieve its production (John Innes Centre, 2014).

Some have argued that anthocyanin has many potential health benefits, but the real improvement is that these tomatoes also last longer after being picked from the vine—a whopping 48 days—more than doubling the 21-day average of non-modified varieties. This means the tomatoes can stay on the vine longer. And longer vine times give flavors time to develop, bringing that fresh-picked flavor to your table.

11.7.9. Pink pineapples

Del Monte holds the patent for a new pineapple with pink flesh, which gets its color by producing an abundance of lycopene—the same chemical that gives tomatoes their red color. The Rosé pineapple was created by adding genes from a tangerine to a Del Monte Gold pineapple. Field trials were completed in Costa Rica and the pink pineapple is approved for import by the USDA (Anon, 2015e; Anon, 2012; Perkowski, 2013).

11.7.10. Fast-maturing farm salmon

Regular farm salmon take about three years to reach maturity. Despite abundant resources in a farm setting, salmon are programed to grow only in the summer time. The AquAdvantage Salmon grows year-round thanks to the help of genetic engineers. The fast-growing fish was engineered by inserting a gene from the Chinook salmon and another from the Pout. They have the same nutritional content, the same fatty acids, and taste the same—but take just 18 months to reach full maturity (Sentenac, 2014).

AquaBounty, the company who pioneered these fish, applied for permission to sell them in the United States back in 1995 (Anon, 2015f). Passed from administration to administration, these salmon have been in regulatory limbo for 19 years at the time of this writing.

11.7.11. Bringing back the mighty chestnut

When settlers first came to America, American chestnuts trees would have been a prominent part of the forest. There were over 4 billion trees which produced billions of pounds of chestnuts every year for both squirrels and people to collect (*Economist*, 2013). The nuts fell from groves in a seemingly limitless supply of shiny brown gems that could be scooped up and crushed to make bread or be roasted or eaten raw. People also desired the trees for their hardwood—they grow faster than oaks and produce a strong wood that is easy to split. Back in 1900, chestnut made up about 25% of all mature timber in the United States (Rosen, 2013).

Their tremendous size earned them the nickname "the redwood of the East." But the giant trees were virtually wiped out by an enemy thousands of times their junior. In 1904, people began to notice that the trees were dying off, and it was discovered that a fungus was to blame. The blight fungus hitched a ride to the new world on immune imported Asian chestnut saplings. Soon the fungus leapt from chestnut tree to chestnut tree and by 1950, the American chestnut was virtually wiped out.

William Powell at the State University of New York and Scott Merkle at the University of Georgia began searching for a genetic engineering solution in 1990. Genomes from both the American and Chinese chestnut tree were mapped, and the genes that seemed to give these trees immunity from the blight were flagged. But the big discovery came from another plant entirely. The team noticed that wheat generates an enzyme that easily detoxifies this particular blight. Powell and his collaborators have now created a number of American chestnut trees that include a few genes from Chinese chestnut trees, other chestnut blight immune trees, and, of course, the special wheat gene. A group of 800 of these precisely engineered trees were planted in 2013 to see how they fare against the blight.

Although anti-GMO activists relentlessly paint GMOs as dangerous, the truth is that patented technology is making food safer and more nutritious for people around the world.

11.7.12. Golden rice

Rice is a staple crop for billions. Although an excellent source of calories and carbohydrates, plain rice lacks the nutrition needed for a balanced diet. Even among populations that have access to enough calories from rice to sustain them, they may still be susceptible to malnutrition. One particularly pervasive form of malnutrition is vitamin A deficiency.

The World Health Organization estimates that between 250,000 and 500,000 children go blind each year for want of vitamin A. About half of those children will be dead within twelve months. Last year, vitamin A deficiency affected "1.7 million children under the age of five and 500,000 pregnant and nursing women," according to the International Rice Research Institute. Vitamin A deficiency is the world's leading cause of preventable blindness and increases the risk of death from disease and infection (WHO, 2015; IRRI, 2015).

That was exactly the problem that scientists Ingo Potrykus and Peter Beyer hoped to tackle with genetic engineering (Potrykus, 2015). These scientists began the search for a way to fortify rice to help with vitamin A deficiency in 1982. In 1999, Potrykus and Beyer had come up with a prototype called "Golden Rice (FAO, 2015; Anon, 2015f; IRRI, 2015a) (Fig. 11.6)." Golden Rice contains beta carotene, which is sometimes called "pro-vitamin A" because it can be converted into vitamin A by the body. Beta carotene is found in a number of vegetables, like spinach and carrots, but not rice.

Fig. 11.6. Golden rice rich in Vitamin A and iron (Right - yellow to orange in color) compared to ordinary rice (Left - white in color)

The two researchers discovered that rice had all the pathways necessary to produce beta carotene, but it was lacking genes to turn that capability on. The new rice was created first by moving two genes from a daffodil into the rice and later by incorporating a maize gene and another gene found in a common soil microorganism (IRRI, 2015b).

The idea is simple: offer the rice for free to any third-world farmers who wish to plant it. By growing Golden Rice in place of beta carotene-poor rice and propagating seed to neighbors, the rice could spread like a golden light preventing blindness.

Sadly, field trials of this rice are regularly destroyed by activist groups, such as Greenpeace. When a mob stormed a field trial in the Philippines in 2014, inventor Potrykus stated that he was "outraged" by the destruction and that it would set the project back months (Kupferschmidt, 2013). The Golden Rice that was growing in that field was going to be eaten by volunteers as part of a scientific test to see how much the rice could help in fighting vitamin A deficiency in the malnourished.

11.7.13. Golden bananas

Bananas are a major food and cash crop in Africa. In Uganda, it is estimated that "a typical adult will eat about three times his body weight [in bananas] in a year (BBC News, 2013)." And in that region of Africa, 30% to 60% of daily calories come from bananas (Tripathi *et al.*, 2009). But bananas, like rice, lack beta carotene.

Researchers at the National Agricultural Research Laboratories in Uganda are testing out a genetically modified banana that has the potential to combat malnutrition. Similar to Golden Rice, researchers have planted test crops of a "golden banana." As this banana produces its own beta carotene, researchers hope the banana will help Ugandans combat blindness and vitamin A deficiency (NARL, 2015).

11.7.14. Non-toxic cotton seeds

Chances are you have eaten a sunflower seed or a pine nut, but have you ever wondered why you have never tasted a cotton seed? Cotton seeds, like other nuts and seeds, are packed with a savory 22% protein, but are also packed with a deadly toxin called gossypol. Gossypol, when eaten, drops blood potassium to dangerously low levels and can severely damage the liver and heart. In other words, eating cotton seeds can kill you.

Researchers at Texas A&M have found a way to remove that toxin from the seeds without making the plant an easy target for insects. The cotton was produced with a technique called "RNA interference," where genetic engineers insert an extra copy of the DNA sequence coding for gossypol right next to the first, but in reverse. The two genes then "interfere" with each other thereby inhibiting gossypol's production entirely (*New Scientist,* 2006).

The cotton already grown worldwide has enough protein in its seeds to feed 500 million people (*New Scientist,* 2006). Cotton is grown mainly in developing countries and by small farmers— these people could benefit hugely from this new variety because they will be able to use the cotton fibers for textiles and also the cottonseed for food, the seed's inventor said (Anon, 2015g).

11.7.15. Cassava

The "cassava," tuber (root) is the main source of nutrition for 290 million people in sub-Saharan Africa (Danforth Center, 2015). It is known as a food-of-last-resort since it can be left growing underground, while other crops are plentiful, for up to three years before it must be dug up. But cassava is prone to diseases, has the lowest protein-to-energy ratio of any staple crop, and lacks adequate levels of beta carotene, Vitamin E, iron, and zinc.

Cassava also contains elevated levels of cyanide, which have no effect when cassava is properly processed and is eaten in small quantities, but during food shortages, when corners are cut and large quantities are typically ingested, it can be toxic.

Using genetic engineering, Ohio State scientists have reduced cyanide levels by 99% in the plant's tuber, and two teams of genetic engineers at Missouri's Danforth Plant Science Center have set out to fix other flaws in the crop. One team is working on a virus-resistant variety and the other on a version that contains adequate zinc, protein, and beta carotene (Wagner, 2003).

11.7.16. Daisy the hypoallergenic cow

In their first year of life, about 2 or 3 in every 100 infants are allergic to a whey protein naturally found in cow's milk called beta-lactoglobulin (BLG). Researchers in New Zealand were able to target and prevent the gene in dairy cows responsible for BLG from expressing itself. They first modified a cow skin cell nucleus then transferred the nucleus into a cow egg using the same methods that resulted in Dolly the sheep. After implanting the fertilized egg in a surrogate mother cow and waiting, "Daisy" was born in 2012—the first cow that produces milk without BLG (Sample, 2012; Jabeda *et al.*, 2012).

11.8. NEW AND FUTURE INITIATIVES

Both public and private sector are actively involved in R and D of GM crops. A working paper on Agricultural Biotechnology by Indira *et al.* (2005) tells about the major players in public and private R and D of GM crops in 2004 (Table 11.6). The public organizations are basically doing primary research, while the research conducted so far by private sector in India involves backcrossing the genetically engineered traits from the imported GM crops seeds into selected local varieties of the crop through standard breeding techniques. For instance, India cultivates 4 species of *Gossypium* (cotton) consisting of 30 varieties and 20 hybrids. The *Bt* cotton commercialized thus is principally a backcross with these local/hybrid varieties.

Table 11.6. R and D Scenario of GM crops in India (2004)

R and D Scenario	Public Sector	Private Sector
Number of crops being researched	19 - Rice, wheat, cotton, potato, banana, tomato, oilseed rape, mustard, coffee, tobacco, brinjal, cabbage, cauliflower, melon, citrus fruits, black gram, groundnut, chickpea and pigeon pea	11 - Cotton, rice, mustard, maize, tomato, pigeon pea, brinjal, cauliflower, cabbage, chilli and bell pepper.
Number of institutions / companies engaged	22 Indian institutions & 2 International centers National: 2 stations of IARI, 7 CPMB's, DRR (Hyderabad), IIHR (Bangalore), Univ. of Agri. Sci. (Bangalore), PAU (Ludhiana), MSSRF (Chennai), CPRI (Shimla), CTRI (Rajahmundry),	0 companies (5 national and 5 joint ventures) Mahyco-Monsanto, Syngenta (Switzerland)-4 subsidiaries, Bayer / Proagro -PGS (Germany), Tata Group/ Rallis (India) and Indo-1American Hybrid Seeds (USA), Rasi Seeds, Navbharat Seeds,

	Assam Agr. Univ (Jorhat), CICR (Nagpur), CCMB (Hyderabad), CRR (Cuttack), Narendra Dev Univ. of Agr. (Faizabad), TERI (Delhi) International: ICGEBT (New Delhi) & ICRISAT (Patancheru, Telangana)	Hybrid Rice International, Ankur and Swarna Bharat Biotechnics
Number of Institutes/ companies working on Cotton	3 Institutes - NBRI (Lucknow), CICR (Nagpur); CPMB; Univ of Delhi, South Campus	6 companies – Mahyco, Monsanto, Syngenta, Ankur seeds, Rasi seeds and Swarna Bharat Biotechnics
Funding	DBT	Self

Source: Indira *et al.* (2005)

Examples of crops and traits that are being focused under public R and D programs are presented in Table 11.7.

Table 11.7. Crops and traits being focused under public R and D programs

Crops	Traits
Rice	Tolerance/resistance to drought and salinity, tungro virus, gall midge, bacterial leaf blight, biofortification.
Wheat	Breeding for quality traits, heat tolerance, biofortification, resistance to leaf and stripe and rust, karnal bunt, powdery mildew.
Cotton	Fiber strength and oil content, gene stacking against insect resistance, herbicide tolerance in Bt cotton.
Maize	Quality protein, biofortification, insect resistance, herbicide tolerance.
Brinjal	Resistance against fruit and shoot borer.
Cabbage, Cauliflower	Insect resistance
Potato	Transgenic dwarf potato, disease resistance, reduction in cold induced sweetening, chip color improvement.
Mustard	Seed yield and oil content, low glucosinolate, aphid resistance.
Soybean	Resistant to yellow mosaic virus
Chickpea	Resistance against pod borers.
Sorghum	Shoot fly resistance
Groundnut	Resistance against TSV Virus

To date, commercial GM crops have delivered benefits in crop production, but there are also a number of products in the pipeline which will make more direct contributions to food quality, environmental benefits, pharmaceutical production, and non-food crops. Examples of these products include: 'Golden Rice' with higher levels of iron and beta-carotene (an important micronutrient which is converted to vitamin A in the body); long life banana that ripens faster on the tree and can therefore be harvested earlier; tomatoes with high levels of flavonols, which are powerful antioxidants; arsenic-tolerant plants; edible vaccines from fruit and vegetables; and low lignin trees for paper making.

11.9. PERSPECTIVES, LIMITATIONS AND ENVIRONMENTAL RISKS

For a GM seed to be released, it has to undergo various examinations by different committees before it can be commercialized. Firstly, there is Institutional Biosafety Committee (IBSC) which is linked to the lab undertaking transgenic work, then the Review Committee on Genetic Manipulation (RCGM) hosted by DBT dealing with small scale trials, Monitoring and Evaluation Committee and lastly the Genetic Engineering Assessment Committee (GEAC) which recommends the final release to the Government. The RCGM consists of expert scientists in areas of molecular biology and biochemistry. Once the RCGM has reviewed the field trial data, approval for large scale trials, seed bulking and commercialization has to be sought by the GEAC. In the view of DBT, the RCGM provides the necessary technical and scientific data and then advices GEAC, to take a final decision.

Time clearance to release a crop in India is minimum three years. This includes one year of multi-locational trials for generating bio-safety data monitored by RCGM, followed to two years of large-scale field trials under the aegis of GEAC. Along with this seeds are also provided to ICAR for assessing its agronomic performance for a period of two years. If the reports are satisfactory then GEAC recommends to the Government to allow the company to proceed with seed multiplication for commercial release.

During field trials, the criteria for assessment of GM crops are extensive. The regulation specifies a range of studies that must be undertaken at the field level like basic agronomic monitoring, pest incidence, pollen flow and in addition to this, toxicity and allergenicity data must be generated. In the case of *Bt* cotton, the preconditions for commercial release were: growing of a refugee area of five rows surrounding each *Bt* cotton plot, early removal of the cotton crop following harvest and continuous inspection throughout the crop period.

Ecological impact related with the introduction of GMOs is always an open debate. The application to agriculture of these new technologies certainly opens interesting perspectives, but also raises potential problems. The risk of crop transgene spreading has been demonstrated. A researcher of Clemson University in South Carolina reported "that in a population of wild strawberries growing within 50 meters of a strawberry field, more than 50% of the wild plants contained marker genes from the cultivated strawberries" (Kling, 1996). A Danish team has shown a possible rapid spread of genes from oilseed rape to the weedy relative *Brassica campestris*.

The introduction of *Bt* gene allows a drastic reduction in the use of toxic chemicals for crop protection. But a poorly controlled use of *Bt*-technology can destroy more effectively the predators than the pests. Or when "many crop plants are transformed with similar effective traits, in such situation, many polyphagous pest species, which by nature are more flexible evolutionarily than those that have a narrower diet, are likely to overcome *Bt* resistance very quickly" (Hokkanen, 1998).

Therefore, before releasing a transgenic plant, any risk has to be weighed against the benefit of the transgenic crops. We must not forget that annually in the world 500,000 acute pesticide poisonings, with 5000 deaths, are observed. Some of these estimated environmental and human health benefits potentially provided by *Bt* crops are presented in Table 11.8.

Table 11.8. Estimated global economic, environmental, and human health benefits of *Bt* transgenic plants (Hokkanen, 1998)

Global market penetration of transgenic plants (% use)	Reduced use of conventional insecticides (US$ millions)	Cost of transgenic plant (US$ millions)	Economic gain (US$ millions)	Estimated environmental gain (US$ millions)	Estimated human health gain (US$ millions)
1	90	45	45	91	16
10	900	47	853	917	157
25	2250	49	2201	2293	394
50	4500	51	4449	4585	787

11.10. FUTURE PROSPECTS AND CONCLUSION

11.10.1. Future prospects

As crop biotechnology faces the third decade of commercialization, new innovations are expected to transform the development of biotech crops and traits. This is manifested by the increasing use of stacked traits, the new generation of biotech crops that does not only address farmers' concerns but also consumers' preference and nutritional needs, and the amplified use of gene discovery in crop improvement and development of new varieties.

Together with conducive and harmonized regulations, crop biotechnology innovations can help double food production to address the needs of the growing global population, especially those in the developing countries.

11.10.2. Conclusion

Genetic engineering is responsible for a massive increase in human well-being, from a cornucopia of safer, more plentiful, and easier-to-grow foods.

Since the first genetically engineered organisms were created forty years ago, the field has experienced an explosion of new and potentially game-changing products. It seems that the sky is the limit when it comes to the technology of genetic engineering.

But unfortunately, activists and regulators are striving to put the gene genie back in the bottle.

The commercial adoption by farmers of transgenic crops has been one of the most rapid cases of technology diffusion in the history of agriculture. There are a number of fascinating developments that are approaching commercial applications in agriculture. Transgenic varieties and hybrids of cotton, maize, and potatoes, containing genes from *Bacillus thuringiensis* that effectively control a number of serious insect pests, are now being successfully introduced commercially. Considerable progress also has been made in the development of transgenic plants of cotton, maize, oilseed rape, soybeans, sugar beet, and wheat, with tolerance to a number of herbicides. The development of these plants could lead to a reduction in overall pesticide use, lower production costs, and environmental advantages.

Dr. Norman E. Borlaug (2000) (Nobel Prize Laureate for Peace in 1970) opines that "Genetic modification of crops is not some kind of witchcraft; rather, it is the progressive harnessing of the forces of nature to the benefit of feeding the human race. The genetic engineering of plants at the molecular level is just another step in humankind's deepening scientific journey into living genomes. Genetic engineering is not a replacement of conventional breeding but rather a complementary research tool to identify desirable genes from remotely related taxonomic groups and transfer these genes more quickly and precisely into high-yield, high-quality crop varieties. To date, there has been no credible scientific evidence to suggest that the ingestion of transgenic products is injurious to human health or the environment. So far, the most prestigious National Academies of Sciences and now even the Vatican, have come out in support of genetic engineering to improve the quantity, quality, and availability of food supplies."

Dr. Borlaug further stated that "The world has or will soon have the agricultural technology available to feed the 10 billion people anticipated by 2050. In order to achieve this target, we will have to nearly double current production. This increase cannot be accomplished unless farmers across the world have access to new biotechnological breakthroughs that can increase the yields, dependability, and nutritional quality of our basic food crops. The more pertinent question today is whether farmers will be permitted to use that technology. Extremists in the environmental movement, largely from rich nations and/or the privileged strata of society in poor nations, seem to be doing everything they can to stop scientific progress in its tracks. It is sad that some scientists, many of whom should or do know better, have also jumped on the extremist environmental bandwagon in search of research funds."

Dr. Borlaug concludes that "Most certainly, agricultural scientists and leaders have a moral obligation to warn the political, educational, and religious leaders about the magnitude and seriousness of the arable land, food, and population problems that lie ahead, even with breakthroughs in biotechnology. If we fail to do so, then we will be negligent in our duty and inadvertently may be contributing to the pending chaos of incalculable millions of deaths by starvation. But we must also speak unequivocally and convincingly to policy makers that global food insecurity will not disappear without new technology; to ignore this reality will make future solutions all the more difficult to achieve."

It is not the case that every new biotech product will necessarily be a winner, but regulatory burdens, anti-GMO activists, and consumer fear have ensured that good products are unfairly shunned, will never see the grocery store shelf, or, worse yet, will never even be created.

Every life-changing invention throughout history has been met with irrational opposition that must be fought with the bright light of science and reason. In our age, anti-biotechnology activists are successfully strangling what could and should have been the "third industrial revolution."

The question is: Are we going to let them?

Chapter - 12

White (Milk Production) Revolution/ Operation Flood

12.1. INTRODUCTION

In pre-Independence India, farmers reared indigenous or native breeds of cattle. Given the relatively low productivity of native breeds, milk production in the country was very low in relation to the huge cattle population present and dairying was confined to traditional pockets in the country. Various projects - technological as well as institutional - have been taken up since 1950 onwards to promote milk production in the country. In the post-Independence period, the Indian dairy sector has undergone a major shift, mainly due to the introduction of new technologies during the implementation of various dairy development programs. These initiatives covered the vital spheres of breeding, nutrition and health of milch animals as well as marketing of milk. After Independence, various programs of dairy development have been taken up, such as the Key Village Scheme (KVS), Intensive Cattle Development Project (ICDP), and Operation Flood (OF).

Despite of holding the number one position in global milk production, the milk productivity in India remains one of the lowest as compared to many leading countries of the world. At national level, milk yield of indigenous cow is about 3 to 3.5 liters, of buffalo 3.96 to 5.39 liters and of crossbred cow between 5.82 to 7.80 liters per day. As per FAO data, productivity of an average milch animal in India is even less than half of the world average. Productivity growth can be enhanced through two pathways – technological progress and technical efficiency improvement (Karanja *et al.*, 2012).

12.2. HISTORICAL

Traditionally the milk markets had been controlled by private middlemen, who imported milk from the surrounding countryside, and by urban producers who kept their cows and buffaloes in the residential areas. The Polsons - a private dairy at Anand established in 1929 - procured milk from milk producers through middlemen, processed it and then sent it mainly to the Bombay Milk Supply Scheme. Bombay was a good market for milk and Polsons profited immensely.

The first dairy cooperative in Gujarat was the result of a farmers' meeting in Samarkha village (Kaira district, Gujarat) on 4th January 1946, called by Morarji Desai on the advice of Sardar Vallabhbhai Patel, to fight rapacious milk contractors. Kaira District of Gujarat was one of India's main dairying areas. It was Sardar's vision to organize farmers, to have them gain control over production, procurement, and marketing by entrusting the task of managing these to qualified professionals, thereby eliminating the middle men. Sardar Vallabhbhai Patel assigned Tribhuvandas Patel the task of "making the Kaira farmers happy and organize them into a cooperative unit". The infant co-operative dairy, Kaira District Cooperative Milk Producers' Union (KDCMPU), - now famous as AMUL - was fighting a battle with the Polson Dairy which was privately owned.

KDCMPU (AMUL) was well organized and soon began regularly supplying milk to Bombay, 425 km away. It was particularly fortunate to have two forceful personalities, Tribhuvandas Patel and V. Kurien, who were the driving force behind the expansion of output and the introduction of modern methods of processing and marketing. The essence of the cooperative's success, however, has been the fact that it is owned by the fanners themselves. They feel that it is their organization and that they have a stake in its future. Although they employ professionals to run the system, producers can participate in decision-making in various ways.

A visit by the Prime Minister, Lai Bahadur Shastri to the scheme in 1964, convinced him that KDCMPU type of cooperative stimulate cooperative effort in the countryside. He sought to have the structure of the Kaira cooperative reproduced in other parts of India, so that as many farmers as possible could share a new prosperity. He expressed the desire for a national-level organization to replicate Anand Pattern dairy cooperatives throughout the country.

12.3. CONSTRAINTS IN DAIRY DEVELOPMENT

India has 2% of the geographical area of the world. It supports about 15% of world bovine population (Wisdom, June 2000) but contributes only 14% of world's milk output.

The main constraints are as follows:

- » Poor management.
- » Majority of stock of poor genetic quality.
- » Inadequate inputs.
- » Poor credit facilities.
- » Lack of veterinary extension services.
- » Insufficient nutrients and green fodders.
- » Poor quality semen.
- » Inadequate and improper breeding.
- » Lack of vaccination facilities.
- » Adverse environment.

12.3.1. Constrains for co-operative development

- » Inadequate support from government.
- » Regional imbalance in the co-operatives.
- » Poor-viability due to losses.
- » Low capacity of utilization of dairy plants.
- » Lack of working capital.
- » Inability to repay the loans.
- » Political interference in co-operatives.

> » New economic policies.
> » Improper management.
> » Competition from private sector.

12.4. WHITE REVOLUTION/OPERATION FLOOD

Operation Flood is the program that led to "White Revolution." It created a national milk grid linking producers throughout India to consumers in over 700 towns and cities and reducing seasonal and regional price variations while ensuring that producers get a major share of the profit by eliminating the middlemen. At the bedrock of Operation Flood stands the village milk producers' co-operatives, which procure milk and provide inputs and services, making modern management and technology available to all the members.

The revolution associated with a sharp increase in milk production in the country is called the White Revolution in India also known as Operation Flood. White revolution period intended to make India a self-dependent nation in milk production. Today, India is the world's largest producer of milk (Fig. 12.1) and Dr. Verghese Kurien is known as the father of the White Revolution in India (Fig. 12.2).

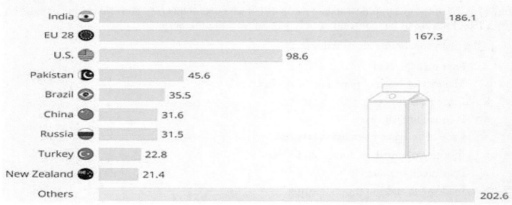

Fig. 12.1. India is the biggest producer of milk in the world.

Fig. 12.2. Dr. Verghese Kurien is known as the father of the White Revolution in India.

12.4.1. Objectives

The main objectives of OF were as follows:

» To monitor over the various development programs by Indian Dairy Corporation set up in Jan. 13, 1970.

» To promote the Co-operate dairy development programs.

» To make the success of operation flood program.

» To increase research and extension work in the field of dairy.

» The main objectives of the cooperative society are the procurement, transportation, storage of milk at the chilling plants.

» To provide cattle feed.

» The production of wide varieties of milk products and their marketing management.

» The societies also provide superior breeds of cattle (cows and buffaloes), health services, veterinary, and artificial insemination facilities.

» To provide extension service.

» The technology of the White Revolution is based on an extensive system of cooperative societies.

» Milk after being collected at a village collection centre, is promptly transported to the dairy plant at the milk chilling centre.

» Timing of collection is maintained by the village society, truck operators, and the quick transport to the dairy plants.

» The chilling centres are managed by producer's cooperative unions to facilitate the collection of milk from producers who live at some distance from the chilling centres and thus, the middlemen are eliminated.

12.4.2. Interventions

In 1965 the National Dairy Development Board (NDDB) was founded with the specific task of encouraging the formation of producers' cooperatives. The mandate of NDDB is to replicate Anand Pattern Dairy Cooperatives in other parts of the country through its now famous Operation Flood (OF) program. Along with those of the Indian Dairy Corporation (IDC) which followed in 1970, its efforts have been dubbed 'Operation Flood', an attempt to boost India's domestic milk production and thereby improve the supply situation in the large urban markets. By 1970, the demand exceeded supply, a gap which was partially met by adulteration with water by traders, whose cleanliness was also very much in doubt. OF planned to circumvent these traders and to establish a new marketing system, with milk supplied by rural producers, organized in Kaira-style cooperatives, to urban processing plants built to hygienic modern standards. At a stroke, it was hoped to eliminate the exploitation of rural producers and to improve both the quantity and quality of supply to city consumers: a bold plan indeed.

In order to overcome this difficulty, exotic cattle such as Jerseys and Friesians have been imported and cross-bred with local races. Fifteen per cent of milch animals had been upgraded by 1984.

Verghese Kurien after returning to India from Michigan State University, where he earned a Master of Science of Mechanical Engineering (with distinction), was posted as a Dairy Engineer at the Government Creamery, Anand, in May 1949. He volunteered to help Shri Tribhuvandas Patel, the Chairman of KDCMPU, to set up a processing plant. At that time, only two village milk cooperatives were involved, representing only a handful of farmers.

The Kaira Union began with a clear goal, to ensure that its producer members received the highest possible share of the consumers' rupee. This goal itself defined their direction. The focus was on production by the masses, not mass production. By the early 1960s, the modest experiment in Kaira had not only become a success, people began to recognize it as such.

During the 1970s, dairy commodity surpluses being built in Europe posed a threat and provided an opportunity to Indian dairy industry. The opportunity was the potential of the European surpluses as an investment in the modernization of India's dairy industry. Assistance of the World Food Program, food aid in the form of milk powder and butter oil was obtained from the countries of the European Economic Community (EEC) to finance the program. It was the first time in the history of economic development that food aid was used as a buffer stock to stabilize market fluctuations as well as to prime the pump of markets that will later be supplied by domestic production.

12.4.3. Three tier system in Anand pattern of dairy co-operatives

The three tiers in the Anand model of dairy co-operatives (Table 12.1) (Fig. 12.3) are:

» Primary Milk Producers co-operative Society at village level
» District Co-operative Milk Producers Union
» State Milk Producers Co-operative Federation

Table 12.1. Three tier structures of Anand pattern co-operatives

Primary Milk Production Co-operative Society	» Providing credit facilities. » Supply of feeds on subsidy rates. » Provision of free veterinary aid and artificial insemination. » Welfare of farmers.
District Milk Producers' Co-operative Union	» Collection » Transport » Processing » Product manufacturing
State Milk Co-operative Federation	» Marketing of products. » Control of dairies. » Control over the marketing of milk and milk products. » Training and research.

Milk producer   Village dairy Co-operative

District Co-operative milk Union

State Co-operative milk   The consumer

Fig. 12.3. Three tier structures of Anand pattern co-operatives

12.4.4. The three phases of Operation Flood (OF)

The OF program was implemented in three phases between 1970 and 1996. Working through the State governments, the OF program created the 3-tier structure of Anand Pattern dairy cooperatives all over India.

12.4.4.1. Phase-I of OF implemented during 1970-79 was financed by the sale within India of skimmed milk powder and butter oil gifted by the EC countries via the World Food Program. As founder-Chairman of the NDDB, Dr. Kurien finalized the plans and negotiated the details of EEC assistance. During its first phase, the main thrust was to set up dairy cooperatives in India's 18 best milksheds, linking them with the four main cities of Bombay, Calcutta, Delhi and Madras, in which a commanding share of the milk market was to be captured. It involved organizing dairy cooperatives at the village level; creating the physical and institutional infrastructure for milk procurement, processing, marketing, and production enhancement services at the union level; and establishing dairies in India's major metropolitan centers. Thus the first phase of OF program not only laid the foundation for India's modern dairy industry but also established the possibility of successfully replicating a robust design-concept.

i) Achievements

» Establishment of 13,000 Dairy Co-operative Societies (DCS) in 39 milk sheds with membership of 18 lakh farmers.

> » Achievement of a peak milk procurement of 34 lakh liters per day and marketing of 28 lakh liters per day.

12.4.4.2. Phase-II, implemented during 1979-85, covered some 136 milksheds linked to over 290 urban markets. The seed capital raised from the sale of WFP/EEC gift products and World Bank loan had created (by end of 1985), a self-sustaining system of 43,000 village cooperatives covering 4.25 million milk producers. Milk powder production went up from 22,000 tons in the pre-project year to 1,40,000 tons in 1989, mainly due to dairies set up under OF program. The EEC gifts thus helped to promote self-reliance.

i) Achievements

> » The numbers of village level Co-operatives reached to 34,500 covering 36 lakh farmers in 136 rural milk sheds.
> » The peak milk procurement increased to a level of 79 lakh liters per day and marketing to 50 lakh liters per day.

12.4.4.3. Phase-III, implemented during 1985-96, ensured that the cooperative institutions become self-sustaining. With an investment of US$ 360 million from the World Bank, commodity and cash assistance from the EEC and NDDB's own internal resources, the program achieved substantial expansion of the dairy processing and marketing facilities; an extended milk procurement infrastructure; increased outreach of production enhancement activities; and professionalization of management in the dairy institutions. Facilities were created by the cooperatives to provide better veterinary first-aid and health care services, breed improvement technologies like artificial insemination, cattle feed, hygienic milk production techniques and modern animal husbandry management techniques to their producer members.

i) Achievements

> » The number of primary Dairy Co-operative Societies increased to 70,000 covering 170 milk sheds with 93.14 lakh farmer memberships.
> » Sum of 26,000 tons of balanced feed, 24,000 tons of bypass protein feed was sold away through co-operatives.
> » Average milk production per day reached to 115 lakh liters per day and total markets of milk is 100 lakh liters per day.

12.4.5. Post-Operation Flood program

The post phase of operation flood was started with the following objectives:

> » To create infrastructure and strengthening of democratic values.
> » Strengthening the co-operatives by providing funds on 50: 50 basis from Central and State Government.
> » Increase the extension work in the fields of co-operatives education, personal training, marketing support, product development, and improving standards.

12.5. IMPACT OF OPERATION FLOOD PROGRAM

The growth of dairy cooperatives during the three OF phases along with the post OF growth is presented in Table 12.2. It is heartening to note that the momentum of growth is maintained during the post OF period. This indicates the solid foundation laid by the OF program.

Table 12.2. Salient features of operation flood and achievements, 1970-96

Features	OF Phases			Post of Phase
	OF - I	**OF - II**	**OF - III**	
Date when started	July 1970	October 1979	April 1985	April 1996
Date when concluded	March 1981	March 1985	March 1996	March 2002
Investments (Rs. crore)	116.5	277.2	1303.1	---
No. of Federations/Apex Milk Unions set up	10	18	22	22
No. of Milksheds covered	39	136	170	170
No. of DCSs set up ('000)	13.3	34.5	72.5	74.3
No. of Members (lakh)	17.5	36.3	92.63	110.6
Average milk procurement (mkgpd)	2.56	5.78	10.99	17.60
Liquid milk marketing (llpd)	27.9	50.1	100.2	126.72
Processing capacity: Rural dairies (llpd)	35.9	87.8	180.9	264.7*
Metro Dairies (llpd)	29.0	35.0	38.8	NA
Milk drying capacity (MTPD)	261.0	507.5	842.0	990.0*
Technical Inputs: No. of AI centres ('000)	4.9	7.5	16.8	22.0
No. of AIs done (lakh/year)	8.2	13.3	39.4	~60.0
Cattle feed capacity ('000 MTPD)	1.7	3.3	4.9	5.2*

Note: mkgpd: million kg per day; llpd: lakh liter per day; MTPD: metric tons per day; *Figures pertain to 1997-98.

Sources: (1) Dairy India 1997; (2) Quarterly & Monthly Progress Reports on Operation Flood, NDDB, Anand, 1996 as quoted in Singh (1999b:205); NDDB Annual Report 2001-02.

The OF program was funded by a World Bank loan, EEC food aid and internal resources of NDDB. The total investments at the end of Phase III of OF program were estimated at Rs. 15.87 billion. By 1996, the higher growth rate attributed to OF program was resulting in an extra 43 million metric tons of milk per annum. Since the start of the faster growth trend, and using a 70 percent rate to compound its value to 1996, the total increment was 1,086 million metric tons. Each ton would require about $310 of imported ingredients if it were to be replaced with recombined milk. If even 2 percent of the observed increase in milk production were due to all investments from World Bank, EEC, and NDDB's own resources, it would return an economic rate of return (ERR) of 10 percent. The returns are phenomenal if most of the increased growth is attributed to OF program. Partly this is due to congenial environment created by OF for members to invest in biological assets - milch animals - that periodically reproduced themselves without major reinvestment and continue to yield regular benefits utilizing crop residues that otherwise do not have much economic value.

The performance of the Indian dairy sector has been quite impressive. Milk production in India increased from 17 million tons in 1950-51 to over 146.3 million tons by 2014-15. During the past one and half decades (2000-2015), milk production has grown at a rate of nearly five per cent per annum *vis a vis* world growth rate of 1.5 per cent. The exports of dairy products increased from Rs. 13.98 million tons in 1990-91 to Rs. 66,424.34 million tons in 2014-15. The per capita availability of milk per day has been increased from 178 grams/day in 1991-92 to 322 grams/day in 2014-15 (Table 12.3). Hence, as the net trade balance of dairy products has changed from negative to positive and the country is now a net exporter of dairy products.

Table 12.3. Milk production and per-capita availability in India

Years	Production (Million Tons)	Per capita availability (Grams/day)
1991-92	55.6	178
2000-01	80.6	220
2010-11	121.8	281
2014-15	146.3	322

12.5.1. Significance of Operation Flood

» The White Revolution in India helped in reducing malpractice by traders and merchants. It also helped in eradicating poverty and made India the largest producer of milk and milk products.

» Operation Flood empowered the dairy farmers with control of the resource created by them. It helped them in directing their own development.

» Connected milk producers with the consumers of more than 700 cities and towns and throughout the country, a 'National Milk Grid' was formed.

» The revolution also reduced regional and seasonal price variations ensuring customer satisfaction at the same time. Also, it ensured that the producers get a major share of the price that customers pay.

» Improved the living standards of the rural people and led to the progress of the rural economy.

12.6. DAIRY FARMING INNOVATIONS FOR PRODUCTIVITY ENHANCEMENT

Farm innovations are the novel practices/products/techniques suitable for particular area, physiological stage of animals and economically viable option to enhance the animals' per diem yield. Low cost and user-friendly dairy farming innovations (technologies) suitable for all kinds of farms, maintained under rural conditions existing in different tropical countries are proved to be useful in enhancing animal productivity and henceforth farmers' socio-economic welfare. As dairying has become a commercial enterprise and needs technology adoption for higher milk yield and lower per unit costs (Hisham El-Osta and Mitchell, 2000); there is a need for innovations applicable for increasing net returns, reducing costs and optimizing production; so that a common dairy farmer as well as consumer can contribute to a more resilient and more sustainable future for all of us.

Though developing countries like India contribute above half in world milk pail, productivity per animal is poor compared to other countries. Low animal productivity might be a result of ineffective breeding, improper feed and fodder management, deficient veterinary care, poor farm management *etc*. Dairy farmer has to improve the amount of milk each animal produces, thereby reducing the amount of feed, water and space needed per liter of milk resulting in less manure production. This provides a big window for different innovation application to enhance productivity in such developing nations where majority dairy farms are small scale and managed on traditional practices. The sustainability studies concluded that market-oriented farms with a high degree of technology adoption was the most economically, socially and ecologically sustainable farms. Various dairy farming issues, animal ailments along with dairy farming innovations and various lab to land approaches are presented jointly in Fig. 12.4.

Strategies to achieve the higher productivity include:

» Selection of superior bulls and their wide spread use.
» Rapid extension in coverage under breeding programs.
» Focus on higher percentage of animals in milk.
» Area wise priority approach.
» Participation of all agencies.
» Quality feeding.
» Self-sustainable inputs delivery.

Major dairy issues (Field)

- Germplasm, housing.
- Ideal body weight.
- Estrus & cyclicity.
- Heat detection, timely AI/mating & PD.
- Record keeping.
- Post-partum care.
- ICI, parturition hygiene.
- Nutrition, balanced-feeding.
- Mastitis, lameness.
- FMD, prophylaxis, parasitism.
- Disease diagnosis, economics, marketing

Common ailments

Repeat breeding, anoestrus, prolapse, torsion, dystocia, infectious diseases, abortion, mastitis, digestive disorders, wounds, etc.

I. Dairy innovations for adoption:

- AI, ETT, sex semon, group calving, farm mechanization, robotic approaches, hoof trimming, oral magnet, herd health softwares, vaccination, deworming, mobile apps, management tools

II. Lab to land approaches

- Trainings

 – Farm literature (magazines, books, pamphlets/ folders/ bulletins)

– Animal health calendars

–Advisory messages (WhatsApp, SM)

– Videos, CDs

– Mass media (radio, TV, news paper)
– ICT tools
- Mobile apps
- Field camps/ visits

Fig. 12.4. Dairy farming issues, innovations and lab to land approaches.

12.6.1. Breeding innovations

Breeding innovations generally known as cross breeding have resulted in profitable dairy farming with serious health and fertility concerns. Selection of good, diseases resistant and climate resilient breed coupled with adoption of scientific breeding innovations laid the strong foundation to the dairy farm to grow in future. Topography, soil type, feed and fodder availability must also be given due consideration while selecting the animals. Highly productive animal requires special care in terms of management, disease control and feeding strategies. Native breeds with quality germplasm would be more appropriate for local climatic conditions. Genetic up-gradation of non-descript animals by using local superior germplasm proves more beneficial in terms of sustainable production. However, introducing exotic germplasm to a certain limit generally known as cross breeding have resulted in profitable dairy farming with serious concerns. Breeding innovations commonly introduced at field level are highlighted below.

12.6.1.1. Artificial insemination (AI) technique: Artificial Insemination (AI) is an Assisted Reproductive Technology (ART) used worldwide to deposit proven sire's stored semen directly into a cow's uterus. The technique is used as a rapid way to improve desired characteristics through intensive genetic selection. Advantages, such as facilitating the use of superior quality semen without the expense and risk of sire's ownership; reduction in the risk of introducing venereal diseases into the herd have achieved with this innovative technique. Being the quickest and most effective means of breeding through AI, developing countries like India could witness position as the top most milk producing country of the world. Not only it exclude the need of keeping a bull for natural service but also helps in exploiting the excellent germplasm up to the fuller extent.

12.6.1.2. Progeny testing: Progeny testing is the practical and best technique, in which bulls are evaluated on the basis of their daughters' performance. When large numbers of animals are spread in many villages for a particular breed in its native tract, these villages can get AI services and progeny produced in this way is evaluated for their performance. Progeny testing is a practical and the best option for achieving genetic improvement in that breed.

12.6.1.3. Embryo transfer technology (ETT): Embryo transfer technology (ETT) is one of the latest tools available for the faster improvement of livestock worldwide particularly for exploiting the genetic potential of high-quality females and the males simultaneously. Prior to the development of this technology, a limited number of off springs were achieved from a superior/high milk producing cow in her life time. Higher cost of technology with low conception rate might be the factors limiting its implementation.

12.6.1.4. Sexed semen: Sexed semen is processed semen of proven bull from where 'Y' chromosomes bearing sperm cells are removed through sorting process. Sexed semen predominant with 'X' chromosomes can ensure birth of female calf. Reduction in economic burden and production of more number of female calves as a future productive cattle are the main advantages popularizing this technology among dairy farmers. However, the higher cost of semen coupled with low conception rate are important factors to be considered before its use and that too in heifers or primiparous animals for better results.

12.6.1.5. Hormonal synchronization/protocols: Different hormone protocols are being adopted for getting group calving or desired calving in a year for efficient and controlled management. Such desired calving matches with market demand and season. It is planned administration of hormones with fixed time AI for specified calving. In addition to this, the advanced reproductive techniques such as Multiple Ovulation and Embryo Transfer (MOET), ovum pick up technique and embryo manipulation (splitting, sexing and cloning *etc.*) offer possibilities for faster multiplication of superior germplasm from highly selected elite donors to achieve the target producing large number of superior bull calves/bulls and their adequate number of quality semen doses.

12.6.2. Feeding innovations (cost-effective feeding strategies)

Steady supply of quality feed and fodder assures productivity enhancement. Feeding constitutes about 60–70% of total cost of milk production in dairying. Feeding management plays a crucial role in exploiting real potential of dairy animals. Balanced feed (green and dry fodder along with concentrate ration) proves beneficial for sustainability as well as profitability of the farm. Fodder both green and dry needs to be grown inside the farm. High yielding fodder varieties like bajra, Napier hybrids, maize, sorghum can be grown in fertile and well irrigated land, while guinea/rye grass can be grown in barren rain-fed land. In draught prone areas, planting of local fodder trees will sustain the animal production during scarcities. Some trees like *Prosopis cineraria*, *Leucaenale ucocephala* and *Moringa oleifera* are gaining popularity among fodder due to their high nutritional value. Further, slight improvement in animal nutritional status with additional supplementation can improve animal productivity with mere addition of cost. Different types of animal feed innovations, easily applicable at every farm are discussed below.

12.6.2.1. Baled silage: Silage is a method of preserving surplus green fodder, predominantly adopted on large dairy farms as far as tropical countries are concerned. It is the product of controlled fermentation of green fodder retaining high moisture content. Many countries are propagating tube silage or bag silage, as one of the innovative technique of silage making, introduced for a marginal dairy farmer possessing one-two dairy animals and limited fodder acreage. Standard plastic tube/polythene bags of recyclable material are available in markets in India with a capacity of producing 500–1000 kg of silage. Baled silage is the latest upgraded innovation of fodder conservation. In this, forage is baled at higher moisture than forage to be stored as dry hay. The sealed airtight plastic bales remain sealed until they are required. The high moisture and lack of air promote fermentation within the sealed bale that preserves forage quality. Such baby corn silage bales of 50 kg are available for sale at a reasonable price on online portals like Indiamart.com.

12.6.2.2. Rumen inert protein (bypass protein): Protein meals are subjected to suitable physical/chemical treatment, energy and nitrogen balance gets improved with only marginal increase in treatment cost. Chemical or heat treatments are the main methods used for protecting proteins. In this technique, part of the protein is not degraded in the rumen and it can be utilized more efficiently in the small intestine. This rumen inert protein commonly known as Bypass protein, that is a misnomer. This protein supplies more essential amino acids at the intestinal level, which can lead to increase in milk yield by 10–15% and growth rate by 20–25%.

12.6.2.3. Bypass fat: Dietary fat, that resists lipolysis and bio-hydrogenation in rumen by rumen microorganisms, but gets digested in lower digestive tract, is known as bypass fat or rumen protected fat or inert fat. Among all forms of bypass fat, calcium salts of long chain fatty acids (Ca-LCFA) has highest intestinal digestibility and act as an additional source of calcium. A simple cost-effective indigenous technology has been developed for the preparation of bypass fat (Ca-LCFA) using vegetable fatty acids. Ration of the high producing animals should contain 4–6% fat, which should include fat from natural feed, oil seed and bypass fat in equal proportions. Bypass fat supplementation has proved beneficial without any adverse effect on the rumen fermentation, feed intake, digestibility of nutrients, and different blood parameters of the dairy animals. Rise in milk is recorded by 5.5–24.0%. Improvement in post-partum recovery and reproductive performance of dairy animals are the added advantages of this innovation.

12.6.2.4. Total mixed ration: The term Total Mixed Ration (TDR) may be defined as, "The practice of weighing and blending all feedstuffs into a complete ration which provides adequate nourishment to meet the needs of dairy cows." Each bite consumed contains the required level of nutrients (energy, protein, minerals and vitamins) needed by the cow. A 4% increase in feed utilization, greater accuracy in formulation and feeding, masking of the flavor of less palatable feeds (urea, lime stone, fats, and some by-pass protein sources) and use of commodity ingredients can be expected while using TMR. While blending all the feeds together in a TMR, over mixing and under mixing of ingredients need to be avoided.

12.6.2.5. Buffers: Dietary changes like shift from hay to silage, feeding high level of grains/concentrate mixture cause increased acidity in rumen which may become detrimental for rumen microorganisms thereby affecting not only digestion but production and reproduction too. Buffers like Sodium bicarbonate, Magnesium oxide neutralize the acids produced by metabolism or fermentation. They are particularly required during hot weather when forage intake is lower and due to less chewing action natural buffer produced i.e. saliva is produced less.

12.6.2.6. Probiotics (prebiotics/synbiotics): Probiotics are the live microorganisms that may beneficially affect the host upon ingestion by improving the balance of the intestinal microflora. *Lactobacillus* sp. is the most prevalent probiotic bacteria, known as lactic acid producing bacteria (LAB). Control of diarrhea in calves, increased milk production and better composition, control of ruminal acidosis, control of growth of pathogens in rumen, reduced pathogen load are the advantages of the technology. The appropriate level of 20 g probiotic per day per animal is found effective. Prebiotic are the ingredients [like Fructo-Oligosaccharides (FOS), Mannan Oligosaccharises (MOS) *etc.*] used to enhance the population of already present good bacteria and synbiotic pertains to combination of pre and probiotic.

12.6.3. Management innovations

Building a hygienic cow shed is another important aspect to be considered among the many factors that lead to the success or failure of dairy farms. Housing systems that require less labor, which provide a comfortable and healthy environment to animals, manage space including storage efficiently and take care of bio-security measures with easy modification and expansions are more profitable than heavy structures with huge capital investment. Sufficient sunlight, proper ventilation, clean, and dry flooring along with sufficient space for lying down and protection from adverse weather conditions are the basic necessities of animal housing. Further, an effective management program has to be developed, so that animals are prevented from falling ill and there is no need for antibiotics/medicines. The direction and orientation of shed plays an important role in keeping the animals healthy as well as reducing laborious work. Considering these factors, loose housing barn with open cattle shed are recommended here, as that can be easily adapted at small as well as marginal dairy farms.

12.6.3.1. Health tracking devices: Digital animal health tracking devices are getting attention now a days as they help farmers in tracking, monitoring and managing animal's health, nutrition, behavior, pregnancy, milking frequency, milk production anomaly and activity level in real-time. These smart animal wearing gadgets can be implanted in the cattle's ears, tail, legs, neck or any part of the body. For tracking the health and early diagnosis of medical condition in dairy animals, GPS-enabled digital chips have been implanted widely in India. A huge database will be generated if these devices are used efficiently. Accuracy in such data will guide in formulating strong and concrete policies for welfare of both human and animals.

12.6.3.2. Heat detection systems: Detection of heat is very important aspect of management for performing timely AI with successful animal conception. Heat Detection System is heat management software which monitors the cow's activity for the whole day, predicting heat on the basis of unrest and hyperactivity along with other features to check milk flow, conductivity for suspected mastitis. It has obvious advantage over visual heat detection which is based on observation, behavior and miss heats or false negatives leading to huge economic losses. Further, these gadgets help the farmers for fertility management to get the target of a 'calf at foot every year'.

12.6.3.3. Robotic milking machines: Innovation of robotic milking machines is useful in eliminating the pressure on physical labor and maintaining a hygienic milking process with remarkable improvement in milk production. These machines have cups with sensors that can be attached individually to cows' teats. The sensors play important role in detecting readiness of teats for milking and also identify impurities, color and quality of milk. Milk not fit for human consumption, is diverted to a separate container. The machines automatically clean and sanitize the teats once the task is over. Few models of low-cost, non-electric milking machines are also developed considering locality and need of dairy farms. Innovation of mobile milk collection unit installed with Robotic milking machines and bulk coolers will introduce a way to produce clean and quality milk from small and marginal farms.

12.6.3.4. Waste disposal and management: Scientific disposal of excreta (dung, urine), other organic waste (aborted fetuses, dead calf/animals, placenta) demands utmost attention. Presently, there is not a clear-cut policy for dung and carcass disposal. In majority of Asian countries, both these are disposed in open, which is a serious concern from zoonotic and infectious diseases point of view. Electric incinerator and community biogas plants can provide the tangible solution. Technology of dung cleaning robot or manure robot is available for barn cleaning and scrapping the dung in slatted floors beneath the barn. Recently manure eating robot has been launched for cow garden cleaning that cleans the barn/cow gardens.

12.6.3.5. Digital farm management: Completion of farm management includes accounting, finance, labor management, and supply chain management. Dairy farm management soft-wares are the innovative tools available in markets for automizing and digitalizing end-to-end production and operations activities. It provides a holistic view for entire farm activities, manage records, generate reports and detect inefficiencies; assuring profitable dairy farming.

12.6.4. Health care innovations

Reduction in milk production is the first sign of animal discomfort and illness; whereas getting back to the production is one of the major challenge and costly affair for small as well as marginal farmer. Also there is reduction in per lactation as well as life time production of that animal. Any kind of disease treatment compels to use antibiotics. This part is of a global conversation about antibiotic resistance, which is a serious public concern shared by animal and human health experts. So, it is always better to prevent the occurrence of diseases rather to treat. This could be possible only through application of health care management innovations.

12.6.4.1. Vaccination: Livestock vaccination is considered an emerging innovation of socio-economic importance in the Indian dairy industry (Rathod and Chander, 2016) and reported more profitable and sustainable than artificial insemination (Lal, 2000). Majority of tropical countries like India are endemic to many diseases that cause severe economic losses due to drastic reduction in the production capacity. Some of the diseases are even highly fatal. Fortunately, vaccines are available for most of these diseases and can be easily controlled if timely vaccination is carried out in a mass scale, covering a large proportion of the susceptible population (at least 80%) (NDDB, 2015). Farmers must stick to the standard vaccination protocol recommended by the Government following all precautions and regularity in inoculations.

12.6.4.2. Teat dip: The teats of all the lactating dairy animals and dry cows (during first 10–14 days of dry period) are dipped regularly after every milking in a germicidal solution. The recommended teat dips are:

» Iodine (0.5%) solution 5 parts + Glycerine 1 part.
» Chlorhexidine (0.5%) solution 1 L + Glycerine 60 ml.

The Iodine teat dip is the best as it treats various types of teat lesions and injuries also. Post-milking teat dipping with 'Iodine-glycerine teat dip' for prevention of new mammary infections is also recommended by many research institutes. Studies have reported that the treatment applying the post-milking teat dip automatically *via* milking machines had the lowest number of new intra-mammary infections (IMI).

12.6.4.3. Mastitis diagnosis kit: Mastitis, one of the expensive diseases, affects economic returns of dairy farms heavily. Farmer has to suffer with huge financial burden due to sub-clinical mastitis (SCM) as it incur heavy losses related to culling, decreased production, decreased fecundity, and treatment costs. Diagnosis of mastitis at sub clinical stage and its management results in milk production rise with quality milk and safety to consumer health (Wattiaux *et al.*, 2011). Innovation of mastitis diagnosis kit includes Sodium Lauryl Sulphate (SLS) Paddle with reagent and Bromothymol Blue (BTB) card. Such innovations can be used by the farmers on their own for early diagnosis and reducing the further incidence of diseases for improving productivity (Nimbalkar *et al.*, 2020).

12.6.4.4. Lameness management: Lameness is reported as the third most economically important disease in world after infertility and mastitis (Enting *et al.*, 1997). It is a major cause of involuntary culling after mastitis. About 90% of lameness in dairy cattle and buffaloes occurs due to foot lesions. Recommended guidelines for prevention of lameness include hoof trimming of all the animals at every 6 months and footbath of size 3 m long, 1 m wide and 15 cm high. Formalin (39–40%) should be preferred for foot bathing as a 4% solution (120 L water +5 L of formalin) in the footbath. Concrete footbaths are best and cheaper. In case there are few animals (unorganized farms), formalin spray (40 ml per liter of water) can be used on 1st, 2nd, and 3rd day of every fortnight along with close monitoring of animal gait at the time of walking.

12.6.4.5. Oral magnet feeding: Hardware disease is a common term for bovine traumatic reticulo-peritonitis, which is usually caused by the ingestion of a sharp, metallic object. Due to industrialization and urbanization, it is commonly found in dairy cattle than any other ruminants. It can be difficult to conclusively diagnose, but can be prevented by the oral administration of a magnet around the time that the animal reaches the age of 1 year. This innovative technology is beneficial to control Traumatic Reticulo-Peritonitis (TRP) that occurs due to the intake of any sharp foreign object such as nails, blades *etc.* along with feed by the animal.

12.6.5. Communication innovations

In this competitive world, farmers are not only looking for various information sources for carrying out their production and marketing tasks efficiently but also for ensuring delivery of safe and quality products to the consumers. Food safety for consumers is at greater risk because of the increasing globalization of food systems. Information and Communication Technology (ICT) has potential to mitigate the needs of both ends by introducing virtual platform for dairy product production and marketing. ICT based information delivery to dairy sector can significantly improve the quality of decision-making in dairy farming system. Mobile phones with internet facility have been one of those successful innovations which benefit a large number of people in the developing world. As worldwide acceptance for mobile phones has improved among all users, it can be used as a major tool for communication and dissemination of information for quality decision making. Different mobile apps, web portals such as epashupalan.com and expert systems are being used by dairy farmers. The mobile application for dairy farmers, named 'Pashu Poshan', is available on both web and android platform, can be accessed by registering on the INAPH portal (http://inaph.nddb.coop). Guru Angad Dev Veterinary and Animal Sciences University, Punjab, India, has launched 'Precision Dairy Farming' mobile application dealing with important aspects of dairy farming including important milch breeds, breeding, feeding and housing management, record keeping, health management, and economics.

12.6.5.1. Product traceability in dairy sector: Traceability is commonly defined as the ability to trace products back and forth throughout the supply chain, from farm or point of production to the end user. The growing complexity of food supply chains, the heterogeneity in food safety regulations across countries, and lack of uniform requirements from one commodity to another are some factors that explain why greater efficiency in food traceability systems has increased in recent years. Block chain technology to give real-time data about the products to customers has been introduced among dairy manufacturers, suppliers and other stakeholders. QR code provided on the packaging of the product can be scanned on personal mobile devices to get information on the origin of the milk. Information about, how and where from the product has collected and packed, how old it is, what kind of transportation and cold milk chain facilities are used, is being provided on internet. However, scattered, diversified and unorganized dairy farming is the major barrier for deep penetration of this innovation at grass roots of the sector. The application can be highly useful in organic milk production as demand for organic milk is increasing in the market. Organic milk is considered as the ultimate milk with almost nil risk of chemicals, drugs and also free from stress factors. However its production is quite cumbersome as it needs a lot of efforts, monitoring and adhering to the organic standards for a branded product. It fetches good value in the market and the product traceability is quite easy as its each and every production component is documented.

12.6.6. Marketing innovations

E-commerce market places have played revolutionary role in input availability and product sales in dairy sector. Modern equipment and advisory services have been made available at the doorstep to farmers and dairy manufacturers on their smart phones through online Business-to-Business (B2B) market places. Many Business-to-Customers (B2C) platforms have also emerged at a rapid pace. They have major role in picking fresh produce from farms and delivering them to the doorsteps of end users. These marketing innovations have reduced spacial barriers for both producers as well as consumers. Online portals like Indiamart.com, amazon.in, reliance fresh at relianceretail.com are the successful examples of innovative online marketing of various dairy products.

12.7. MAJOR THRUST AREAS

The major thrust is on the following critical areas:

» Rapid genetic upgradation and conservation of germplasm of cattle and buffaloes and improvement in the delivery mechanism of breeding inputs and services to farmers.

» Extension of dairy development activities in non-operation flood, hilly and backward areas, including clean milk production.

» Promotion of fodder crops and fodder trees to improve animal nutrition.

» Provision of adequate animal health services with special emphasis on creation of disease-free zones and control of foot and mouth disease.

» Improvement of small ruminates and packs animals.

» Provision of credit facility to farmers for viable activities.

» Development of reliable database and management information system.

12.7.1. Launch of National Milk Vision

As envisaged in the India vision 2020, particularly for MILK we need to launch a "National Milk Vision" in an integrated manner in the following fronts.

» Cattle breeding.
» Feed and nutrition.
» Cattle healthcare.
» Farm management.
» Milk procurement and transportation.

When we launch National Milk Vision, as a variant of White Revolution, we will empower the farmers with the implementation of above missions will certainly put India on the high yielding milk map and also bring sustainable development in the rural area enriching the agriculture, animal husbandry and food processing creating value added employment opportunities to the 60 million families. The second green revolution will be hastened up due this National Milk Vision, because it will act as feeder channel and result into an economic multiplier for the nation especially during the recession.

12.8. CONCLUSION

To claim that Operation Flood is a panacea for the development ills of India, or even that it is the best approach for developing a dairy industry, would be short-sighted; however, it has undeniably impacted India's rural development in a positive way. Lessons from Operation Flood are vital, given the recent focus on livestock products as a means of addressing protein and micronutrient deficiencies in developing countries. Fortunately, in many countries, demand for livestock products is rapidly growing, and this livestock revolution not only allows smallholder farmers to benefit economically from expanding markets but also provides their families with energy-dense calories and micronutrients. Failure to act could promote that this revolution be of such a manner to not promote poverty alleviation, enhanced nutrition and health and environmental preservation (Delgado, 2003). Specifically, market-oriented milk production has proven to be a key income-generating livestock activity available to poor and marginal households. It generates a steady flow of income and also has been shown to play a role in capital accumulation among resource-poor households, which fosters their investments in education as well as other productive activities and assets. Oftentimes, women of the household are the ones accruing this income and a subsequent positive investment in child welfare and nutrition can be observed (Staal *et al.*, 2008).

Overall, regarding the lessons gleaned from Operation Flood, the message to the world is one of guarded optimism (Scholten and Basu, 2009, Hindu Business Line, 2009). New challenges, such as rising competition from investor-owned firms will continue to emerge and must be addressed. However, at the end of the day, Operation Flood established a reliable, profit-generating market for smallholder farmers that engendered confidence and increased investment in the dairy sector: the result was an expansion of production and improved productivity to meet the growing demand for dairy and in turn, enhance the dietary quality of millions of Indians who could not consume greater quantities of milk and milk products.

Chapter - 13

Blue Revolution
(Fisheries Production)

13.1. INTRODUCTION

The fisheries sector of India is immensely contributing to the economy of the country by way of food production, nutritional security, provides valuable foreign exchange, and employment generation for millions of people. At the same time, it is an instrument of livelihood for a large section of economically backward population of the country. More than 7 million fishers in the country depend on capture fisheries and aquaculture for their livelihood. The fisheries sector has grown from a traditional livelihood activity in the fifties and sixties to science and technology led commercial enterprise in the past four decades (Ayyappan, 2012; Ayyappan *et al.*, 2013). Fishery sector occupies an important place in the socio-economic development of the country.

Aquaculture is one of the fastest growing food production sectors globally. The seafood (fish and other aquatic foods cultivated both in freshwater and marine environment) production is growing at the rate of over 7% per year and accounts for almost half of all seafood destined for human consumption (Edwards *et al.*, 2019). India is the second largest producer of fish and offers a vast potential for aquaculture, contributing significantly to the economy and nutrition of millions of people. Indian aquaculture encompasses a range of species and cultivation systems which is spread across the freshwater and marine ecosystems exhibiting spectacular growth (Katiha *et al.*, 2005; Munilkumar and Nandeesha, 2007; Nair, 2014; Jayasankar, 2018). The rich aquatic genetic resources of the country ranging from deep seas to the Himalayan lakes and rivers support about 10% of the global biodiversity in terms of fish and shellfish species whose number has currently reached 3137 with an additional 462 exotic species (Lal and Jena, 2019).

Today, India ranks second in global fish production registering an annual growth rate of over 7% with a production of 13.4 million tons in 2018-19 which includes 3.7 million tons and 9.7 million tons from marine and inland sector, respectively (Fig. 13.1). The sector contributes 6.58% to the gross domestic product (GDP) of the agriculture sector and 1.03% to the total GDP of the country. Fish and fishery products exports have emerged as the largest group in agriculture exports and in value terms, accounted for US$ 6.73 billion in 2018-19. The fisheries potential of the country has been estimated at 22.31 million tons with 5.31 and 17 million tons from marine and inland sector, respectively (GoI, 2020b). Andhra Pradesh is the frontrunner in inland aquaculture production in the country with a boom in fish culture, first observed with Indian major carps (IMCs) and then Pangasius, resulting in expansion of pond area to 142,000 ha and subsequent massive increases in inland farmed fish production to 1.5 million tons (Belton *et al.*, 2017). The technological advances and environmental changes have made it necessary to seriously address the issues of sustainability (Christopher *et al.*, 2015; Taryn *et al.*, 2020).

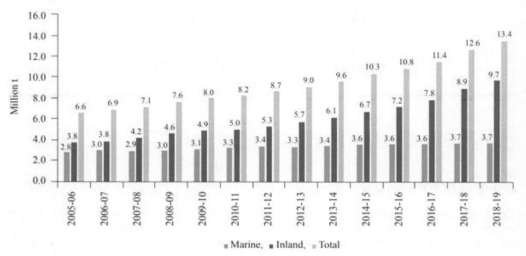

Fig. 13.1. Fish production in India from 2005-06 to 2018-19 (*Source:* Ministry of Fisheries, Animal Husbandry and Dairying, Govt. of India)

13.2. AREA AND PRODUCTION

The share of various states in the total fish production is presented in Table 13.1 (GoI, 2018). Fish production in India for the period 1980-81 to 2019-20 is presented in Table 13.2. The share of Inland fisheries in 1950-51 to the total fish production of the country was 29% which increased to 72% in 2018-19 (Fig. 13.2). Gujarat leads the marine fish landings followed by Andhra Pradesh, Tamil Nadu, Maharashtra, Karnataka, and Kerala (Fig. 13.3). Andhra Pradesh is the largest fish producer in the inland sector followed by West Bengal and together they contribute more than 40% of the total inland fish production in the country (Fig. 13.4).

Table 13.1. Major state-wise fish production in India during 2019-20 (In Lakh Tons)

State	Fish production during 2019-20 (lakh tons)		
	Inland	Marine	Total
Andhra Pradesh	36.10	5.64	41.74
Telangana	3.00	0.00	3.00
West Bengal	16.19	1.63	17.82
Gujarat	6.99	1.58	8.57
Uttar Pradesh	6.99	0.00	6.99
Odisha	6.60	1.58	8.18

Bihar	6.41	0.00	6.41
Chhattisgarh	5.72	0.00	5.72
Karnataka	2.29	4.03	6.32
Kerala	2.05	4.75	6.80
Maharashtra	1.18	4.43	5.65
Tamil Nadu	1.74	5.83	7.61
Goa	0.04	1.01	1.05
Assam	3.73	0.00	3.73
Other States	10.75	1.36	12.11
TOTAL	104.57	37.27	141.84

Table 13.2. Fish production in India for the period 1980-81 to 2019-20

Year	Fish production (in million tons)		
	Marine	Inland	Total
1980-81	1.555	0.887	2.442
1990-91	2.300	1.536	3.836
2000-01	2.811	2.845	5.656
2010-11	3.250	4.981	8.231
2019-20	3.727	10.437	14.164

Source: Department of Fisheries, States Government / UTs Administration

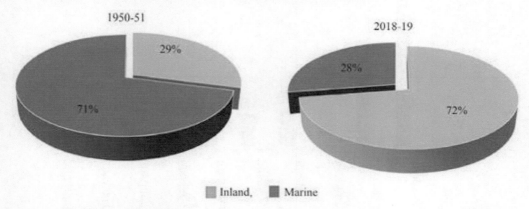

Fig. 13.2. Percentage contribution of inland and marine fisheries during 1950-51 and 2018-2019. (Source: Ministry of Fisheries, Animal Husbandry and Dairying, Govt. of India)

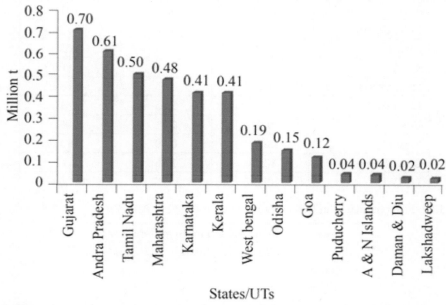

Fig. 13.3. Major inland fish producing states during 2018-19 (Source: Ministry of Fisheries, Animal Husbandry and Dairying, Govt. of India)

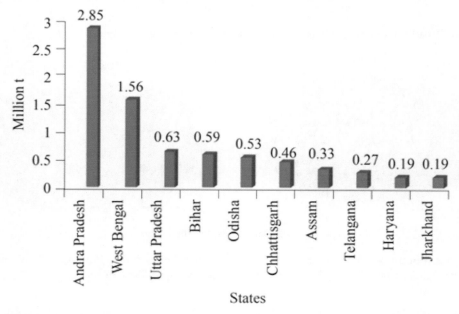

Fig. 13.4. Major marine fish producing states during 2018-19 (Source: Ministry of Fisheries, Animal Husbandry and Dairying, Govt. of India)

13.3. FISHERIES RESOURCES

Varieties of fish commonly cultured in India include;

- » Freshwater fish
- » Brackish water fish

13.3.1. Freshwater fish

Varieties of freshwater fish include: Catla, Rohu, Mrigal, Silver Carp, Grass Carp, and common carp (Fig. 13.5)

Fig 13.5. Freshwater fish varieties.

13.3.2. Brackish water fish

Varieties of brackish water fish include: Seabass, Pompano, Milkfish, and Cobia (Fig. 13.6)

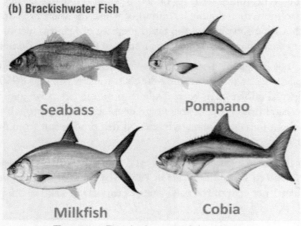

Fig. 13.6. Brackish water fish varieties

13.3.3. Resources

India with a coastline of 8118 km and the exclusive economic zone (EEZ) of 2.02 million km^2 (0.86 million km^2 on the west coast, 0.56 million km^2 on the east coast and 0.60 km^2 around the Andaman and Nicobar Islands) provides huge potential for marine capture and culture fisheries (Fig. 13.7). The inland fishery resources broadly include 0.38 million km stretch of rivers and canals, 2.70 million ha of reservoirs, 2.47 million ha of ponds and tanks, 0.43 million ha of beels, 0.34 million ha derelict water bodies/oxbow lakes and 0.96 million ha brackish water area. The diverse cold-water resources are distributed in the upland streams, rivers, lakes and reservoirs located at medium to high altitudes, in the Himalayan and North-Eastern region.

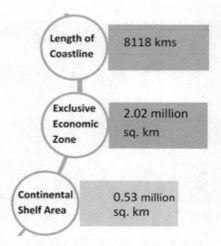

Fig. 13.7. Fisheries resources in India (2019-20)

13.4. BLUE REVOLUTION

In the 1960s, India made headlines with its Green Revolution, using high-yielding varieties and improved technology to more than double its output of wheat between 1965 and 1972. Today, India is pushing ahead with a Blue Revolution, the rapid increase of fish production in small ponds and water bodies, a boon to small farmers, the nation's nutrition and its gross domestic product.

Fish production has increased more than five-fold since India's independence and is a major industry in the coastal states. It rose from only 800,000 tons in the year 1950 to 4.1 million tons in the early 1990s. Special efforts have been made to promote extensive and intensive inland fish farming, modernize coastal fisheries, and encourage deep-sea fishing through joint ventures. These efforts led to a more them four-fold increase in coastal fish production from 520,000 tons in 1950 to 2.4 million tons in 1990.

The increase in inland fish production was even more dramatic, increasing almost eightfold from 218,000 tons in 1950 to 1.7 million tons in 1990. The value of fish and processed fish exports increased from less than 1 per cent of the total value of exports in 1960 to 3.6 per cent in 1993.

13.4.1. Objectives

The Blue Revolution aimed to enhance the economic condition of India through the augmentation of fisheries and thus contributing towards the food and nutritional security. The utilization of the water resources for the development of fisheries was done by the Neel Kranti Mission in a sustainable manner.

The objectives of the Blue Revolution mission are mentioned below:

» Completely tapping the total fish potential of India on both islands as well as in the marine sector and to triple the production by the year 2020.

» Transforming the fisheries sector into a modern industry through the utilization of new technologies and processes.

» Doubling the income of the fishers through increased productivity and improving the post-harvest marketing infrastructure including e-commerce, technologies, and global best innovators.

» To ensure the active participation of the fishers and the fish farmers in income enhancement.

» Tripling the export earnings by the year 2020 with a major focus on the benefits covering the institutional mechanisms.

» Developing the nutritional and food security of the nation.

13.4.2. Strategies

The strategies to achieve 15 million tons fish production are presented in Table 13.3.

Table 13.3. Strategies to achieve 15 million tons fish production

Ponds	Reservoirs	Brackish waters	Coastal waters
• Area expansion • Production enhancement • Productivity 2.33 tons/ ha to 3.90 tons / ha	• Production enhancement • Productivity 100 kg/ ha to 170 kg/ ha • Cage culture	• Infrastructure • Seed production • Productivity 3.52 tons / ha to 6.45 tons / ha	• Sea cage farming of Finfish • Seaweed cultivation • Mariculture of Shellfish
Wetlands	**Cold Water**	**Deep Sea**	**Species**
• Community participation • Increased utilization • Productivity 220 kg/ha to 1000 kg/ ha	• Conservation • Rainbow Trout Raceway farming • Mahseer sport fishing	• Exploitation of deep-sea resources • Export of Sashimigrade Tuna	• Minor carps • Ornamental fishes • Tilapia, Pungassius • *P. (F) indicus* • Standardisation of breeding technology

13.4.3. Highlights of Blue Revolution

» Blue Revolution has been attained by enhancing fish production from 0.75 million tons in 1951 to 5.4 million tons in 1997. India has emerged as the second largest fish producing country in the world in freshwater aquaculture.

» Phenomenal growth of marine products export.

» Indigenous design of fishing craft and gears.

» Developed national standards for fish inspection and quality control.

» Nutritional evaluation of major fish species and fishery products.

» Created national collection centre for characterization and storage of important marine microorganisms.

» Developed value-added fishery products for export market.

» Technology development of Retortable Pouch Process as a substitute for canning fish.

» Commercial production of chitin and chitosan from shrimp head and shell.

» Hatchery technology for shrimp.

» Semi-intensive shrimp farming.

» Fattening of lobsters and crabs.

» Artificial feed for shrimp farming.

» Technology package for brood stock management, production of fingerlings and grow-out systems for major finfish, shellfish and molluscs.

» Culture and utilization of sea weeds.

» Technology package for mass culture of one species of micro-algae.

» Production of ornamental fish under hatchery conditions.

» Breeding of sea bass under controlled conditions.

» Commercial production of cultured pearls from pearl oysters.

» Induced breeding of major carps, catfishes and other fin fishes.

» Production of freshwater pearls.

» Development of vaccines and formulation of drugs for fish diseases.

» Genetically improved, rohu, CIFAIR-I.

» Commercialization of fish feeds for inland aquaculture.

» Production of mahseer and snow-trout in hatcheries.

» Conservation of endangered species.

» Cryopreservation of milts of consumable important fish species.

» Enhanced fish productivity of reservoirs.

13.5. MARINE FISHERIES

The marine fisheries resources of the country are in terms of 8.129 km long coastline. 0.5 million sq. km of continental shelf and 2.02 million sq. km of exclusive economic zone. The annual exploitation from the marine sector was 3.56 million tons in 2019 against the potential of 5 million tons (CMFRI, 2020). Marine capture fisheries play a vital role in India's economy, providing employment and income to over two million people. The total number of marine fishing fleet is estimated to be 2,80,491 consisting of traditional crafts (1,81,284), motorised traditional crafts (44,578) and mechanised boats (53,684) and the share of traditional, motorised and mechanised sector in the catch is estimated at 9, 26 and 65% respectively (Shinoj *et al.*, 2020). India's marine fish harvest mostly centres around coastal waters up to 100 m depth and about 90% of the catch comes from within 50 m resulting in overexploitation of species in the nearshore waters. Coastal resources up to 100 m depth are subjected to intense fishing pressure and is expected at levels close to or exceeding optimum sustainable limit. While the inshore waters have been almost exploited to the maximum sustainable yield (MSY) levels, the contribution from the deep sea has been insignificant, hitherto directed at shrimps only. The country is presently harvesting about 12% of the potential and the balance can be optimally harvested using a judicious mix of technology, infrastructure, and human resources development. Hence, after having almost reached a plateau in production from the coastal waters, the scope for increasing fish production from marine sources now lies in the exploitation of the deep sea. In order to harvest deep sea resources, diversification of the existing deep-sea fishing fleet and introduction of resource specific vessels for long lining, purse seining and squid jigging are favorable options attracting more attention.

To fish those resources for increasing fish production from the marine sector, the industry needs ocean-going vessels and sophisticated on-board facilities which are capital-intensive. The strategies proposed for marine fisheries management are: regulated and diversified fishing, targeting the under-exploited and non-conventional resources of the EEZ, identification of potential fishing zones, stock enhancement through sea ranching, installation of fish aggregating devices and artificial reefs, community based resource management, responsible fishing including closed seasons and mesh regulations, assessment arid exploitation of resources available around islands and infrastructural support in terms of deep sea vessels, on-board and onshore facilities.

The marine environment provides an immense biodiversity that is being catalogued for commercial uses. These include several microorganisms, algal forms, and invertebrates, that could serve as- potential sources of bioactive substances including antimicrobials, anaesthetics, anticarcinogens *etc.* as well a wealth of valuable genetic material for transgenics and thus it presents a huge opportunity for both Food and Drugs from the seas. Identification of suitable sites along the Indian coastline of over 8000 km, hatcheries and grow-out systems for finfish, shellfish, and other organisms, possibilities of cage culture in island eco-systems are the strategies for realizing these potentials. Research thrusts in the next five years pertain to studies in the shelf, slope and oceanic realms of the EEZ to assess and map the resource potential, upgradation of mariculture technologies, sociotechno-economic aspects of marine fisheries and brackish water aquaculture, design and fabrication of modern fuel-efficient fishing vessels, development of cost-effective and responsible fish harvesting systems, diversification and value addition for utilization of low value fish, quality assurance, and management systems.

13.6. INLAND FISHERIES

13.6.1. Fresh water aquaculture

India is blessed with vast area of inland open water in terms of rivers and canals, estuaries, reservoirs, ponds and tanks, flood plains, backwaters and lagoons, beels and wetlands and the inland saline and brackish-water which are invaluable aquatic resources of the country offering tremendous scope for capture and culture fisheries development.. The vast and varied cold water fishery resources extend from north-eastern to north-western Himalayan Ranges and parts of Western Ghats in peninsular regions encompassing about ten states. These cold-water resources are spread as upland rivers, streams, high and low altitude natural lakes, and reservoirs. The cold-water fisheries resources include 258 indigenous, exotic, cultivable and non-cultivable fish species belonging to 21 families and 76 genera (Debajit Sarma *et al.*, 2011).

It is estimated that freshwater aquaculture contributes to over 95% of the total production. The three Indian major carps (IMCs) namely Catla [*Labeo* (=*Catla*) *catla*], rohu (*Labeo rohita*) and mrigal (*Cirrhinus mrigal*) are the key species contributing to about 80-90% of the freshwater aquaculture production. The other species which contributes to about 10-20% of the production include minor carps, catfish, silver carp, grass carp and common carp. After the historic breakthrough in induced breeding of IMCs in 1957, the major expansion of Indian aquaculture happened in the 1970s; wherein the technology of composite fish farming of three Indian and three exotic carps was developed and disseminated to the farmers leading to a paradigm shift in carp farming, greatly enhancing the production and productivity (Sinha *et al.*, 1973). Subsequently, with the adoption of polyculture, semi-intensive and intensive carp culture practices, a production ranging from 5-15 t ha^{-1} has been demonstrated. The recent years have witnessed successful culture technology development towards diversification of species including new candidates such as minor carps [*Labeo calbasu, L. fimbriatus, L. gonius, L. bata, Sytomus (=Puntius) sarana* and *Amblypharyngodon mola*], catfishes [*Clarias magur, Heteropneustes fossilis, Pangasius, Wallago attu, Sperata (=Mystus) seenghala, S. aor* and *Ompok pabda*]; murrels (*Channa striata, C. marulius*); *Anabas testudineus* and tilapia (*Oreochromis mossambicus, O. niloticus*). The freshwater prawn farming once a boon in Andhra Pradesh, West Bengal and Kerala has been on the decline since 2006 due to inbreeding depression, disease problems and reduced growth. The genetic improvement program initiated at ICAR-Central Institute of Freshwater Aquaculture (ICAR-CIFA), Bhubaneswar, in collaboration with World Fish, Malaysia, is expected to solve the problems of reduced growth and improve the disease resistance in *Macrobrachium rosenbergii*. The freshwater prawn farming has huge potential in India for food and nutritional security (New *et al.*, 2010). Commercial farming of the exotic catfish, *Pangasianodon hypophthalmus*, under monoculture systems with production levels of 15-50 t ha^{-1} yr^{-1} have become popular among the entrepreneurs especially in the states of Andhra Pradesh, West Bengal and Chhattisgarh. In recent years, trout culture is also being undertaken in the hill states of Jammu and Kashmir, Ladakh, Himachal Pradesh and Sikkim. The states of Andhra Pradesh, West Bengal, Bihar, Assam, Panjab, Haryana, Chhattisgarh and Jharkhand have shown impressive growth in freshwater aquaculture in recent years (Jayasankar, 2018).

13.6.2. Brackish water aquaculture

The brackish water fish farming traditionally practiced in the Bheries of West Bengal and Pokali paddy fields of Kerala has transformed into a commercial aquaculture enterprise from 1980 onwards. In fact, shrimp farming emerged as the fastest-growing food producing sector in 1990s contributing substantially to the Indian economy and exports. The brackish water aquaculture production in India touched 0.75 million tons valued at US\$ 7 billion during 2017-18 and the key cultivable shrimp species included *Peneaus monodon* and *P. vannamei*. The fish species namely *Chanos, Lates calcarifer, Mugil cephalus, Etroplus suratensis* and *Trachinotus blochii* have emerged as new candidates for diversification. It is reported that out of 3.9 million ha of the brackish water area comprising estuaries, coastal lagoons, lakes, backwaters, tidal creeks, canals, mudflats and mangroves, only 11% is currently utilized for brackish water aquaculture (NFDB, 2019). In addition, about 9 million ha of salt affected inland saline soils in the states of Haryana, Punjab, Rajasthan, Uttar Pradesh and Gujarat offers immense scope for inland saline water aquaculture (NFDB, 2019). A recent breakthrough in the successful commercial farming of the white legged shrimp (*P. vannamei*) in the inland saline waters of Haryana and Punjab has demonstrated the potential of converting waste into wealth (Lakra *et al.*, 2014). The brackish water aquaculture sector is expected to contribute significantly in the future because of its vast potential. Major deterrent in brackish water aquaculture, particularly in shrimp farming, is the availability of specific pathogen free (SPF) seed and disease management during the culture period. Shrimp production in the country touched 0.7 million tons in 2017-18, valued over 3000 crores (US\$ 30 billion). Presently, about 90% of the brackish water aquaculture production is of the exotic white shrimp, *P. vannamei*. Considering the disease and production cost issues in *P. vannamei* farming, the Indian white shrimp, *P. indicus* is being intensely researched for domestication as an alternative indigenous species to the shrimp farmers (Vijayan, 2019).

13.6.3. Mariculture and cage farming

Global fish consumption is predicted to increase over 20% by 2025 as both human population and economic development would rise in coming decades (FAO, 2020). Culture of fish in cages and pens offers great scope for enhanced production, meeting the rising demand for seafood. Globally, fish provides about 3.2 billion people with 20% of their animal protein intake as per capita consumption has increased from 9 kg in 1961 to 20.5 kg in 2018 (FAO, 2020). Mariculture research and development in India was systematically initiated during 1980s by ICAR-Central Marine Fisheries Research Institute (ICAR-CMFRI), Kochi. The National Institute of Ocean Technology (NIOT) and Marine Products Export Development Authority (MPEDA) have also contributed in the development of this sector. The technological advances include captive maturation, seed production and developing technologies for farming of various finfishes (cobia, pompano, seabass, groupers, snappers, breams and ornamental fishes); shellfishes (mussels, oysters, clams, lobsters, shrimps and crabs) and seaweeds (Syda Rao *et al.*, 2014; Ranjan *et al.*, 2018; Kaladharan *et al.*, 2019). However, the marine aquaculture is still in its infancy and commercial farming is yet to take off despite its huge potential to enhance seafood production in the country. The projected mariculture potential of the country based on the resources available in the maritime states and union territories and islands is 8-16 million tons, while the current mariculture production is around 0.05 tons. Hence, there is an urgent need for technology upscaling, establishment of hatcheries in various states for seed production of candidate species. Identification of suitable sites all along the Indian coast with financial, technological capacity building and policy support to the stakeholders for adopting mariculture activities as a commercial enterprise as well as an alternate livelihood activity involving traditional fishermen is strongly advocated. It is reported that more than 30% of India's EEZ lies around the islands where immense opportunities exist for mariculture of fish and shellfish (NFDB, 2020a).

13.7. FISHERIES POTENTIAL/ PRODUCTION

The country has a long coastline of 8,129 km and a vast area under estuaries, backwaters, lagoons *etc.*, highly amenable for developing capture as well as culture fisheries. After declaration of the Exclusive Economic Zone (EEZ) in 1977, the marine area available to India is estimated at 2.02 million sq. km. comprising 0.86 million sq. km on the west coast, 0.56 million sq. km on the east coast and 0.60 million sq. km around the Andaman and Nicobar islands. With the absolute right on the EEZ, India has also acquired the responsibility to conserve, develop and optimally exploit the marine living resources within this area. The harvestable potential of marine fishery resources in the EEZ has been revalidated by a Group of Experts constituted by the Government of India, Ministry of Agriculture at about 3.93 million tons (October, 2000) consisting of 2.02 million tons of demersal, 1.67 million tons of pelagic and 0.24 million tons of oceanic resources. In the Inland Sector, the resources potential has been estimated at 4.5 million tons which considers the production from both capture and culture fisheries.

While the inshore waters have been almost exploited to the MSY levels, the contribution from the deep sea has been insignificant. The thrust of the deep-sea fishing industry has hitherto been directed at shrimps only, notwithstanding the other resources. As of today, the deep-sea fishing industry is almost a 100% shrimp-oriented enterprise, faced with overexploitation of the available shrimp resources as well as the fierce competition from the smaller class of vessels.

The development of deep-sea fishery industry is of concern to the entire marine fishery sector because it would have considerable impact on the management of near-shore fisheries, shore-based infrastructure utilization and post-harvest activities, both for domestic marketing and export. Similarly, the upgradation of the small mechanized sector to support the entrepreneurial interest in the sector will be given high priority.

13.8. HARVEST, POST-HARVEST PROCESSING AND SEAFOOD TRADE

13.8.1. Harvesting

There is an increasing need in the country for responsible harvesting of fishery resources based on eco-sustainability and equitable access and utilization. The harvest sector in fisheries has been enriched by several technological innovations particularly in the marine sector. This includes modernisation of fishing craft and vessels, standardisation of nets and trawlers for deep sea fishing, technologies on sustainable practices, introduction of turtle excluder device (TED) and selective fishing. Similarly, eco-friendly harvest technologies such as optimised gillnets, lines or traps for inland fisheries are other innovations in the sector.

13.8.2. Processing

The research and developments in the areas of processing, by-products and packaging of fish or fishery products are progressing in the country. The focus is on complete utilization of harvested fish through value addition, waste utilization and appropriate packaging. Several special fishery products viz. fish mince, surimi, balls, cutlets, fingers, patties, burger, coated products and shrimp-based products have been standardised. The value-added fish products include canned products, ready to eat products, extruded products and the recent focus is mostly on restructured products and nutritional protein concentrates. Interestingly, seaweeds have also been used for making cookies, seaweed jelly and yogurts on experimental scale.

Important biomolecules from the processing waste such as chitin, chitosan, fish oil, collagen, gelatin and pigments have been generated converting waste to wealth. In the packaging front, advanced technologies like vacuum packaging, modified atmosphere and active packaging technologies have been developed (Biji et al., 2015). Seafood safety and hygiene assurance have also been addressed by implementing the HACCP (Hazard Analysis Critical Control Point) and EU norms. The technology of DNA barcoding has been introduced to authenticate seafood labelling (Nagalaxmi et al., 2015). However, there are issues and constraints related to raw material, traceability, promotion of domestic marketing and online markets with adequate cold chain facilities.

13.8.3. Export

India is a leading country in seafood export (Fig. 13.8). It is reported that almost 70% of the fish caught is marketed fresh and the remaining is used for processing, drying, smoking and fish meal production. The key markets for India include USA, South-east Asia, European Union (EU), Japan and Middle East (Table 13.4). The major exports are in frozen and chilled form with only 10% or so in the value-added form.

Table 13.4. Market-wise export in value and quantity (2019-20)

Country	Quantity (Million tons)	Value (Rs. crores)
USA	3,05,178	17,904.37
European Union	1,65,773	6,136.71
China	3,29,479	9,617.44
Japan	78,507	2,920.28
Middle East	57,387	2.079.12
South East Asia	2,23,398	4,929.90

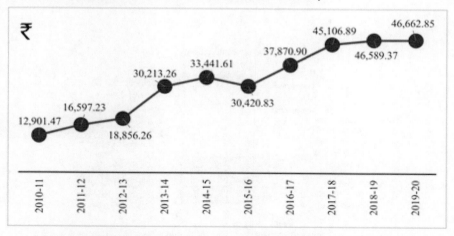

Fig. 13.8. Trend of export of fish and fish products in value (Rs. Crore).

13.9. NUTRITIONAL BENEFITS

13.9.1. Brain food

» Helps in the development and function of brain.

13.9.2. Heart food

» Lowers risk of heart attacks and strokes.

13.9.3. Health food

» Reduces risk of autoimmune diseases, including type-1 diabetes.
» Prevents and treats depression, making a happier person.
» Helps prevent asthma in children.
» Lowers risk of cancer, blood pressure, Alzheimer's disease, *etc.*

13.10. CHALLENGES

The two major constraints impacting the marine resources and fisheries are poor governance and overexploitation due to overcapacity. The overexploitation has been recognised as one of the key threats to global biodiversity (Trindade-Santos *et al.*, 2020). Other important factors include vagaries of climate change, traditional fishing and harvesting practices, poor infrastructure especially of fishing harbors and landing centres, lack of sufficient cold chain and processing facilities for value addition, post-harvest losses and shortage of skilled manpower (Fig. 13.9).

The aquaculture sector despite its fast growth, is facing the challenges of quality seed, feed, disease and improved breed. Diversification of fish species has provided some options but mass scale seed production of several of these new candidates still remains elusive. The use of genetically improved seed of Jayanti rohu in the freshwater aquaculture is providing rich dividends, but no such choices are available in the brackish water and mariculture sectors for improved indigenous species.

A largely, low input culture practice with low productivity reflects low adoption of modern technology and scientific farming in the inland aquaculture.

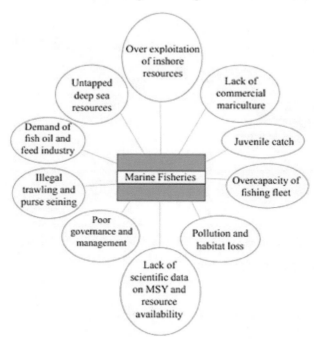

Fig. 13.9. Factors responsible for stagnating marine fisheries.

Lack of skills and institutional credit in aquaculture including both freshwater and marine are other impediments to faster growth and enhanced production. The aquaculture sector is also confronted with reduced availability of water in the inland areas, disease outbreaks in fish and shrimp farming, cumbersome institutional credit and increasing input costs especially of feed. Therefore, addressing all these issues through research and development as well as policy formulation is of utmost importance for realising the goals of blue revolution under a mission mode approach.

13.11. EFFECT OF CLIMATE CHANGE

Climate change and its impact on Indian fisheries and aquaculture has been reviewed comprehensively (Mohanty *et al.*, 2017). Rise in sea surface temperatures (SST), ocean acidification and coral bleaching are increasingly witnessed which are affecting the marine fisheries. Coastal communities and their livelihoods are seriously threatened by cyclones, storms and extreme weather conditions. Changes in oceanic weather systems such as SST, pH, salinity and EL Nino Southern Oscillation (ELSO) are becoming evident as a result of climate change. It is predicted that SST is likely to increase by 2.0 to 3.5°C by the end of the century in Indian Ocean. Unnikrishnan *et al.* (2015) reported that the net sea level has increased by 1.09 to 1.75 mm year^{-1} in the Indian seas in the last 55 years with faster rise observed along the east coast. Similarly, the number of cyclones occurring in the Bay of Bengal is four times than that in the Arabia Sea (Shaji *et al.*, 2014). Ocean warming and increased stratification of the upper ocean caused by global climate change will lead to decline in the dissolved oxygen with implications on ocean productivity, nutrient cycling, carbon cycling and marine habitat. Inland fisheries and aquaculture are also severely affected by climate change especially the river systems and their fisheries. The major river systems of India affected by climate change are the Himalayan glacier fed rivers such as Ganga, Brahamaputra and rivers of Deccan plateau including Narmada, Tapti, Godavari and Mahanadi. These rivers will experience more reduced water flow or flooding due to climate change. The changes are likely to affect fish production and biodiversity depending on the severity of the situation.

It is opined that an active partnership between fishermen, managers, scientists and policy planners will provide a better understanding of the climate impacts on fisheries and to evolve adaptation options and mitigation strategies. A mega awareness campaign on the lines of Swachh Bharat Abhiyan is required about the impact, vulnerability, adaptation and mitigation related to climate change among the stakeholders in the fisheries sector. The climate change resilience as an adaptation strategy will help addressing the livelihood security of the stakeholders. It is well established globally that effective fisheries management including the solutions to the problems of over-exploitation, habitat degradation, pollution and climate change could be achieved by adopting an ecosystem-based approach. The ecosystem approach will build resilience to the ecological and fisheries effects of climate change.

13.12. CAPACITY BUILDING AND TRAINING OF STAKEHOLDERS

The major aim for skill development in agriculture and allied sectors is to double the farmers' income by doubling production and reducing input costs. In fisheries sector, Government of India has entered into several memoranda of understanding (MOUs) with Government institutes, state agricultural and fisheries universities to take forward the skill development programs. Shortage of private players as training partners necessitates the need to open up the same to Non-Governmental Organisations (NGOs) and corporates having recognised presence in fisheries and aquaculture. It is a challenge to convert the mindset of the institutes and capacity building providers, orienting them to skill-based training program modules. Presently, majority of them indulge only on knowledge based theoretical learning with minimum focus on applied skills. Most of the institutes lack quality trainers, infrastructure and environment for long-term skill-based training. There is a need to collaborate with industry and private sectors for these purposes including identification of potential partners, training the trainers, master trainers, curriculum and content development, for speeding up the skill development programs in the fisheries sector and for meeting the increasing skilled manpower requirement in the country.

There are immense entrepreneurship opportunities available in aquaculture. Traditionally, all fisheries capacity building and extension programs were handled either by State Fisheries Department or through the ICAR Fisheries Institutes and Fisheries Universities and Colleges under State Agricultural Universities. A centrally sponsored scheme on Fisheries Training and Extension aided various central and state government organisations towards Farmers' Training Centre, publication of handbooks, training manuals, organisation of workshop/symposia/seminar, production of documentary films on fisheries and establishment of awareness centres.

13.13. FUTURE THRUST AREAS OF RESEARCH AND DEVELOPMENT

The main objective of the new schemes namely, PMMSY, Fisheries and Aquaculture Infrastructure Development Fund (FIFA) and Blue Revolution launched in the country is to enhance production and productivity along with modern infrastructure development, increased employment generation for youth and women, improved socio-economic conditions of fisherfolks, augmentations of exports and adopting an integrated approach to marine and inland fisheries towards responsible and sustainable fisheries and aquaculture development. It is directed at comprehensive development of the sector through innovative technology applications and policy interventions addressing the critical gaps in knowledge, technology and governance.

The marine capture fisheries is plagued with declining catches due to excess capacity of fishing fleets, juvenile catch besides aquatic pollution and climate variability. An ecosystem approach to fisheries management and the state and region-specific management plans are advocated to be effectively implemented by central and state governments towards providing solutions to the depleting resources and alternate livelihood options for the coastal communities. The exploitation of deep-sea resources and the increased application of remote sensing and space technology in locating new fishing grounds holds a great promise. The recent initiatives of the Government of India namely Sagarmala and PMMSY are mainly directed towards enhancing the production and productivity, strengthening the value chain including infrastructure, post-harvest management, quality control and marketing.

There is immense scope in aquaculture sector through both horizontal expansions and vertical intensive farming using advanced technologies of recirculating aquaculture systems (RAS) and integrated multi-trophic aquaculture (IMTA). Hi-tech aquaculture may enhance vulnerability to diseases but a coordinated approach between the central and state agencies will help to timely mitigate and control diseases in aquaculture. A national program on aquatic animal disease surveillance involving more than 20 national organisations, 16 states and 3 Union Territories is already underway under its second phase. A fisheries agency at central and state level is proposed to be created to address standards and diseases in aquaculture.

Cage culture in sea and reservoirs, integrated fish farming, ornamental fish culture and seaweed farming are the sectors of immense potential which have been prioritised to be upscaled and promoted with appropriate technology and diversification of species and associated support for large scale expansion creating new employment and alternate livelihood options to fishers and fish farmers.

Domestic fish marketing infrastructure has been upgraded in selected pockets in the country in recent years. However, development of new domestic markets across the country with adequate levels of hygiene remains elusive and it needs to be addressed on priority. Similarly, India is one of the leading countries in seafood export and strengthening and modernising the value-chain to increase shelf life, reduction in post-harvest losses and production of more value-added products for the domestic and international markets are on the priority agenda under the New India Vision. Efforts are being made on value addition of export items so as to bring higher acceptability, value per unit and meeting the target of foreign exchange to the tune of 1,00,000 crores (US$ 1000 billion) by 2025 as envisioned by the Government.

The goals of sustainable fisheries and aquaculture development in India are targeted to be achieved through appropriate fisheries policy framework, improved management and governance and strong partnership amongst the stakeholders. The environmental, economic and social sustainability of aquaculture systems have to be addressed based on quantitative indicators (Valenti *et al.*, 2018). A sectoral mission mode approach is suggested for realising the full potential of the sector towards blue revolution encompassing the principles of blue economy and blue growth initiative.

Considering the vast potential of fisheries resources and to achieve the targeted production and productivity, following thrust areas of research and development have been identified which need attention on priority basis:

» Stock assessment and monitoring of commercially important marine fish species.

» Clarification on development of fuel-efficient crafts and gears for deep sea fishing.

» Prevention of post-harvest losses, development of improved transport, storage and processing of fish.

» Development of value-added fishery products for export.

» Extraction, production and evaluation of biomolecules from marine organisms and plants for industrial and pharmaceutical applications.
  • Mariculture of fish/shellfish and culture of ornamental fish.
  • Introduction of HACCP in seafood processing.
» Marine biodiversity conservation and management
  • Rural aquaculture and integrated fish farming, increasing availability of fish seed and feed.
  • Application of molecular biology in aquaculture.
» Genetic mapping of important fish species through DNA fingerprinting.
» Fishery informatics and databases.
  • Fish nutrition and feed development.
» Diversified aquaculture.
» Breeding and culture of new fish and shellfish species in fresh and brackish water environments.
» Environment monitoring of aquatic ecosystems.
  • Development of eco-friendly and sustainable fish/ shellfish farming.
  • Development of hill fishery resources and management.
» Human resource development in emerging areas.
» Bioactive substances from aquatic biota.
» Coastal Zone Management and environment impact assessment.

13.14. RECOMMENDATIONS

The following specific recommendations are provided to further transform the fishery sector of India:

» The economic, environmental and biological sustainability of aquatic resources is critical to fisheries management. Therefore, the National Fisheries Policy (NFP) should be strongly aligned with sustainability of resources and livelihoods of the traditional fishing communities. The fishing communities should be involved in responsible actions in the resource conservation, rather than simply involving beneficiaries of the state and central government schemes.

» Special emphasis is required to be given to increase the domestic consumption of fish and shrimps through innovative marketing strategies based on the success stories of poultry and dairy sector. This will address the marketing issue of shrimps during pandemics wherein international exports are badly affected for months together affecting the whole industry, there by incurring great losses.

» The umbrella scheme of blue revolution and PMMSY should have a major focus on blue economy encompassing target-oriented programs on deep sea fishing, marine aquaculture and biotechnology. The capacity development in the exploitation of the deep-sea resources, open sea cage farming including seaweed farming and marine ornamental fish culture should be taken up with quantitative targets. The captive breeding and mass scale seed production of potential marine cultivable fish species along with feed mills are required to be undertaken all along the Indian coast. Marine biotechnology has immense potential for cutting edge research in bioprospecting and green growth.

» A strategy on the job mentoring of farmers and fishers by scientists/experts on regular basis at their fields has proved very effective in several countries including China. A country with third largest scientific manpower in the world, should therefore consider this and mandate the research and development organizations to develop action plans for farmer mentors in the aquaculture sector.

» A National Academy of Fisheries and Aquaculture (NAFA) should be established as a platform for policy advocacy and support to the farmers and to strengthen national and international interface and collaborations.

» The role of coastal and marine fishers in the fisheries sector is well recognized for their significant contribution to food security. Representation of these communities and their wards in the fisheries higher education is almost negligible. A special scheme needs to be formulated on the lines of the civil service coaching to Scheduled Caste/Scheduled Tribe communities in the country, with an objective to facilitate a minimum representation in degree courses for the wards of fishing communities in all fisheries colleges.

» The coastal security and safety at sea has assumed global significance in view of increased threats from pirates and terrorists. Youth from coastal villages should be motivated and prepared for national services especially by special recruitment drives in navy, coast guard, coastal police and forest services.

» A global partnership in fisheries research and development is the need of the hour. The recent National Education Policy and the administrative reforms implemented in the federal ministries and structures brings several opportunities for the scientific and academic community to develop partnership with reputed organizations in terms of joint degree programs, exchange of faculty and students and sharing of advanced knowledge.

» Diversification in aquaculture through the establishment of regional hatcheries of new candidate species whose seed production technology has been standardised. This needs to be accomplished with funding from Blue Revolution/ PMMSY schemes which will facilitate mass seed production and its availability to the farmers on a sustainable basis for both freshwater and marine aquaculture.

» Formation of fish farmer producer organisations (FPOs) holds great promise. The schemes of NABARD and Government of India provide several opportunities for these beneficiaries covering the complete value chain from production to market.

» A comprehensive third-party review of the central scheme on Blue Revolution implemented through the state/union territories covering all the states and beneficiaries for its impact on the sector is required in terms of success stories, challenges and lessons learnt. This will provide new insights for effective implementation of PMMSY towards improved management of the fisheries sector.

13.15. CONCLUSION

The fisheries and aquaculture sector in the country is poised to play a major role in the lives of people in the coming decades, with increasing population pressure on land and alternate food production system being increasingly projected from the aquatic resources. The research and development activities as indicated above in the frontier areas of fisheries sector are urgently required on priority basis to meet the new challenges in fisheries sector and to make the whole system sustainable and eco-friendly. Such research and development support through various organizations will not only boost fish production and productivity but also ensure nutritional and food security, employment opportunities and socio-economic upliftment of the poorest of the poor.

Chapter - 14

Silver (Poultry Egg and Meat Production) Revolution

14.1. INTRODUCTION

Poultry is one of the fastest growing segments of the agricultural sector in India today. While the production of agricultural crops has been rising at a rate of 1.5 to 2.0% per annum, that of eggs and broilers has been rising at a rate of 8 to 10% per annum. As a result, India is now the world's fifth largest egg producer and the eighteenth largest producer of broilers. The potential in the sector is due to a combination of factors – growth in per capita income, a growing urban population and falling real poultry prices. Poultry meat is the fastest growing component of global meat demand, and India, the world's second largest developing country, is experiencing rapid growth in its poultry sector. In India, poultry sector growth is being driven by rising incomes and a rapidly increasing middle class, together with the emergence of vertically integrated poultry producers that have reduced consumer prices by lowering production and marketing costs. Integrated production, market transition from live birds to chilled and frozen products, and policies that ensure supplies of competitively priced maize and soybeans are keys to future poultry industry growth in India. There are a number of small poultry dressing plants in the country. These plants are producing dressed chickens. In addition to these plants, there are five modern integrated poultry processing plants producing dressed chicken, chicken cut parts and other chicken products. These plants will manufacture egg powder and frozen egg yolk for export.

The poultry segment in India has experienced a change in perspective in structure and activity which has been its change from a backyard activity into a significant business agri- based industry over a time of four decades. The constant efforts in upgradation, modification and application of new technologies paved the way for the multi-fold and multifaceted growth in poultry and allied sectors. Development of high yielding layer (310- 340 eggs) and broiler (2.4-2.6 kg at 6 weeks) varieties together with standardized package of practices on nutrition, housing, management and disease control have contributed to spectacular growth rates in egg and broiler production in India during the last 40 years. Currently the total poultry population in our country is 851.81 million numbers out of which 317.07 million is backyard poultry and egg production is around 103.93 billion during 2018-19 (BAHS, 2019). Backyard poultry production has shown a tremendous growth rate of 45.79% from last census as compared to commercial poultry which has increased only by 4.5%. The current per capita availability (2018-19) is around 79 eggs per year (BAHS, 2019). The poultry meat production is estimated to be 4.06 million tons (BAHS, 2019) which is around 50% of total meat production. Share of poultry meat in India is highest as compared to other species (Fig.14.1) (BAHS, 2019). However, it is lower than the suggested level of consumption of 180 eggs and 10.8 kg poultry meat per person per annum by Indian Council Medical Research. The poultry husbandry has involved a vital position both in giving work just as in contributing a significant extent to the national GDP. It has been seen that the interest for the animal protein source is on the rise in developing countries (Raveloson, 1990).

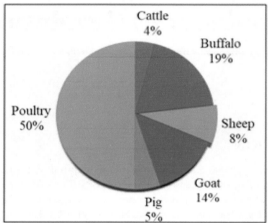

Fig. 14.1. Share of poultry meat in India is highest as compared to other species.

The top 5 States in terms of poultry population are presented in Table 14.1.

Table 14.1. Top 5 States in terms of poultry population

S. No.	State	Poultry population (in million)
1	Andhra Pradesh	161.33
2	Tamil Nadu	117.35
3	Maharashtra	77.79
4	Karnataka	53.44
5	West Bengal	52.84
6	All others	266.45
	TOTAL	729.21

Poultry can be defined as the rearing and breeding of domesticated fowls such as chicken, geese, ducks, turkey and some varieties of pigeon for their meat and eggs. Poultry farming is a method where the birds are raised domestically or commercially for their meat and eggs and also for their feathers. There are some important birds that are commonly farmed in poultry like chicken, turkey, geese and ducks.

Most non-vegetarian people find poultry products as a cheap source of protein from animals. Thus, it is one of the fastest-growing segments in animal husbandry. The meat-producing poultry birds are called broilers, and the egg-laying female birds are called layers. Poultry birds benefit humans in many ways as they are not only efficient converters of agricultural by-products, but also provide egg, feathers, and rich manure.

14.2. POULTRY PRODUCTION SYSTEMS

FAO classified poultry production systems into four categories based on the volume of operation and level of biosecurity:

» Village or backyard production,
» Commercial production with low biosecurity
» Large scale commercial with high biosecurity
» Industrial and integrated production systems.

14.2.1. Village or backyard production

14.2.1.1. Objectives

» Improve village poultry production and productivity in a sustainable manner.
» Adapt village poultry production to the changing environment.
» Maintain the diversity of local bird types.
» Improve access to markets and the supply of poultry and poultry products to remote locations that do not attract commercial poultry producers.
» Build the capacity of stakeholders involved in village poultry production.
» Raise awareness and influence livestock policy to promote village poultry production.

India has nearly 70% of its population living in rural areas. Small scale semi-commercial backyard poultry production can be advantageously promoted in rural areas. It can be used as a powerful tool for alleviation of rural poverty, eradication of malnutrition and creation of gainful employment in vast rural areas (Sharma and Chatterjee, 2009; Rajkumar *et al.*, 2010).

The most basic and simple backyard production system involving a few hens and a cockerel is essentially a closed system. Home-produced fertile eggs are hatched to provide replacements, birds feed by scavenging or are provided with household scraps and crop by-products; there are virtually no veterinary inputs and the remaining eggs and meat produced are consumed within the household.

Producers with even slightly larger flocks, generate cash income from the sale of eggs and birds within the local community. Village or backyard production systems are widely distributed and exist in both rural and urban areas. It is estimated that today in India, about 15 percent of total poultry output is derived from "backyard" production (Landes *et al.*, 2004). In areas that are less densely populated by poultry, "backyard" systems are likely to contribute a larger proportion of total poultry production. In the village or backyard sector, production is generally based on traditional local, native breeds, producing both eggs and birds for meat. In the recent past, improved backyard varieties (like Vanaraja, Gramapriya, Srinidhi, Giriraja *etc.*) developed mostly by public sector and a few by private sector (like Kroiler, Rainbow Rooster) are substantially contributing to the total chicken egg and meat production of the country. Nonetheless, village or backyard production can make a useful contribution to dietary protein intake and incomes of resource poor households (Acamovic *et al.*, 2005, Rajkumar *et al.*, 2010). Furthermore, given the lower opportunity costs of resources and the higher market prices offered for local poultry, backyard systems are likely to yield a positive economic return, despite increasing competition from the commercial sectors.

Suggestions for obtaining high returns from backyard poultry farming include:

» Protect birds from predators, to avoid predator's poultry can be kept on roof or inside house.

» Provide additional concentrate feed if available.

» Provide clean and fresh drinking water.

» Provide optimum space to avoid overcrowding and disease spread.

» Proper vaccination, de-worming and veterinary care.

» Regular disinfection of poultry house and surrounding.

» Need for flexibility in communication and extension strategies to take account of differences (e.g. between districts, villages and groups; a 'one size fits all' approach is not appropriate).

» For appropriate flock size, scavengeable feed resource base method should be applied.

14.2.2. Commercial poultry production with low biosecurity

This sector is based on commercial production, but it retains some characteristics of the traditional, backyard systems, particularly in selling live birds in wet markets or directly to retail shops. Production units are generally intermediate in scale between backyard systems of up to 200 birds and commercial systems of 10,000 to 50,000 birds. Levels of biosecurity are low, in that birds are often not permanently housed, mixed flocks of chickens and waterfowl may be kept, birds are generally marketed live, and a range of different markets, un-monitored for health risks, are used for produce sales and input supplies (Upton, 2007). The flocks are generally reared either for broiler meat production or for egg production. Feed is generally purchased either as premixed rations or as raw materials for home milling and mixing.

In India, it is suggested that relatively small-scale producers are at a disadvantage in facing high feed and transport costs, limited access to vaccines and veterinary services, and shortage of credit (Upton, 2007).

14.2.3. Large-scale commercial with high biosecurity

This sector consists of the generally larger scale (50,000 to 1.00 lakh birds) commercial flocks of broilers, layers or breeding birds. Only relatively wealthy individuals or commercial joint-stock companies have the necessary investment funds or can raise sufficient credit for these larger-scale investments. Biosecurity levels are defined as high, as birds are continuously housed, strictly preventing contact with other flocks or with wildlife. Despite this, many outbreaks of HPAI appear to have started in large-scale commercial flocks. Inputs are generally supplied and products marketed through formal market agencies. The scale and intensity of production is substantially higher in the commercial and industrial sectors than in backyard systems. Advantages are derived from economies of scale, providing scope for specialization and division of labor between the different stages in the production process, leading to automation of operations and labor-cost savings. These advantages add to those derived from the use of highly productive commercial hybrid chicks and improved technologies such as the evaporative cooling or air-conditioning of poultry houses. The four southern states, where poultry densities and flock sizes are high, together contribute 57 percent of the nation's egg production (FAO, 2007).

14.2.4. Industrial and integrated production

This sector consists of the largest and most industrialized (more than 1.00 lakh birds) enterprises in the poultry industry. The various stages in the value chain are vertically integrated into a single industrial company. The broiler or layer components are either fully integrated as part of the parent company, or are separate production units operating under contract to the parent company, it has been assumed that although the whole process, from chick breeding and hatching through to distribution and retailing is integrated in a single organization, feed milling remains as a separate business enterprise. In many instances, the feed and poultry production activities are integrated, together with 'horizontal' links to other sectors. In other cases, vertical integration is partial – from breeder down to broiler grower, or from market distributor up to broiler producer. Vertical integration yields financial benefits by reducing the operational costs at different stages of the value chain. The vertical integration of the different stages of the breeding, production, processing and marketing of poultry produce is a rational economic response, which should increase efficiency and reduce unit costs. In India, substantial numbers of integrated poultry production companies have been established, particularly in the four southern states (Landes *et al.*, 2004).

The introduction of improved, exotic, genetic material is an important first step in the growth and development of the commercial poultry sector. Generally, the new strains are less hardy and less resistant to endemic diseases than indigenous birds. The greater productive potential cannot be attained without complementary inputs of specially compounded concentrate feeds, and improved housing, management, and veterinary care. Nonetheless, the introduction of new genetic material is the foundation on which other technological improvements are added.

14.3. EGG PRODUCTION

14.3.1. Objectives of increasing egg production

» To enable doubling of farmers income.
» To fulfil the objective of protein enriched food requirement of the growing population of the country and prevent malnutrition in one of the highest malnourished children population in the world.
» To achieve 2% of world egg market trade through exports.

The egg production in the country has increased from around 88 billion in 2016-17 to 114.38 billion during 2019-20. The per capita availability of egg has increased from 69 in 2016-17 to 79 eggs per year during 2018-19 (BAHS, 2019).

14.3.2. Areas of egg production

Overall, Tamil Nadu counts for maximum egg production. In Telangana, Hyderabad is the city with maximum poultry and hatcheries. Besides Telangana, Andhra Pradesh (Vishakapatnam and Chittoor), Karnataka, Tamil Nadu, Maharashtra, Gujarat, Madhya Pradesh, Odisha and North-Eastern States are the major egg contributors.

14.3.3. Egg production- growth rate

Growth rate of egg production from 2011-12 to 2016-17 is presented in Table 14.2.

Table 14.2. Egg production growth rate from 2011-12 to 2016-17

Year	Egg production (in billion numbers)	% annual egg production growth rate
2011-12	66.45	5.40%
2012-13	69.73	4.94%
2013-14	74.75	7.20%
2014-15	78.48	4.99%

| 2015-16 | 82.93 | 5.66% |
| 2016-17 | 88.13 | 6.28% |

14.3.4. Per capita egg availability

Per capita egg availability from 2011-12 to 2016-17 is presented in Fig. 14.2.

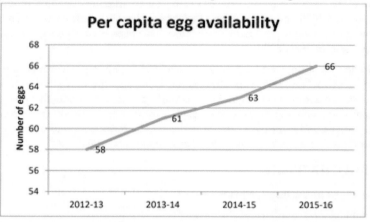

Fig. 14.2. Per capita egg availability from 2011-12 to 2016-17

14.3.5. Gap analysis: Egg production

By 2050, it is expected that the population in India would increase by 34% and to fulfil the dietary recommended levels of the livestock products by Indian Council for Medical Research (ICMR) for a population of 1.7 billion people, the livestock sector should produce 306 billion eggs per annum from the current level of egg production, it has to increase by 4.7 folds, respectively. Fulfilling the feed demand for this huge livestock from same resource base of land and water is going to be a huge challenge (NIANP, 2013).

As per nutritional requirement, half an egg a day is optimal for an average healthy person, which translates into 180 eggs/person/annum and present availability is around 69 only. Therefore, there is a huge gap in demand and supply. However, the limiting factor for growth is the prohibitive prices of important feed ingredients for the hen, namely, maize and soya. Therefore, we propose doubling of egg production assuming that adequate feed would be available at reasonable price.

Egg production is currently having a 5-6% CAGR. However, with newer scientific advances like 500 eggs in 100 weeks compared to present 320+ eggs in 72 weeks we can expect an achievable substantial increase in egg production by 2022-23, provided suitable policy support is provided to poultry industry. Government can limit its fiscal intervention to rural backyard poultry which is about 29% of total egg production.

14.3.6. Nutritive value of eggs

Egg is a wholesome, nutritious food with high nutrient density because, in proportion to their calorie count, they provide 12% of the Daily Value for protein and a wide variety of other nutrients like vitamins, essential amino acids and minerals such as vitamin A, B_6, B_{12}, folate, iron, phosphorus, selenium, choline and zinc *etc.* along with various other important ingredients so crucial for growth and good health. Protein in the nutrition is one of the most important health indices that affect children's growth and development. Lutein and zeaxanthin are two newly-recognized nutrients that have put eggs in the "functional foods" category. A functional food is one that provides health benefits beyond its basic nutrient content. Recent studies have shown that consuming lutein and zeaxanthin can significantly lower risk of age-related macular degeneration (AMD), a leading cause of blindness affecting people over the age of 65. In addition, there is a less likelihood of cataracts.

14.3.7. Need for enhancing of egg production

One of the great challenges currently posing us is how to feed an estimated 9 billion people by the year 2050 world-wide, 40% more than presently inhabit the planet, even more formidable is the challenge to achieve this without damaging the environment. The challenge is how to increase the food supply, particularly food of animal origin, in light of increasing global demand from predominantly urban populations with increased purchasing capacity.

Concomitantly, the rural backyard poultry systems play a pivotal role in achieving nutritional security of the country in rural areas. In village poultry systems, the production of poultry meat and eggs is extremely efficient in terms of feed and water inputs. These nutritious products can supplement household grain-based diets. Family poultry have a special place as they are under the control of women, require low investment, assist in pest control and provide manure. Improvements in their production can meet the nutritional demand in the household and in the community by increasing their social standing and financial autonomy.

Major objective of livestock/poultry production is to provide safe and healthy animal food/protein for the growing population. However there are many serious challenges cropping up on its sustainability.

14.4. BROILER PRODUCTION

14.4.1. Poultry meat production

In 2020, production of poultry meat in India was 3.6 million tons. Production of poultry meat in India increased from 95,088 tons in 1971 to 3.6 million tons in 2020 growing at an average annual rate of 8.00%. The growth in the broiler segment is expected to remain strong due to consumer preference for poultry, increasing income levels, and changing food habits. The live market sales of broiler meat still constitute more than 90-95 percent of total volume of sales; the processed chicken meat segment comprises only about 5% of total production. Broiler production is mainly concentrated in the states of Tamil Nadu, Andhra Pradesh, Maharashtra, Uttar Pradesh, and Telangana.

Indian Broiler Industry experiences the rapid growth driven by increase in per capita consumption. The impressive growth in the poultry sector in general and broiler industry in particular is the result of technological breakthroughs in breeding, feeding and health, and sizeable investments from the private sector. Broiler industry is growing with the backward integration system providing opportunities for the rural masses with all the technical inputs and assured remunerations.

14.4.2. Nutritive value of poultry meat

Poultry meat is a good source of protein, minerals, such as iron, selenium, zinc, and B vitamins. It is also one of the main sources of vitamin B_{12}. It has several advantages as half of the fat from chicken meat is made up of the desirable monounsaturated fats, and only one-third of the less healthy saturated fats. There are much higher proportions of saturated fats in most cuts of red meat, which also vary considerably in total fat. Chicken meat is therefore seen as a healthy meat. It does not contain the trans-fats that contribute to coronary heart disease. Poultry meat is rich in the omega-3 fats and is an important provider of the essential polyunsaturated fatty acids (PUFAs), especially the omega (n)-3 fatty acids. Scavenging chickens are a particularly good source because of their varied diet. The amounts of these important fatty acids can be increased more easily in chicken meat than in other livestock meats; so too can some trace minerals and vitamins. Poultry meat can be enriched with several of the important dietary nutrients like selenium whose deficiency is becoming more widespread in humans because soils are becoming depleted and the foods grown on them are therefore lower in selenium.

14.5. SWOT ANALYSIS

14.5.1. Strengths

>> Low cost protein in the country.

>> Good growth rate- CAGR is around 5% for eggs and 7% for poultry; The CAGR of GVA for last 5 years for egg and poultry is nearly 13% and 15% respectively. Rising economy & growing emphasis on poultry products.

>> Livestock contributes nearly 12% to rural household monthly income; poultry alone can contribute nearly half of the same coping up with captive production of soya bean and maize.

>> Consolidation of integrated operations would strengthen poultry supply chain

14.5.2. Weaknesses

>> Lack of infrastructure facilities for value addition such as poultry processing, warehousing, cold storage, refrigerated vehicles.

>> High maize and soya price fluctuation leading to availability issues of poultry feed at reasonable prices.

>> Small farms, losing out on economies of scale and biosecurity.

>> Lack or undefined standards leading to impending cheaper imports.

14.5.3. Opportunities

>> 95% raw/ wet market – can transform.

>> Work on developing alternate breeds and LIT birds for upgraded family poultry.

>> Untapped potential for the export and value-added chicken products.

14.5.4. Threats

>> Avian influenza and other emerging/re-emerging diseases.

>> Calamity.

14.6. GROWTH DRIVERS AND EMERGING TRENDS FOR EGGS AND POULTRY

>> In India, poultry sector growth may be attributed to many factors like rising incomes and a rapidly expanding middle class, together with the emergence of vertically integrated poultry producers that have reduced consumer prices by lowering production and marketing costs.

>> Integrated production, market transition from live birds to chilled and frozen products, and

policies that ensure supplies of competitively priced corn and soybean are keys to future poultry industry growth in India. Further, disease surveillance, monitoring and control will also decide the fate of this sector.

» Concurrently, India's unorganized and backyard poultry sector is also one of the potent tool for subsidiary income generation for many landless/ marginal farmers and also provides nutritional security to the rural poor.

» These achievements and growth rates are still being sustained despite the ingress of avian influenza which was a severe setback for the industry, showing the resilience of the subsector, perseverance of the private sector and timely intervention by the Government.

» To assess the future trends, we have to review the past planning and present scenario to extrapolate the future. The externalities and variables are often unprecedented and sudden. Both empirical and statistical methods need to be accounted for while making any predictive assumptions. The 'Livestock Revolution 2020' study by Delgado *et al.* (2001) gained such acceptability and popularity because variables and factors like economic crises and increase in consumption of meat products *etc.* were integrated in the study.

14.7. TECHNOLOGIES TO ENHANCE POULTRY PRODUCTION

14.7.1. Breeding for low-input technology variety of chickens

» One of the most effective ways of improving heat tolerance / temperature modulation is through the incorporation of single genes that reduce or modify feathering, such as those for naked neck (Na), frizzle (F) and scale less (Sc), as well as the autosomal and sex-linked dwarfism genes, which reduce body size.

» One of the major breeding strategies in India is based on crossing Aseel breed males with CARI Red hens to produce crossbred CARI Nirbheek hens. Kadaknath is an indigenous breed whose flesh is black is considered not only a delicacy but also of medicinal value and their crosses like CARI-Shyama are also popular. Some of the stocks developed for the purpose are Chabro, Kalinga Brown, Kaveri, Vanaraja, Gramapriya, CARI-Gold, Hitcari, Upcari, Cari-Debendra, Giriraja, Girirani, Krishipriya, Swarnadhara, Nandanam 99 and Rajasri.

» New varieties have also come up like Srinidhi, Jharsim, Kamrupa and Pratapdhan. A few private sector players like Keggfarms, Khandsa, Gurgaon; New Dr. Yashwant Agritech Pvt. Ltd., Jalgaon; Indbro Research and Breeding Pvt. Ltd., Hyderabad; Shipra Hatcheries, Patna; are also producing stocks like Kuroiler, Satpuda-desi, Rainbow Rooster and Shipra in this segment.

14.7.1.1. High yielding breeding stocks: India has the technical know-how in maintaining the high-yielding stocks and some companies have entered into franchisee agreement with pure line/ Grandparent stock providers. Commercial birds with laying capacity of around 320+ in case of layers and broilers with FCR 1.6 are now common in this segment. Some of the stocks available are as follows:

i) Layers

» Pure lines: Babcock, CARI Gold layer
» Grandparents: Bovans, Hyline, Lohmann

Most of the GP stocks are imported e.g. Bovans (Netherlands), Hyline (USA/ Germany) etc. Major market share is of Babcock in layer segment.

ii) Broilers

» Pure lines: Cobb, Hubchicks, CARIbro, Indbro, Marshall

» Grandparents: Ross, Hybro, Hubbard, Lohmann broiler

Most of the GP stocks are imported e.g. Ross (UK), Hybro, Hubbard, *etc.* Major market share is of Cobb in broiler segment.

14.7.1.2. Cutting-edge technology in poultry breeding

» Biotechnological and immunological tools have to be adopted in combination with breeding methods to develop robust stocks having higher production level like QTLs through genome wide scan.

» Microarray analysis for elucidating biological pathways.

» Identifying the genes involved in particular biological processes.

» Mining of allele for identifying useful alleles affecting phenotype.

» The genetic modifications like transgenesis, knocking down a gene and RNAi, use of CpG motifs, proteomics, nanotechnology, epigenetics, aptamers, *in-vivo* approaches and even CRISPR gene editing technology holds immense potential.

» Breeds/ strains having high immune competence will be another priority area for research due to adaptability of future stocks to changing farming systems and climate.

» For smallholder systems, creep-upgrading or nucleus crossbreeding, community-based breeding programs and strategies to generate sustained replacement stocks in systems where crossbreds are the best option may be explored further.

14.7.2. Poultry nutrition

14.7.2.1. Feed & feed supplements

i. **Feed resources:** Feed accounts for 65- 70% of broiler and 75-80% of layer production cost. Maize is the popular cereal used in combination with protein meal like soybean meal which generally determines the cost of compounded feed. Production of maize increased from 9.65 million tons in 1989-90 to only 24.4 million tons in 2015. Similarly, soybean meal production increased to 11.35 million ton in 2015 from 3.52 million tons in 1999-2000. Average increase in maize availability has been 3.8% per annum which is far below the growth rate of egg or meat production. Thus, there is a need to increase the production of maize and soybean or explore the usefulness of other alternate energy and protein rich feedstuffs to maize and soybean meal, respectively, in poultry diets.

In view of the large gap between the demand and availability of feedstuffs for poultry production, a holistic approach is needed to meet the demand of ever-growing poultry industry. Some of the approaches in these respects are

ii. **Identification of newer feed resources:** The escalation in the cost of feed ingredients and consequently the cost of eggs and meat situation can be corrected by developing strains that need less feed input. However, alternate feed ingredients that are not related to human consumption and available in plenty should be identified and their suitability should be tested including the economic aspects.

iii. **Utilization of structural carbohydrates and phytate phosphorus:** The reduction in dependency of poultry on the storage plant carbohydrate, protein or other nutrient and to allow them to make greater use of structural carbohydrates and other nutrients. Hence the dimension from research should change from as such providing feed than technologies that utilize feed better. There are many components of feed such as ß-glycans, pentosans, mannans, cellulose, lignin and phytic acid which cannot be digested by poultry normally. These non-digestible feed ingredients frequently generate digestive stress in poultry with a consequent reduction in nutrient utilization and wet litter problems. These problems could be largely alleviated by use of feed enzymes.

» Feed represents the major cost of poultry production, constituting up to 70 percent of the total. Of total feed cost, about 95 percent is used to meet energy and protein requirements, about 3 to 4 percent for major mineral, trace mineral and vitamin requirements, and 1 to 2 percent for various feed additives.

» The predominant feed grain used in poultry feeds worldwide is maize. The plant protein source traditionally used for feed manufacture is soybean meal, which is the preferred source for poultry feed. Feed supplements like probiotics, vitamins, minerals, amino acids, mold inhibitors, enzymes, preservatives, coccidiostats, antioxidants *etc.* are mostly imported.

» The total feed requirement of organized poultry sector is nearly 23 million tons and nearly the whole of it is in compounded form.

14.7.2.2. Cutting-edge technology in poultry nutrition

» Single-cell protein products such as algae, bacteria, and yeasts are now showing promise to meet the demand. Technology has made it feasible to produce transgenic feeds with high protein and amino acid contents (quality protein maize with high lysine and tryptophan), low anti-nutritional factors (Canola meals with low erucic acid, tannins, and glycosinolates) and with high vitamin activity (yellow sorghum with high beta-carotene activity), *etc.*

» The synbiotics (probiotics and prebiotics) have been considered as suitable substitute for antibiotics, which are slowly being phased out, especially the gut-acting ones.

» Micro-organisms have been selected and optimized by classical biotechnological methods to produce amino acids in fermentation process to produce the limiting amino acids in particular in large quantities for the feed industry.

» Production of trace mineral proteinates (organic minerals) utilizing yeast (*Saccharomyces cerevisae*) has become feasible in augmenting availability of various trace minerals including zinc, manganese, chromium, selenium, copper, *etc.* Using Dried Distillers Grain Solubles (DDGS) which is left over after corn is turned into ethanol and other such alternatives can help alleviate stagnant growth of maize.

» There exists enormous potential in India for investing in research and development (R&D) to enhance maize productivity, including increased use of high-yielding, production risk-reducing improved maize varieties/hybrids.

» Currently, the limiting amino acids in poultry diets – methionine, lysine, threonine and tryptophan – are added to poultry feed. However, development and distribution of bio-fortified maize – Quality Protein Maize (QPM) and High Methionine Maize (HMM), containing enhanced levels of limiting amino acids – offers some prospects for the poultry industry by reducing the requirement for synthetic amino acid supplements (Krishna *et al.*, 2014; Lopez-Pereira, 1993; Prasanna *et al.*, 2001).

14.7.3. Poultry health

14.7.3.1. Disease management: Management of diseases in poultry plays an important role for the progress of the industry. Birds in the commercial farms are reared in open-sided houses and maintained under optimum management conditions. Birds are reared under veterinary supervision. Vaccination is regularly practiced to protect the bird against diseases.

To minimize the occurrence of disease in poultry, the three most important components of disease control are biosecurity, vaccination, and medication. Biosecurity refers to all measures taken to secure prevention of all types of pathogens in poultry farms. Effective bio-security and implementation of successful hygienic procedures are increasingly dependent on Hazard Analysis Critical Control Point (HACCP) approach. The principles of HACCP such as hazard analysis, critical control points, critical limits, correction, recording, and verification should be strictly followed for analyzing risk assessment and risk management. Vaccination should be practiced regularly following the regulatory procedures.

Priorities for effective disease management in making the poultry industry a sustainable enterprise include:

- » Trans-boundary disease – Many of the diseases which are not endemic to India (Avian Influenza, VVND) may enter through germplasm and biologicals. This needs strict quarantine measures.

- » Establishing and strengthening surveillance and monitoring system – The surveillance and monitoring system should be carried out in established laboratories. There is need to establish a National Avian Disease Laboratory with all modern facilities for surveillance and monitoring .of infectious diseases in poultry.

- » Diagnoses through genomic approach – Efforts may be made to develop new diagnostics and biologicals using genomic approaches for rapid and accurate diagnosis and effective control of poultry diseases.

14.7.3.2. Veterinary infrastructure

- » World class laboratories providing disease diagnostic service will play a crucial role in assessing epidemiological profile mapping.

- » Laboratories engaged in quality control and monitoring of vaccines and biologicals.

- » There are a number of ancillary as well as specialized Veterinary Organizations and Institutes, which by employing qualified veterinarians undertake works related to vaccination, treatment, and welfare of animals.

- » Animal quarantine and certification service's objective is to prevent ingress of livestock diseases into India by regulating import of livestock and livestock related products.

14.7.3.3. Cutting-edge technology in poultry health

- » Epidemiology, economics, and impact assessment: Studying the evolution of pathogenic infectious agents with varying infectivity, virulence, transmissibility and adaptations over time to re-emerge; and analysis of social factors responsible for transmission of pathogens, studying genetic resistance factors.

- » Technology development and improvisation: For example, development of tools for diagnosis, management, control and prophylaxis of diseases; training, infrastructure and information sharing for responding to emerging diseases; combating outbreaks of avian influenza and strengthening Sanitary and Phyto-Sanitary measures to deal with exotic agents; development of effective and convenient biosecurity; and establishment of Compartments / Disease-Free Zones *etc.*

» Innovation tools that consider effective use and application of new technologies: For example, participatory epidemiological tools, GIS techniques *etc.* to help effective need-based input and service delivery.

14.7.4. Poultry processing and value addition

14.7.4.1. Processing: Live and fresh dressed broilers account for the bulk of sales and sale of processed meat is limited (below 10%). However, acceptance of processed chicken is on the rise, particularly in the urban markets. Thus, the sale of slaughtered chicken is expected to increase. Hence, there is a need to develop processing facilities. Hence, there is an urgent need of many chicken processing plants in the near future and sale of processed chicken to increase both to cater domestic as well as export markets.

A few plants for processing eggs have been installed using state of the art machinery in some states with an average daily turnover capacity of 0.7 - 0.8 million eggs. Whole egg powder, yolk powder, egg weigh powder, lysozyme *etc.* are being produced under high standards of operation. Egg powder from India is well accepted in EU, Japan, and Far-east. However, to tap the international market, there is a need to establish many more egg processing plants. It has been told that India is geographically ideally located to cater to the Middle East and far eastern countries for shell eggs. Therefore vast scope exists to increase the export of shell eggs from India to these countries.

14.7.4.2. Value addition: Value addition in poultry plays in important role in increasing the profits. The value addition may be through nutritional manipulations, processing and transgenesis. Omega-3 enriched eggs and meats are available in the market for premium price developed by nutritional approaches. Feeding the chicks with rich sources of omega-3 fatty acids will aid in increasing the levels of omega-3 fatty acids in eggs and meat of the birds. Experiments on fortification of Zinc and Vitamin B_{12} in chicken egg and meat through dietary manipulation for enhanced value addition and shelf life are going on and need commercialization. The second one is through biotechnological approaches, where in the gene (inter species) responsible for specific trait can be made through transgenesis. However, this approach is still in primitive stage wherein research is being carried out. The commonly utilized method for value addition is processing of the poultry products. By value addition low valued meats and by products can be processed into a highly nutritious finished products adding to the returns.

14.7.4.3. State of the art post-harvest technologies

» Egg processing sector is still in infancy stage in India in spite of commendable production. Installation of about half-a-dozen egg processing units, rapid urbanization and industrialization and proliferating fast food parlours, *etc.* over the last decade have given some impetus to the growth of egg processing sector.

» The country has, thus, begun exporting table eggs, egg powder and frozen egg products on a limited scale in recent years.

» At present, hardly 5% of eggs produced are processed into dehydrated/frozen products, primarily for export purpose or used in bakeries and other food and non-food industries.

» Low-cost processing technologies have been developed for both cottage and large industries.

» India is leading in the wet market share compared to other Asian markets. Live broilers are more than 95% of total consumer sales. Small birds 1.8-2.0 kg dressed weights are the norm. Skinless raw poultry products are preferred by many buyers.

» There is huge preference for freshly slaughtered chicken which is slaughtered in local meat shops or municipal slaughter houses. The reasons behind this preference may be many. Indian consumer is price conscious.

» Industry must come forward to create awareness about processed products. This will not only help in the improvement of production lines but will also promote consumption of healthy, safe, and hygienic meat products among Indian population.

» Chilled poultry is said to be gaining more rapidly than frozen, but both are a very small share of the total market.

14.7.5. Infrastructure and poultry equipment

14.7.5.1. Modern poultry equipment in India

» The demand for modern poultry equipment in India is fuelled by an increase in the size of the poultry farms. Previously poultry farms had production of only a few hundred birds (200-500 chickens) per cycle. However presently, poultry units with fewer than 5,000 birds are an exception with the majority of the farms breeding more than 500,000 birds. Similar is the case with layer poultry farms.

» Further, with increasing demand for quality poultry products, the adoption of better machinery to ensure quality has become very important to the Indian poultry units. Barring a few items like egg graders, poultry processing equipment *etc.* most of the equipment are available/ produced in the country. India is almost self-sufficient in indigenous production of most of the basic equipment like hatchers/ incubators, feeders, poultry houses, even environmentally controlled and pre-fabricated houses *etc.*

14.7.5.2. Some modern farm house innovations

» Automated public access control system with automatic showers, concrete flooring between houses to reduce vegetation, pad cooling with easy cleaning and disinfecting even when birds are present.

» Chain-feeder technology promotes efficient feed distribution by accurately measuring feed and providing uniform nutrition for every bird.

» Fluid LED light level control, flicker free lighting system, with multiple light level settings.

» Air quality monitor is designed to sample the air within the building every two minutes, and display the following air quality information CO_2/ ammonia/ humidity/ temperature.

» Water system designs to keep water uncontaminated by preventing dirt, faeces, and other pollutants from entering the automatic drinking system.

» Innovative waste management methods: Manure belt systems in egg production.

» Palletisation of dried manure further stabilizes the material, reducing dust. Some countries are using Black soldier fly (BSF) larvae are an alternative system for manure treatment.

» Remote access livestock monitoring: Our livestock monitoring system allows poultry farmers the ability to view their broiler sheds internally from their smartphones, tablets and personal computers, in great detail they can view feed and drinker lines, hoppers, bird spread, all without the need to enter the houses as regularly as they normally would.

14.8. IMPACT OF CLIMATE CHANGE

Poultry production accounts for 8 percent of global anthropogenic greenhouse gases (GHG) emissions

(Steinfeld *et al.*, 2006). Poultry (chicken) are more vulnerable to climate change because birds can only tolerate narrow temperature ranges. Poultry are not well adapted to high ambient temperatures because they lack sweat glands. The internal body temperature of chickens (41-42°C) is higher than that of mammalian livestock and humans (36-39°C). Poultry have considerably less threshold to heat stress as compared with other animals.

Birds subjected to heat stress conditions spend less time for feeding, more time for drinking and panting, as well as more time with their wings elevated, less time moving or walking, and more time resting. Increased temperature had significant adverse effects on both broiler and layer production. Heat stress impairs overall poultry meat and egg production by modifying the bird's neuro-endocrine profile both by decreased feed intake and by activation of the HPA axis. This leads to reduced feed intake, excessive panting to maintain thermo regulation and diverting more energy towards homeostasis instead of growth and production. The chronic heat exposure negatively affects fat deposition and meat quality in broilers, in a breed-dependent manner.

Maintaining the in-house temperatures is very important for sustaining the productivity from the birds. In extreme summers, heat management through spreading the paddy/wheat straw on roofs and sprinkling water on the roofs maintains the temperatures. Effective cross ventilation, use of coolers and foggers is also recommended. In winters, generally gunny bags are used to cover the sides especially in high raised poultry houses.

Nutrition management can also allow improvement to feed conversion ratios through optimal diet balancing and feeding regimes, and improvement to feed digestibility. Feeding the antioxidants like vitamin E, plant extracts, and trace minerals like selenium, chromium, zinc *etc.* reduces the stress condition and improves the heat tolerance in birds. Many researchers formulating feeds that closely match the nutritional requirements of birds in their different production and growth stages to reduce the amount of nutrients excreted.

14.9. FOOD SAFETY

There is a worldwide concern to minimize the use of antibiotics in poultry because of disease resistance and antibiotics residues in food chain. In such case, suitable alternatives need to be explored, which could be beneficial and cost effective. Many products of such nature like probiotics, gut acidifiers, immunomodulators, eubiotics, organic acids *etc.* are available in the market, but need further research. Ensuring safe food is paramount for the protection of human health and for enhancement of the quality of life.

Over the last decades, the food chain approach has been recognized as an important step forward to ensure food safety from production up to consumption. This approach requires the commitment of all players in the food chain, involving producers, traders, processors, distributors, competent authorities as well as consumers.

14.10. MARKETING SYSTEM

There is a need to develop and strengthen the reliable and stable market chain round the year for marketing of poultry products. Some of the approaches in this direction include:

»   Development of reliable and stable market chain round the year for marketing of poultry products.
»   Facilities for hygienic slaughter and preservation of eggs should be made available at market places in both urban and rural areas.

» Formation of producer co-operatives/ associations and rural market yards will help in proper marketing.

» National Egg Coordination Committee, a farmers' cooperative agency has been contributing to the improvement in marketing of eggs. However, more systematized marketing strategy and the state's involvement in minimizing the channels are required for making poultry farming remunerative and cost-effective in the years to come.

» Because of the location of farms in urban and peri -urban areas that too concentrated in few states, availability of eggs and chicken meat are high in these areas only, but in rural areas and rest of the country the availability is low. Thus, there is a vast scope to tap the rural markets and remote areas of the country where availability is low.

14.11. EXPORTS

The country has exported 2,55,686.92 MT of poultry products to the world worth Rs. 435.53 crores during the year 2020-21. Major export destinations during 2020-21 include Oman, Maldives, Indonesia, Vietnam Socialistic Republic, and Russia. At present mainly table eggs (UAE, Kuwait and Oman), hatching eggs (UAE, Oman and Kuwait) and egg powder (Japan, Poland, Belgium and UAE) are exported from India. Our major markets Middle East and Asia. Egg powder is exported to Japan and EU. India has infrastructure to export eggs including all primary packaging mechanism and cold chain to deliver top quality produce to customers.

14.11.1. Export of egg and poultry / products

» Major items exported are table eggs, egg powder, hatching eggs, SPF eggs, live birds, and poultry meat. The current export value of poultry products is to the tune of around Rs. 435.53 crores in 2020-21. The strength of exports mainly lies in the competitive cost of production, proximity to international markets and successful regaining of freedom from Highly Pathogenic Avian Influenza (HPAI).

» Although some efforts have been made to increase poultry exports from India, the trade is very small in comparison to the global trade. At present the exports are mainly in table eggs, hatching eggs, frozen eggs, egg powder, and to a small extent for live poultry.

» Our major markets are Middle East and Asia. Egg powder is also sent to Japan and in EU. Now we have extended exports to many African countries. We have infrastructure to handle egg exports and we also have primary packing materials and the full logistics cold chain to deliver top quality fresh eggs to all our customers.

Interestingly, it is noted that though the volume of exports was commensurate with the value of exports, it is desirable to have low-volume high-value products to have more profitability. We were exporting more raw material than processed product, which explains that either we have less processing capacity or value-addition was negligible.

14.11.2. Scope with value added products

» Whole egg powder, brined and pickled eggs, egg roll, egg strips, egg soufflé, egg cutlet, egg crepe and waffles, albumen flakes/ rings, yolk powder, natural yellow pigment from yolk, lecithin, conalbumin and avidin from eggs used in pharmaceutical industry.

» Lysozyme, di-calcium phosphate from shell and shell membranes, cured and smoked chicken, chicken patties, intermediate moisture diced products with long shelf-life, battered and breaded enrobed products, chicken chunkalona, chicken soup, chicken essence, nuggets, kababs, meat spreads, marinated breast fillet, hotdogs, frankfurters *etc.*

»   Giblets, liver, and liver extract, deboned meat for airline industry, chicken gizzard pickle, Feather meal, poultry by-product meal from inedible portions as a source for poultry feed *etc.*

»   High demand for various forms of egg powder and hatching eggs including Specific Pathogen-Free (SPF) eggs.

The exports are not equitable across the globe and are concentrated in certain clusters like Middle-East, South-East and immediate neighbors but as stated earlier, the expansion is taking place across newer territories like Africa.

14.12. CHALLENGES

In spite of rapid growth, the poultry industry suffered many setbacks in recent times due to rising cost of feed, emergence of new or re-emerging of existing diseases, fluctuating market price of egg and broilers, *etc.* which need to be addressed to make the poultry sector as a sustainable enterprise. Issues relating to animal welfare and environmental pollution by poultry units have been of increasing concern. Some of the challenges facing poultry industry include:

»   A major constraint affecting the growth of the poultry industry in India is the lack of basic infrastructure such as storage and transportation, including cold chain. As a result, there are wide price fluctuations in the prices of poultry products, i.e., eggs and broilers.

»   An inefficient marketing system- The presence of so many market intermediaries' harms both the producer and the consumer.

»   The price and availability of feed resources- Maize or corn plays a major role in broiler production, as it constitutes 50 to 55 percent of broiler feed. As the broiler industry is growing at the rate of 8-10 percent per annum, the demand for maize and soya is thus likely to increase.

»   Emerging and re-emerging diseases of poultry-Mutations in viral genomes leading to new variants in viruses and developing resistance to vaccines and antibiotics. Avian Influenza outbreaks occurring in parts of India, is a very good example.

The policy measures that are required to improve the poultry industry must involve:

»   Improving infrastructure facilities, which will help not only to stabilize the price of poultry products in the domestic market, but will also make them available in remote areas.

»   Creating an efficient marketing channel that will help provide remunerative prices to producers (in other words, India's marketing set-up should also grow along professional lines).

14.13. CONCLUSION

The poultry production in India continues to exhibit spectacular growth in spite of several challenges encountered over the years. With increasing demand for chicken egg and meat, the poultry production in India foresees further expansion and industrialization. Adoption of small-scale poultry farming in backyards of rural households will enhance the nutritional and economic status of the rural people. With the advent of knowledge and new discoveries in different fields of poultry, the future challenges will not be a hindrance and thus sees a bright future for poultry production in this country.

India has almost doubled its meat consumption during the past decade spurred by domestic economic growth and consumption dynamics. Still, the average Indian only consumes about 4.5 kg

(10 pounds) of meat per year, reflecting the country's low-income status and preference for non-animal protein sources. Poultry occupies a crucial place in India and chicken is the most widely accepted meat in India – helped by religious taboos around beef and pork. Many Indian families in urban areas have begun to accept eggs as a regular supplementary part of their vegetarian diet. The domestic demand for poultry meat and eggs in India is expected to continue to grow at a brisk pace.

A key factor under-pinning India's poultry industry is the availability of animal feed, particularly maize. Maize production in India is expanding and changing rapidly in response to the growth in the poultry industry. Meanwhile, maize value chains are growing more sophisticated and their changing structure provides investment opportunities for public and private sector actors. Maize thereby helps drive India's agricultural and economic growth especially through its role as feed for the flourishing poultry industry. An appropriate institutional and policy environment should enable India's poultry revolution to continue into the future – with due attention for feed market development and the environment.

India's poultry revolution has already made its mark on global poultry production and trade. The overwhelming majority of the demand in India was and will continue to be met by domestic production – whereby traditional poultry exporters such as the United States and Brazil have largely missed out on India's burgeoning poultry market. Furthermore, India is likely to become a more important player on the export market especially in the Middle East and thereby presents an emerging competitor in the global poultry trade arena.

Chapter - 15

Red (Ruminant Livestock Meat Production) Revolution

15.1. INTRODUCTION

Livestock production is a vital activity in rural areas, which provide employment, livelihood and income for farmers, rural poorer and weaker sections of society. Meat sector plays an important role in India as it not only provides meat and by-products for human consumption but also contributes towards sustainable livestock development and livelihood security for millions of men and women from weaker sections. Among agriculture produce, meat occupies a significant place as about 80% of Indian population is non-vegetarians. Although, India is bestowed with huge livestock resources with enormous potential but Indian meat industry has not attained its due status because of obvious reasons. Major portion of meat from sheep, goat, pig and poultry is primarily used for domestic consumption in the form of hot meat. Certain portion of meat from buffaloes, cattle and sheep is exported.

India has the largest livestock population throughout the country. It has 512.0 million of animal population excluding poultry. India is the largest exporter of buffalo meat and third largest exporter of meat after Brazil and Australia. It accounts for about 58% of the world buffalo population and 14.7% of cattle population. There are about 300 million bovines, 65.07million sheep, 135.2 million goats and about 10.3 million pigs and 729.2 million poultry in the country as per 19th Livestock Census.

Meat and meat products are crucial part of the man's diet. Meat is highly demanded food items of human being due to presence of plentiful proteins, minerals and all the B-complex vitamins with excellent digestibility and well-balanced composition of essential amino acids. Population growth, urbanization, changed life styles and increased per capita income are fuelling a massive increase in demand for food of animal origin all around the world. As per World Bank projection, worldwide demand for food will increase by 50% and for meat by 85% by 2030. Governments and industries must prepare for meeting demand of meat in the country with long run policies and investments to satisfy ever rising consumer demand, improve nutritional status, generation of income opportunities and alleviate environment stress. Government of India has already recognized livestock and poultry as an important sector for the socio-economic development of the country.

15.2. LIVESTOCK RESOURCES

The livestock resources in India include (Table 15.1 and Fig 15.1):

- » World's highest livestock owner at about 535.78 million
- » First in the total buffalo population in the world - 109.85 million buffaloes
- » Second in the population of goats - 148.88 million goats
- » Second largest poultry market in the world
- » Third in the population of sheep (74.26 millions)
- » Fifth in in the population of ducks and chicken (851.81 million)
- » Tenth in camel population in the world - 2.5 lakhs

Table 15.1. Livestock population (2019 Livestock census)

Species	Number (in millions)	Ranking in the world population
Cattle	192.49	Second
Buffaloes	109.85	First
Total (including Mithun and Yak)	302.79	First
Sheep	74.26	Third
Goats	148.88	Second
Pigs	9.06	---
Others	0.91	---
Total livestock	535.78	---
Total poultry	851.81	Seventh
Duck	-	Fifth
Chicken	-	
Camel	0.25	Tenth

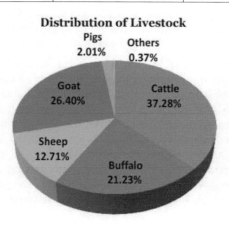

Fig. 15.1. Species-wise distribution of livestock in 2018-19.

15.3. CONSTRAINTS

There are many constraints for the slow growth of the Indian meat industry, including lack of scientific approach to rearing of meat animals, unorganized nature of meat production and marketing, socio-economic taboo and inadequate infrastructure facilities, and poor post-harvest management.

There are many reasons for slow growth rate of meat industry in India:

» Myths about meat consumption and half-truths in the media against meat consumption.
» Insanitary and unhygienic conditions of the slaughterhouses and the meat shops.
» Lower carcass weight and dressing percentage of food animals due to the slaughter of spent / aged animals.
» Indian consumer prefers fresh meat, resulting in less demand for frozen meat.
» Diseases such as Foot and Mouth Disease (FMD) are a major concern.
» Subsidies in developed countries, stipulation of Sanitary and Phyto Sanitary (SPS) measures, and increasing cost of production and inputs as compared to competing nations.
» Non-availability of good quality livestock in the open market.
» Overcrowding of food animals during transport resulting into inferior meat quality.
» The lack of a sufficient cold chain infrastructure.

15.4. CONTRIBUTION OF LIVESTOCK

The livestock provides food and non-food items to the people.

15.4.1. Food items

The livestock provides food items such as meat for human consumption. India is number one milk producer in the world. It is producing about 7.70 million tons of meat in a year. The value of output from livestock sector at current prices was Rs 9,17,910 crores at current prices during 2016-17 which is about 31.25% of the value of output from total agricultural and allied sectors.

15.4.2. Income

Livestock is a source of subsidiary income for many families in India especially the resource poor who maintain few heads of animals. Animals like sheep and goat serve as sources of income during emergencies to meet exigencies like marriages, treatment of sick persons, children education, repair of houses *etc.* The animals also serve as moving banks and assets which provide economic security to the owners.

15.4.3. Employment

A large number of people in India being less literate and unskilled, depend upon agriculture for their livelihoods. But agriculture being seasonal in nature, could provide employment for a maximum of 180 days in a year. The landless and marginal farmers depend upon livestock for utilizing their labor during lean agricultural season.

15.5. CONSUMPTION OF RUMINANT LIVESTOCK PRODUCTS AROUND THE WORLD

Products produced from ruminant livestock are an important source of energy, high-quality protein and micro-nutrients, including vitamins like A, B_{12}, and B_2 and minerals like calcium, iron and zinc. Obtaining sufficient quantities of these nutrients from plant-based foods uniquely is challenging (Mottet *et al.*, 2017; FAO, 2018). Seventeen per cent of calories and 33% of protein consumed worldwide comes from animal sources (FAO, 2018).

Low intakes of animal products are associated with malnutrition leading to serious consequences globally (Adesogan *et al.*, 2019; FAO, 2018). Health problems include anaemia and risks to pregnancy from lack of B_{12}. Even if calorie requirements are met, insufficient consumption of animal-sourced nutrition in the form of meat or milk by pregnant and lactating women, babies and young children can result in stunting. Children from households where women own livestock have better nutritional results than households without livestock (Adesogan *et al.*, 2019).

As is apparent from Fig. 15.2, the importance of ruminant meat and dairy as a protein source varies by region. Populations in Sub-Saharan Africa and India get the majority of their daily protein from crops, while in other parts of the world, animal proteins are the largest source of daily protein intakes. In Latin America, both ruminant meat and dairy are important sources of protein, whereas in Europe dairy is proportionally more important. In India, dairy proteins are important too.

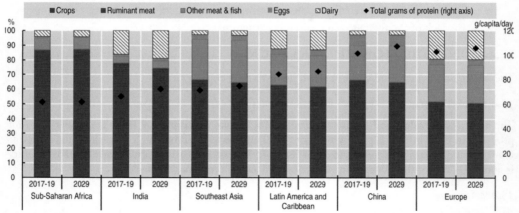

Fig. 15.2. Contribution of ruminant meat and other protein sources to total daily per capita availability [Note: Bars refer to the share of the food group in total daily per capita protein intake (left axis); Dots represent the total quantity daily per capita protein intake (right axis); Crops include arable food crops (cereals, edible oilseeds, pulses, roots and tubers, sugar)] *Source:* OECD/FAO (2020).

Consumption of (all types of) meat in Africa, Asia, and the Pacific is low (Fig. 15.3). In terms of beef consumption, developed countries currently consume approximately three times as much as developing countries. While growing in developing countries, in particular Asia, the rate of beef consumption is on the decline in many OECD countries albeit from high consumption levels. This is particularly the case in the European and Central Asian region (OECD/FAO, 2020). For example, per capita consumption of beef in New Zealand decreased from 23.3 kg per capita in 1992-94 to 15.7 kg in 2012-14 and is predicted to decrease by 11.3 kg in 2028-30. Beef consumption in Canada shows a similar decline, from 23.3 kg in 1992-94 to 19.7 kg in 2012-14, and it is predicted to decline to 16.8 kg per capita by 2028-30.

15.6. RED REVOLUTION

Red revolution is a term used to denote the technological revolutions in the meat and poultry processing sector. In India, the current meat utilization per capita is around 6 grams per day and will enhance to 50 grams a day in the next decade. India's bovine meat consists only of buffalo meat (carabeef) due to cultural reasons. With 58% of the world's buffalo population, India is home to the world's largest population of cattle and buffaloes. The bovine meat industry plays a significant part in employment generation in the agricultural sector. About 10% of the rural labour force is employed in livestock rearing and related occupations, which constitutes around 26% of the total agricultural value added. According to Meat and Livestock Australia, a meat industry research company, the cost of production of bovine meat in India is much lower compared to its competitors like Brazil and Australia. Meat production in India is generally a by-product of livestock rearing. In other competing nations such as Brazil, cattle are reared specifically for the purpose of meat production. This makes their bovine meat industry expensive compared to its Indian counterpart. Given India's geographical location, it can easily cater to markets in the Gulf as well as in the East Asian countries. With rising incomes in the developing world and an expanding youth population, food preferences are shifting towards a protein rich diet. In such a scenario, India stands to gain from the rising demand for meat products. There has been no report of adulteration in the Indian bovine meat industry. Moreover, restrictions had been imposed on imports from EU nations on account of Bovine Spongiform Encephalopathy (BSE) or what is popularly called mad cow disease.

15.6.1. Demand for meat and meat products

Meat serves as the principal source of animal origin protein for mankind. In addition meat also contributes macro- and micro-nutrients required for the growth and health. The rate of increase in per capita consumption of animal products was found to be higher in the developing countries compared to the developed countries. The meat production has registered a healthy growth from 2.3 million tons at the end of Tenth Five Year Plan (2006- 07) to 5.5 million tons at the end of the Eleventh Five Year Plan (2011-12). Meat production in the beginning of Twelfth Plan (2012-13) was 5.9 million tons which has been further increased to 7.0 million tons in 2015-16. The per capita animal protein consumption in India is 10.4 g per day compared to world average of 25 g. Minimum requirement of 20 g per capita /day for animal protein can be fulfilled from milk 10 g, meat 4 g, fish 4 g, and 2 g eggs. As per Indian Council of Medical Research (ICMR) recommendation, minimum amount of 30 g of meat /day/head should be taken which makes 10.95 kg of meat/head/annum. At present availability of meat is only 5.5 kg/head/annum. From the above figures, it is clear that there is a wide gap between demand and availability. During 1993-2020, the average growth rate (weighted) for the total domestic demand for mutton and goat meat was 13.7%, for beef and buffalo meat 3.5%, and for chicken 4.8% (Dastagiri, 2004) indicate bright prospects for meat industry in India.

In India, meat production and consumption has increased remarkably in recent years. Meat eating habits of the consumers are slowly changing from fresh to frozen and to processed meat and meat products. Demand for quality meat and meat products are increasing due to growing awareness about nutritional and sensory characteristics of such meat products. Changing socio-economic status also contributed towards enhancing consumption of processed and convenience meat products. Though, the meat industry in the country is yet to transform into an organized sector, still its contribution to the national GDP at present is significant.

15.6.2. The factors favoring meat sector development

» Low cost of production of meat type of animals to a desired age of 2 years.

» Leanness of Indian meat: Contains less fat and the present international trend is favorable for low fat meat. Average fat content of Indian beef is around 4% compared to 15-20% in most of the developed countries.

» Green fodder feeding, absence of animal protein (carcass meal; meat and bone meal) in the ration are favorable factors for Indian meat industry.

» Price structure of various meats in international market. Beef price is the highest followed by pork, mutton, and chicken.

» The absence of hormones, antibiotics and growth promoters' in the feed, the Indian meat is considered not only lean but also clean and organic.

» There is no incidence of Bovine Spongiform Encephalopathy (BSE) in India.

» Close proximity to promising Middle East market.

15.6.3. Need of modern or semi-modern abattoir

Escalating global market opportunities for the Indian meat sector have significantly generated a plenty of private investment in meat processing through state-of-the-art technology of integrated plants. Many corporate firms are coming up and have set up modern integrated plants with state-of-the-art facilities for slaughtering and dressing of animals, carcass deboning, packing, chilled and frozen storages, by-products processing, effluent treatment *etc*. There are already ten established state-of-art mechanized abattoirs-cum-meat processing plants and four integrated poultry meat-processing plants, which are adopting international standard sanitary and phyto-sanitary (SPS) measures. These plants are eco-friendly, where all the slaughterhouse by-products are utilized in the production of carcass meal, meat-cum-bone meal, tallow, bone chips and other value-added products. Several more plants are coming up due to large potential for export of meat. At present, there are 70 meat processing plants and 29 APEDA approved Indian abattoirs-cum-meat processing plants in India.

15.6.4. Indian meat industry perspective

The FAO has outlined four steps that should be taken if India has to achieve a red revolution.

» Setting up state of the art meat processing plants.

» Developing technologies to raise male buffalo calves for meat production.

» Increasing the number of farmers rearing buffalo under contractual farming.

» Establishing disease-free zones for rearing animals.

15.6.5. Government policies to promote Red Revolution

» 100% FDI is allowed in the sector.

» No excise duty or income tax is charged in the meat and poultry sector.

» No restrictions on the export of poultry and poultry products.

» In order to keep a check on the amount of meat wasted, the quality standards, deterioration and contamination of produce; the government has launched a comprehensive scheme for the modernization of slaughterhouses across the country.

» The Agricultural and Processed Food Products Export Development Authority (APEDA), has approved 70 integrated abattoirs, slaughterhouses, and meat processing plants across the country.

» The National Research Centre on Meat based in Hyderabad is established to conduct basic and applied research in the area of meat quality control and regulations and is aimed at improving the quality and safety requirements for both the domestic markets and the export industry.

15.7. MEAT FOR HUMAN NUTRITION

Chemically meat is composed of four major components including water, protein, lipid, carbohydrate and many other minor components such as vitamins, enzymes, pigments and flavor compounds (Lambert *et al.*, 1991). The relative proportions of all these constituents give meat its particular structure, texture, flavor, color, and nutritive value. Meat is a nutritious, protein-rich food which is highly perishable and has a short shelf-life unless preservation methods are used. Animal food is considered as good source of quality nutrients viz. fat, protein, carbohydrates and minerals. Digestibility of animal source protein is 90-97% while vegetable origin proteins have 75-99%. However, proteins of animal origin are more completely digested and nutritionally superior than those of plant origin.

Animal source foods are energy-dense and an excellent source of high-quality and readily digested protein (Layrisse *et al.*, 1990; Bender, 1992). The proteins in these foods are considered the highest quality available, as they contain a full complement of essential amino acids and most resemble the proteins of the human body in their amino acid composition. Animal source foods are also an efficient source of micronutrients. The main micronutrients offered in abundant and bioavailable form by animal source foods are iron, zinc, and vitamin A from meat, and vitamin B_{12}, riboflavin and calcium from milk (Bender, 1992; Murphy *et al.*, 1992). Additional nutrients that are supplied by meat in abundant amounts are riboflavin, taurine, selenium, and the long-chain polyunsaturated fatty acids, pentaenoic and hexaenoic acids, all of which are increasingly being recognized as important for optimal human health. The main advantage of animal source foods, particularly meat, is the high content and bioavailability of micronutrients; that is, there is a high level of absorption and utilization by the body because of the presence of heme protein found only in meat. Meat contains iron and zinc bound to heme protein, which is readily incorporated into blood cells of the body. The presence of heme protein in a meal enhances the absorption of zinc and iron from cereal and other plant sources (Gibson, 1994; Ferguson *et al.*, 1993).

15.8. RESEARCH AND DEVELOPMENT

15.8.1. Breeding

The use of conventional livestock breeding techniques (such as breed substitution, crossbreeding and within-breed selection) have been largely responsible for the increases in yield of livestock products that have been observed over recent decades (Leakey *et al.*, 2009). Cross-breeding, widespread in commercial production, exploits the complementarity of different breeds or strains and makes use of heterosis or hybrid vigor (Simm, 1998). Rates of genetic change have increased in recent decades in most species in developed countries for several reasons, including more efficient statistical methods for estimating the genetic merit of animals, the wider use of technologies such as artificial insemination and more focused selection on objective traits such as milk yield (Simm *et al.*, 2004). This has been achieved through the widespread use of breed substitution, which tends to lead to the predominance of a few highly specialized breeds, within which the genetic selection goals may be narrowly focused.

There are considerable opportunities to increase meat productivity in developing countries.

Breed substitution or crossing can result in rapid improvements in productivity. There is much more potential in the use of crosses of European breeds with local Zebus that are well-adapted to local conditions. In the future, many developed countries will focus livestock breeding on other attributes in addition to production and productivity, such as product quality, increasing animal welfare, disease resistance and reducing environmental impact. The tools of molecular genetics are likely to have considerable impact in the future. Genomic selection should be able to at least double the rate of genetic gain in the dairy industry (Hayes *et al.*, 2009), as it enables selection decisions to be based on genomic breeding values, which can ultimately be calculated from genetic marker information alone, rather than from pedigree and phenotypic information. Trade-offs are likely to become increasingly important, between breeding for increased efficiency of resource use, knock-on impacts on fertility and other traits, and environmental impacts such as methane production.

15.8.2. Nutrition

The nutritional needs of farm animals with respect to energy, protein, minerals and vitamins have long been known, and these have been refined in recent decades. A considerable body of work exists associated with the dynamics of digestion, and feed intake and animal performance can now be predicted in many livestock species with high accuracy. There is a need for information on the robust prediction of animal growth, body composition, feed requirements, the outputs of waste products from the animal and production costs. Advances in genomics, transcriptomics, proteomics and metabolomics will continue to contribute to the field of animal nutrition and predictions relating to growth and development (Dumas *et al.*, 2008).

Poor nutrition is one of the major production constraints in smallholder systems, particularly in developing countries. Research work has been carried out to improve the quality and availability of feed resources, including work on sown forages, forage conservation, the use of multi-purpose trees, fibrous crop residues and strategic supplementation. There are also prospects for using novel feeds from various sources to provide alternative sources of protein and energy, such as plantation crops and various industrial (including ethanol) by-products. Given the prevalence of mixed crop–livestock systems in many parts of the world, closer integration of crops and livestock in such systems can give rise to increased productivity and increased soil fertility (McIntire *et al.*, 1992).

Another key driver that will affect livestock nutrition is the need (or in countries such as the UK, the legal obligation) to mitigate greenhouse gas emissions. Improved feeding practices (such as increased amounts of concentrates or improved pasture quality) can reduce methane emissions per kilogram of feed intake or per kilogram of product, although the magnitude of the latter reduction decreases as production increases. Many specific agents and dietary additives have been proposed to reduce methane emissions, including certain antibiotics, compounds that inhibit methanogenic bacteria, probiotics such as yeast culture and propionate precursors such as fumarate or malate that can reduce methane formation (Smith *et al.*, 2007).

15.8.3. Disease

The last few decades have seen a general reduction in the burden of livestock diseases, as a result of more effective drugs and vaccines and improvements in diagnostic technologies and services (Perry and Sones, 2009). At the same time, new diseases have emerged, such as avian influenza H5N1, which have caused considerable global concern. A critical area where progress needs to be made if disease diagnostics, monitoring and impact assessment are to be made effective and sustainable. In future, the infectious disease threat will remain diverse and dynamic, and combating the emergence of completely unexpected diseases will require detection systems that are flexible and adaptable in the face of change (King *et al.*, 2006b).

Potentially effective control measures already exist for many infectious diseases, and whether these are implemented appropriately could have considerable impacts on future disease trends. Recent years have seen considerable advances in the technology that can be brought to bear against disease, including DNA fingerprinting for surveillance, polymerase chain reaction tests for diagnostics and understanding resistance, genome sequencing and antiviral drugs (Perry and Sones, 2009). There are also options associated with the manipulation of animal genetic resources, such as cross-breeding to introduce genes into breeds that are otherwise well-adapted to the required purposes, and the selection via molecular genetic markers of individuals with high levels of disease resistance or tolerance.

15.8.4. Nanotechnology

The next few decades may well see nanotechnology applied to various areas in animal management. Nanosized, multipurpose sensors are already being developed that can report on the physiological status of animals, and advances can be expected in drug delivery methods using nanotubes and other nanoparticles that can be precisely targeted. Nanoparticles may be able to affect nutrient uptake and induce more efficient utilization of nutrients, for example. One possible approach to animal waste management involves adding nanoparticles to manure to enhance biogas production from anaerobic digesters or to reduce odors (Scott, 2006).

15.8.5. Artificial meat (more correctly, *in vitro* meat)

Meat 'grown in vats' could be made healthier by changing its composition and made much more hygienic than traditional meat, as it would be cultured in sterile conditions. *In vitro* meat could potentially bypass many of the public health issues that are currently associated with livestock-based meat. The development and uptake of *in vitro* meat on a large scale would unquestionably be hugely disruptive to the traditional livestock sector. It would raise critical issues regarding livestock keeping and livelihoods of the resource-poor in many developing countries, for example. On the other hand, massive reductions in livestock numbers could contribute substantially to the reduction of greenhouse gases, although the net effects would depend on the resources needed to produce *in vitro* meat.

15.8.6. Effect of climate change

Climate change may substantially affect the global livestock production systems (Table 15.2) (Thornton *et al.*, 2009), and changes in productivity are inevitable. Increasing climate variability will undoubtedly increase livestock production risks as well as reduce the ability of farmers to manage these risks. At the same time, livestock food chains are major contributors to greenhouse gas emissions, accounting for perhaps 18 per cent of total anthropogenic emissions (Steinfeld *et al.*, 2006). Offering relatively fewer cost-effective options than other sectors such as energy, transport and buildings, agriculture has not yet been a major player in the reduction of greenhouse gas emissions.

Table 15.2. Direct and indirect impacts of climate change on livestock production systems (adapted from Thornton and Gerber, 2010).

Grazing systems	Non-grazing systems
Direct impacts Extreme weather events Drought and floods Productivity losses (physiological stress) owing to temperature increase Water availability	**Direct impacts** Water availability Extreme weather events
Indirect impacts Agro-ecological changes: Fodder quantity and quality Host–pathogen interactions Disease epidemics	Indirect impacts Increased resource price, e.g. feed and energy Increased cost of animal housing, e.g. cooling systems

More extensive adaptation than is currently occurring is needed to reduce vulnerability to future climate change, and adaptation has barriers, limits and costs (IPCC, 2007). There are several options related to livestock, including grazing management and manure management. Global agriculture could offset 5 –14% (with a potential maximum of 20%) of total annual CO_2 emissions for prices ranging from \$20 to 100 per ton CO_2 eq. (Smith *et al.*, 2008). These mitigation options also have adaptive benefits, such as growing agro-forestry species that can sequester carbon, which can also provide high-quality dietary supplements for cattle. Such carbon payments could represent a relatively large amount of potential income for resource-poor livestock keepers in the tropics.

15.9. PROCESSING OF MEAT

Poultry meat is mostly sold by slaughtering the live birds in the presence of the consumers. However, there are a few modern processing plants where poultry is slaughtered; chilled and frozen birds are sold in big cities. The export of poultry meat sector is very poor or insignificant. The reasons are that mortality of birds is high, overhead inventories are abnormal. The greatest disincentive is that exporters find the prices quoted in importing countries are not favorable.

There is very little processing, hardly 1% of the total meat produced in the country and remaining meat sold in fresh or frozen form. Pork and poultry meat are used for production of ham, sausages, patties *etc.*, for the elite market. The meat processors like Venky, Alchemist foods, Darshan foods, Government Bacon Factories *etc.*, produce these products. Meat from small ruminants, namely, sheep and goat is also used for production of traditional Kebabs (Seekh and Shami Kebab). Buffalo meat is basically used in the household for preparation of curries and Kebabs. It is also mixed with vegetables like potatoes, cabbages, turnips, sugar beet to make delicious dishes, to name a few, besides the irresistible Biriyani, which is a mix of meat and rice. Buffalo steaks are also a delicious product. Both Seekh and Shami Kebabs are delicacies prepared from buffalo meat only, which are liked by all classes of people in India. The buffalo meat has a great water holding and binding properties, and is, therefore, used for industrial purposes in the production of sausages, patties, nuggets, corn beef, ham *etc.*

15.10. CONCERNS

» The FAO has estimated that approximately 18 percent of global greenhouse gas emissions come from livestock production.

» To produce one calorie from animal protein, 11 times as much fossil fuel is required than to produce one calorie from plant protein.

» Energy is devoured by growing feed, transporting feed, transporting animals, processing animals, packaging meat, transporting meat and keeping meat cold.

» In India, 873 liters of water is used to produce one kilo of chicken meat, and 1,471 liters of blue water is used to produce beef in industrial systems.

» The horrors of industrial food animal production facilities include confined and concentrated large animal populations in small areas who experience short lived, poor quality lives.

» The meat production facilities can also pose significant risks to human health and the environment.

» Animal waste may also be a serious source of contamination and pollution of groundwater and air.

» The concentration of parasites, bacteria and chemical contaminants in animal waste can have drastically detrimental effects on ecosystems and communities living near waste disposals.

» According to the United Nations, 30 percent of the earth's landmass is devoted to raising animals to become meat which includes land that is used for grazing and for crop growth which is used as feed.

» India is said to be home to 40 percent of the population favoring a vegetarian diet which means that the resources India is devoting to meat production, which is not even domestically consumed, is high.

15.11. EXPORT

The export of different livestock products during the year 2020-21 is presented in Table 15.3.

Table 15.3. Export of livestock meat from India during 2020-21

Livestock meat type	Qty in tons	Value (Rs. Crores)	Major destinations
Buffalo meat products	10,85,619.93	23,460.38	Hong Cong, Vietnam Socialist Republic, Egypt Arab Republic, Malaysia, Indonesia.
Sheep and goat meat	7050.55	329.96	UAE, Qatar, Kuwait, Saudi Arabia, Oman.
Other meat (pigs, horses, donkeys, rabbits, camels, primates.	894.04	18.06	Bhutan, Nepal
Processed meat	774.11	11.92	Hong Cong, Qatar, Bhutan, Myanmar, Lao, Congo
Animal casings	13,887.74	416.54	Hong Cong, Vietnam, Malaysia, Myanmar, Cambodia.

In 2014, India surpassed Brazil and Australia to become the largest bovine meat exporting country in the world. India's meat production is top agricultural export item ahead of basmati rice in 2014-15. According to the United States Department of Agriculture (USDA) The country has exported 1.24 Million tons of buffalo meat products to the world for the worth of US$ 3.61 billion during 2018-19 (Table 15.4). India ranks first in production of buffalo meat with share of 42.60%. Uttar Pradesh, Andhra Pradesh, Maharashtra and Punjab are major meat production centers. These abattoirs conform to international hygiene standards, with zero hand-touch-processing and regular visits by halal inspectors from the importing countries. Major export destinations during 2014-15 were Vietnam, Malaysia, Egypt, Thailand, and Saudi Arabia.

Table 15.4. Buffalo meat exports from India 2018-19

Country	Qty in tons	Rs. Crore	US$ Million
Vietnam Socialistic Republic	565854.0	11914.5	1711.7
Malaysia	124413.2	2574.6	369.6
Indonesia	94500.0	2267.0	324.5
Iraq	67514.6	1192.8	171.0
Myanmar	45745.4	883.5	124.5
Philippines	44709.1	838.9	120.1
Egypt Arab Republic	47128.0	821.6	117.7
Saudi Arabia	33232.3	785.9	112.8
United Arab Emirates	37852.2	756.0	108.5
Algeria	17286.0	354.9	50.9
Russia	14428.0	330.1	47.6
Oman	13597.1	293.9	42.0
Jordan	10250.2	202.7	28.9
Angola	15584.0	189.9	27.4
Georgia	8929.0	164.2	23.1
Other Countries	95615.3	1597.9	228.5
Total	1236638.4	25168.3	3608.7

Source: DGCIS

15.12. CHALLENGES AND OPPORTUNITIES

15.12.1. Setting up of State of Art-Abattoir cum meat processing plants

In India, there are only 10 most modern sate of art mechanized abattoir cum meat processing plants in various states for slaughtering of buffaloes and sheep. These plants are eco-friendly as the by-products are utilized for production of MBM, tallow, bone chips *etc.* In addition, establishment of Effluent Treatment Plant for waste water treatment from abattoir and lairage, with the water discharged having BOD values ≤ 30 ppm. These plants follow SPS measures prescribed by the International Animal Health Code of O.I.E. There are four fully integrated poultry meat processing plants of world standard too.

15.12.2. Raising male buffalo calves for meat production

In India, 15 million calves are removed from the buffalo production system due to intentional killing by the farmers to save dam's milk in the wake of nonremunerative cost of raising them. Male calf therefore is unfortunately not cared for and resultantly does not survive. A major potential exists for male calf rearing for meat purposes. These calves can be salvaged for meat production thereby improving the economic condition of the farmers and also meat production for domestic and export market. Male calves can be reared without the use of hormones, antibiotics, and growth promoters. They can then be slaughtered scientifically for meat production. The GOI has proposed a financial outlay of $250 million to fund the program 'Salvaging and Rearing of Male Buffalo Calves' for the purpose of increasing meat production during the XI plan (GAIN report, 2008).

15.12.3. Rearing buffaloes

Contractual farming as backward integration to modern abattoir: A strong need has been felt to establish a production base around each modern abattoir to produce quality and disease-free animals as per SPS requirements of O.I.E. The success story of broiler farming with contract farming can be employed here to safeguard the interest of small and marginal farmers by providing them the feed, medicine and marketing of finished product and ascertaining a fixed remuneration to farmers.

15.12.4. Establishing disease free zones for rearing animals

India is fortunately free from most of the trade related diseases and BSE has not been reported. However, FMD is still prevalent in an endemic form in some states. The establishment of 3 zones with 56 districts to control FMD in the Xth five-year plan was the first step by the Government of India towards the establishment of disease-free zones. The World Organization of Animal Health (OIE) declared India "Rinderpest free" in May 2006.

15.12.5. By-product utilization

For the profitability of the meat industry, proficient utilization of meat by-products is important. It has been estimated that 11.4% of the gross income from beef and 7.5% of the income from pork, come from the by-products.

15.12.6. Upgradation of municipal slaughter houses for export purpose

Government should make efforts to upgrade municipal slaughter houses to semi- modern abattoirs with all basic requirements and minimum hygienic standards. In the present context of WTO agreement on sanitary and phyto-sanitary (SPS) measures, the hygienic standards of existing meat plants should be improved at par with the best in the world.

15.12.7. Modernizing the quality control laboratories of State Governments

The State Government laboratories are not well equipped and also lacking in skilled staff to conduct various examinations of meat. It is therefore need of an hour to upgrade the quality control laboratories in terms of modern testing facilities and skilled manpower in order to produce safe, clean and wholesome meat.

15.12.8. Need for strict laboratory inspection of meat and meat products

In most of the slaughter houses, basic facilities required for hygienic meat production are lacking so ante-mortem and post-mortem inspection are not followed.

15.12.9. Training program for meat workers

Regular training should be conducted by Government institutions, SAUs for meat workers, supervisors and managers on importance of scientific slaughtering and dressing. This will help in realization of significance of hygiene and sanitation in slaughter houses, meat processing plants *etc.*

15.12.10. Marketing of meat and meat products

Most of the meat production and marketing practices in India are traditional. Well-integrated marketing system for meat and meat products is lacking in India. The main reasons are monopoly of meat trader, lack of coordination between production and demand, too many middlemen in the trade and inefficiency in the management in slaughterhouse. There is a dire need to modernize the meat production and marketing system. Government of India is keen to improve the marketing system so that the consumers would get the quality meat and meat products at reasonable prices.

15.12.11. Setting up cold storages

Meat is nutrient dense food which makes it perishable commodity. In order to improve keeping quality of meat, cold chain is of crucial importance during transport as well as storage till it reaches to consumers. The Government should support setting up cold storages, supply/value chain and 100% export-oriented slaughterhouses in the country.

15.13. SWOT ANALYSIS OF THE INDIAN LIVESTOCK SECTOR

15.13.1. Strengths

» Vast livestock population, with adaptability to wide range of agroclimatic conditions, is a vital asset for the country and offers scope for diversified animal agriculture.

» Abundant crop residues and common property resources ensure adequate availability of roughages for animals.

» Low cost of production compared to the most other parts of the world, strengthens the possibility of reaping the benefits of comparative advantage.

» Animal protein consumption is regular part of the diet of the people and hence there is presence of large market.

» Increasing number of processing plants is expected to boost value addition on the livestock sector.

» Considerable number of educated youths/non-livestock-based companies and organizations venture into livestock which is a strength to improve the quality of the produce.

15.13.2. Weaknesses

» Though cross breeding programs have improved animal productivity, at least in cattle, generally the country is still largely dominated by low yielding non-descript animals.

» Lack of cold-chain and poor support infrastructure, e.g. roads and erratic power supply remain a major challenge for procurement and supply of good quality raw animal products.

» Inadequate knowledge and low adoption of scientific livestock farming and clean meat production practices.

» Non-maintenance of records by the farmers constrains the availability of comprehensive and reliable production data for proper planning.

» Investments in livestock research is not commensurate with returns and potential.

15.13.3. Opportunities

»   Purchasing power of the consumers is on the upswing with growing economy and continually increasing population of middle class.

»   Expanding market will see creation of enormous job and self-employment opportunities.

»   Demand for livestock products is income elastic. Continued rise in middle class population will see shift in the consumption pattern in favor of value-added products.

»   Untapped potential of improved technologies in certain areas leaves ample scope for improving productivity.

15.13.4. Threats

»   Excessive grazing pressure on marginal and small community lands has resulted in degradation of land.

»   Indiscriminate crossbreeding for raising productivity could lead to disappearance of valuable indigenous breeds and germplasm.

»   Export of quality feed ingredients, viz., cakes, molasses, etc. is making the domestic producers rely on low energy fodders.

»   With intensive industrialization of livestock sector in response to market forces, the small producers will find it increasingly difficult to compete with the industrial sub-sector and thus risk losing a significant means of livelihood.

15.14. FUTURE PROSPECTS

India is having a good potential for meat production because of large livestock population. Measures should be taken to increase the meat production efficiency of different species of animals using the improved management practices. Adoption of improved shelter management practices can reduce the environmental stress. New breeds should be developed for meat production with higher feed conversion efficiency, faster growth and disease resistance. Health management practices should be followed for prevention of diseases and economic loss to the farmers. Regular prophylactic health measures should be carried out against infectious diseases. Regular screening of animals should be carried out against diseases such as tuberculosis, brucellosis, salmonellosis *etc.*

The livestock market yard should have basic facilities for feeding, watering and holding animals for a few days. By vertical integration with meat processing industries, the middlemen can be eliminated, which will ultimately increase the profit of farmers. Need for modernizing the quality control laboratories of the State Government for strict laboratory inspection of meat and meat products. Training programs for meat workers regarding hygiene and sanitation need to be organized regularly. Modernization of abattoirs, setting up of rural abattoirs and registration of all slaughterhouses in cities/towns are essential for quality meat production. The setting up of large commercial meat farms have been recommended to address the traceability issues necessary for stringent quality standards of *CODEX Alimentarius.*

Strategies for the prosperity of Indian meat industry are as follows:

»   In a report entitled "Indian Meat Industry Perspective," the FAO outlined the following four steps that should be taken if India's food industry is to successfully go pink:

·   Setting up state of the art meat processing plants.

·   Developing technologies to raise male buffalo calves for meat production.

- • Increasing the number of farmers rearing buffalo under contractual farming.
- • Establishing disease-free zones for rearing animals.

» Production of good quality animals for slaughter is must for production of good quality meat. Hence, farmers' cooperative can play a major role in the field of production and marketing of quality animals, extension education and encouragement of backward integration / contract farming as in poultry industry for intensive and semi-intensive system of rearing small ruminants.

» Food safety at all stages of production, processing, packing, storage and marketing of meat and meat products, maintenance of standards such as SPS, HACCP certification and others which are prescribed by the importing countries.

» Consumer awareness: Priorities must be given to address the myths prevalent among the public regarding meat consumption and risk of diseases (Meat consumption and risk of cancer and coronary heart disease due to its fat and cholesterol content) with proper extension programs.

» Meat processing and value addition are key for the prosperity of meat industry. The awareness regarding the processed meats and the convenience to the consumers and households should be improved.

» Packaging of meat and meat products: Most of meat is sold in India is in unpacked form. Meat is packed only in some organized meat factories and in bacon factories. For safe delivery of the meat and various value-added meat products through the various stages of processing, storage, transport, distribution and marketing packaging is of utmost importance.

» Breeding strategies: Meat scientists and animal geneticists should collaborate their research for developing a potential cross bred buffalo for meat purpose. Meat production potential under extensive and intensive system should be adequately exploited through cross breeding of selected local breeds/ non-descript breeds with specific exotic and improved breeds.

» At present buffalo meat is obtained as a by-product of buffalo milk production. There is vast scope for increasing carabeef export consequent to cattle slaughter ban act, availability of male buffalo calves and the steady demand for the same from the importing countries. It is high time to consider growing/fattening of male buffalo calves for veal production. Sufficient nutrition and improved levels of hygiene and sanitation at meat handling will enable India a quantum jump in meat production by utilizing the surplus male calves.

15.15. CONCLUSION

Several assessments agree that increases in the demand for livestock products, driven largely by human population growth, income growth and urbanization, will continue for the next three decades at least. Globally, increases in livestock productivity in the recent past have been driven mostly by animal science and technology, and scientific and technological developments in breeding, nutrition and animal health will continue to contribute to increasing potential production and further efficiency and genetic gains.

Since last decade, due to stringent SPS requirements imposed by the importing countries has brought in a sea change in the meat industry This has forced the establishment of state of art abattoirs and modernization of local slaughterhouses by the Ministry of Food Processing Industries, GOI, for export. India had already witnessed revolution in food grain and milk production, now its prime time for the revolution in meat production in up-coming years. Indian farming community has been very progressive by need based adaptation. Animal agriculture has been favorable and so has become popular.

The major role played by buffaloes in strengthening dairy industry in India can well be repeated in meat industry. The non-tariff barriers (NTB) such as SPS should be based strictly on real health and environmental standards. Collective and planned efforts from liaison offices, meat industries, meat corporations, scientists and nodal bodies like APEDA and Directorate of Marketing and Inspection (DMI) are required for exchange of information, overall improvement and thus transformation of this escalating industry.

At present, the Indian meat industry is gradually expanding in all aspects of fresh marketing. Value addition to processed meat, development of specific meat products like designer meat, functional meat *etc.* for specific groups of consumers, online marketing of meat and meat products of different Indian regions are now getting more sophisticated as per public demands. Gradually improving health-conscious consumer awareness has shaped today's Indian meat market. Finally the upcoming modified government rules and regulations related to food safety and security have designed a strong prospect of the Indian meat industry in the present era.

Chapter - 16

Green (Renewable) Energy Revolution

16.1. INTRODUCTION

Energy is one of the major parameters for establishing growth and progress of agriculture in a country, which depends directly upon the per capita energy consumption. Fossil fuel-based energy generation is the current practice in agriculture in India which is expensive and causes greenhouse gas (GHG) emissions enhancing climate change process. Gasoline, coal, natural gas, diesel, and other commodities derived from fossil fuels are non-renewable. There is a need to shift current fossil fuel-based energy generation to renewable based energy generation in order to mitigate climate change and reduce GHG emissions. Non-fossil fuel based renewable energy sources include solar, wind, biomass, hydroelectric and geothermal (Fig. 16.1).

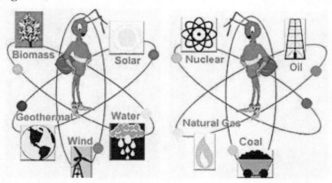

Fig. 16.1. Renewable (left) and non-renewable (right) energy sources

16.2. RENEWABLE ENERGY PRODUCTION IN INDIA

As of 30th April 2016 India's cumulative grid interactive or grid tied renewable energy capacity (excluding large hydro) reached about 42.85 GW, surpassing the installed capacity of large-scale hydroelectric power in India for the first time in Indian history. 63% of the renewable power came from wind, while solar contributed nearly 16%.

Ministry of New and Renewable Energy (MNRE) renewable electricity targets have been upscaled to grow from just under 43 GW in April 2016 to 175 GW by the year 2022, including 100 GW from solar power, 60 GW from wind power, 10 GW from bio power and 5 GW from small hydro power. The ambitious targets would see India quickly becoming one of the leading green energy producers in the world and surpassing numerous developed countries. The government intends to achieve 40% cumulative electric power capacity from non-fossil fuel sources by 2030 (Table 16.1).

Table 16.1. Installed grid interactive renewable energy capacity in India as on July 31, 2016 (RES MNRE)

Source	Total installed capacity (MW)	2022 target (MW)
Wind power	27,441.15	60,000.00
Solar power	8,062.00	1,00,000.00
Biomass power	4,860.83	10,000.00
Waste-to power	115.08	
Small hydro power	4304.27	5,000.00
Total	44,783.33	1,75,000.00

16.3. TYPES OF RENEWABLE ENERGY

Renewable energy technologies are ones that consume primary energy resources that are not subject to depletion. Examples of renewable technologies include solar energy, wind energy, biomass energy, hydro energy, and geothermal energy. Unlike fossil fuels, a renewable resource can have a sustainable yield. Renewable energy is energy that, in its production or consumption, has minimal negative impacts on human health and the healthy functioning of vital ecological systems, including the global environment. It is an accepted fact that renewable energy is a sustainable form of energy, which has attracted more attention during recent years. A great amount of renewable energy potential and environmental interest, as well as economic consideration of fossil fuel consumption and high emphasis on sustainable development for the future will be needed.

The promotion of renewable resources such as solar, wind, biomass, geo-thermal, small-scale hydro, biofuels and wave-generated power have a huge potential for the agriculture industry. The farmers should be encouraged by giving subsidies to use renewable energy technology. Wind, solar, biomass, and geothermal energy can be harvested forever, providing farmers with a long-term source of income. Renewable energy can be used on the farm to replace other fuels or sold as a "cash crop." The characteristics of different forms of renewable energies and their uses in agriculture are presented in Table 16.2.

Table 16.2. Characteristics of different forms of renewable energies and their uses

Energy source	Conversion to	Most applied technologies and applications	Remarks
Solar energy	▪ Heat ▪ Mechanical energy ▪ Electricity	▪ Photovoltaic (PV) driven pumps for irrigation ▪ Crops, drying of fruits / spices, ice making and cold storage (through absorption or heat driven refrigeration)	▪ PV systems are limited to agricultural activities that require little power input only. ▪ FAO provides an inventory of PV applications

Wind energy	▪ Mechanical energy ▪ Electricity	▪ Direct use: grinder, mills, mechanical water pumps ▪ Electrical water pumps	▪ Option for energy intensive processing activities
Biomass energy	▪ Heat ▪ Electricity ▪ Liquefied biofuels ▪ Biogas	▪ Dryer (fruits, herbs, spices) ▪ Fermenter (tea) ▪ Combustion motor or electric motor (fuels like ethanol and biodiesel for transportation) ▪ Anaerobic digester: biogas for lighting, cooking and heating and industrial biogas for decentralized electricity	▪ Biomass is organic material used to generate electricity, to produce heat or biofuels for transportation. ▪ Bioenergy is derived from wood, agricultural crops, residues, animal by-products, agro-industrial by-products.
Hydro energy	▪ Mechanical energy ▪ Electricity	▪ Hydro plants in derivation schemes ▪ Hydro plants in existing water distribution networks	▪ Size: Medium-small

16.4. SOLAR ENERGY

Solar energy is the energy derived directly from the Sun which is the most abundant source of energy on Earth. The fastest growing type of alternative energy, increasing at 50 percent a year, is the photovoltaic cell, which converts sunlight directly into electricity. The Sun delivers yearly more than 10,000 times the energy that humans currently use.

For many agriculture needs, the alternative is solar energy. Modern, well-designed, simple-to-maintain solar systems can provide the energy that is needed at the given location and for the given time period. These are systems that have been tested and proven around the world to be cost-effective and reliable, and they are already raising levels of agricultural productivity worldwide (Gustav *et al.*, 2008).

India has tremendous scope of generating solar energy. The geographical location of the country stands to its benefit for generating solar energy. The reason being India is a tropical country and it receives solar radiation almost throughout the year, which amounts to 3,000 hours of sunshine. This is equal to more than 5,000 trillion kWh. Almost all parts of India receive 4-7 kWh of solar radiation per sq. meter. This is equivalent to 2,300–3,200 sunshine hours per year. Since majority of the population lives in rural areas, there is much scope for solar energy being promoted in these areas.

There is a need for promoting the use of renewable energy systems for sustainable agriculture, e.g. solar photovoltaic water pumps and electricity, greenhouse technologies, solar dryers for post-harvest processing, and solar hot water heaters (Fig.16.2). In remote agricultural lands, the underground submersible solar photovoltaic water pump is economically viable and also an environmental-friendly option as compared with a diesel generator set. If there are adverse climatic conditions for the growth of particular plants in cold climatic zones, then there is a need for renewable energy technology such as greenhouses for maintaining the optimum plant ambient temperature conditions for the growth of plants. Food dryers, irrigation systems, greenhouse environment control, electric fencing, and many other specific applications are being developed to capture the sun's energy and converting it to power equipment.

Fig. 16.2. Solar photovoltaic (PV) panels providing green energy for agricultural growth.

Solar farming uses power generated from solar energy to operate agricultural or farming tools. It is simple, cost effective, reliable and long lasting. Most common agricultural tools such as tractors, watering systems, rotator, roller, planter, sprayers, and broadcast seeder work on battery power and fuel oil. In solar farming, the battery power is replaced with solar power, so that the usage of electricity from grid-power and non-renewable sources can be reduced. Therefore, the solar energy technologies are likely to play an important role in the near future through a variety of thermal applications and decentralized power generation and distribution systems.

In general, there are two types of solar systems – those that convert solar energy into direct current power and those that convert solar energy into heat (to heat water or air). Both convert sunlight into usable energy and both have many applications in agricultural settings to aid farmers and ranchers in satisfying the energy requirements of their operations. Both types have many applications in agricultural settings, making life easier and helping to increase the operation's productivity (Gustav *et al.*, 2008).

16.4.1. Solar electric (Photovoltaic) systems

Photovoltaic has been derived from the combination of two words; "Photo" means light and "Voltaic" means electricity. It is a technology that converts light directly into electricity. PV devices generate electricity directly from sunlight via an electronic process that occurs naturally in certain materials. Solar energy frees electrons and induces them to travel through an electrical circuit, powering an electrical load. PV devices can be used to power anything from small electronics such as calculators and road signs to homes and large commercial buildings.

The basic building block of photovoltaic is a round or square cell that converts sunlight into direct current (DC) electricity. Cells are wired together to form a module; multiple modules are arranged together to form a panel; and multiple panels produce a PV array. In general, the larger the area of a module or array, the more electricity will be produced. The cells and modules can be wired (in series and/or parallel electrical arrangements) to create a wide range of voltage and current combinations (Fig. 16.3). The majority of applications in smaller projects (< 200 W) are for 12 to 24-volt outputs with the amperage depending on how much power is required.

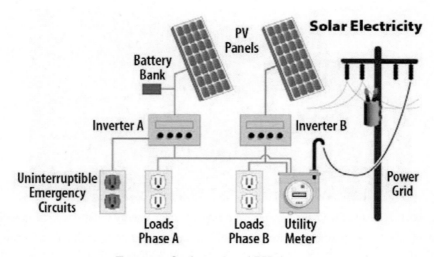

Fig. 16.3. Grid-connected PV electricity

PV systems produce DC power. When energy is needed to operate alternating current (AC) equipment, the DC output is converted to AC with an inverter. Most household appliances require AC electricity, but DC-powered appliances can be ordered.

16.4.1.1. Types of PV modules: Currently, the 2 most common types of PV modules include:

» Crystalline silicon (mono - crystalline and multi - crystalline) (Fig. 16.4).

» Thin film (amorphous – silicon and cadmium –telluride) (Fig. 16.4).

Fig. 16.4. Examples of multi- crystalline modules (left) and amorphous – silicon thin film modules (right)

a) Crystalline silicon modules

i) Advantages

» According to Solarbuzz, an international solar energy research and consulting company, 82% of PV modules manufactured in the world are crystalline silicon, making them easy to find in the market.

» Module efficiency of crystalline silicon modules is higher than thin film (12 to 20% versus 3 to 11%), so fewer modules are required and the system uses less space, which can be of importance when used on high-valued agricultural land.

» The multi-crystalline modules have been demonstrated to last over 30 years and warranties up to 20 years are offered. According to Vick *et al.* (2003), thin film modules have only been around since 1988, and early modules demonstrated problems with performance degradation over time.

» Crystalline silicon modules demonstrate only a slight decline in power output over time (~1% per year) while amorphous-silicon (a-Si) thin film modules experience about a 20% initial decrease followed by a 1% annual decrease thereafter) (Osborne, 2003). Normally, solar-PV companies installing a-Si modules expose the modules to the sun prior to installation so power fluctuation will not vary significantly for customer.

» Tempered glass makes multi-crystalline modules less likely to break (thin film modules currently require untampered glass). While thermal cracking occurred with a-Si modules prior to 2005, most manufacturers are able to either strengthen glass without tempering or using a stronger non-glass material like teller.

» Crystalline silicon modules are non-toxic and can be disposed off in landfills (e.g., unlike cadmium-telluride, according to EPA, which cannot be disposed in landfills due to toxicity of cadmium).

b) Thin-film modules

i) Advantages

» Amorphous silicon modules use less than 1% the amount of silicon that crystalline silicon uses which decreases the manufacturing cost.

» Thin film modules can generate higher voltage than crystalline silicon modules, which is important in applications with power requirements from 200 watts to 2 kilowatts.

» Generally, the price per Watt for thin film modules is cheaper for large PV (Megawatt and larger size) installations.

» The power loss with increased module temperature for a-Si modules is ~0.25%/ °C compared to crystalline silicon modules (King *et al.*, 2001).

» Efficiency improvements have been demonstrated by a-Si modules over crystalline silicon modules in cloudy conditions (Wu and Lau, 2008).

» Flexibility of a-Si modules allows them to more easily be integrated into buildings (e.g., building integrated PV, BIPV).

It is a simple fact that photovoltaic modules produce electricity only when the sun is shining, so some form of energy storage is necessary to operate systems at night. One can store the energy as water by pumping it into an overhead tank while the sun is shining and distributing it by gravity when it is needed after dark. For electrical applications at night, one will need a battery to store the energy generated during the day. Photovoltaic is a well-established, proven technology with substantial applications in agriculture. The photovoltaic systems are increasingly more cost-effective compared with either extending the electrical grid or using diesel generators in remote locations.

Photovoltaic systems are very economical in providing electricity in remote locations on farms, ranches, orchards and other agricultural operations. PV systems can be much cheaper than installing power lines and step-down transformers in applications such as electric fencing, area or building lighting, and water pumping either for livestock watering or crop irrigation. In fact, water pumping is one of the simplest and most appropriate uses for photovoltaic. From crop irrigation to stock watering to domestic uses, photovoltaic-powered pumping systems meet a broad range of water needs. Most of these systems have the added advantage of storing water for use when the sun is not shining, eliminating the need for batteries, enhancing simplicity and reducing overall system costs.

Powering electric fans for air circulation is another application of photovoltaic. Modern pig and poultry farms double and even triple production by raising the animals in enclosed buildings. Another good use of photovoltaic is for lighting in agricultural buildings and enclosures. Electric lighting using solar photovoltaic in these buildings can significantly extend working hours and increase productivity. This is especially true for those who use precious evening hours for equipment repair and maintenance. The photovoltaic systems can be more economical choices than conventional battery-powered fixtures, flashlights and fuel lamps. In addition, they provide more and higher quality light, and emit no smoke or fumes.

16.4.1.2. Potential of photovoltaics for farm use: Solar energy can supply and/or supplement many farm energy requirements (Table 16.3). Motor energy generation is the primary use for PV on farms. Water pumping, one of the simplest and most prevalent uses of PV, includes irrigation in fields, watering livestock, pond management, and aquaculture. Portable or ground-mounted PV systems can be used to pump water from underground wells or from the surface (e.g. ponds, streams). PV water pumping systems can be the most cost-effective water pumping option in locations where there are no existing power lines. When properly sized and installed, PV water pumps are very reliable and require little maintenance. Environmental benefits can include keeping cattle and other livestock out of wetlands and waterways. PV systems are very cost effective for remote livestock water supply, small irrigation systems, and pond aeration.

Table 16.3. Farm applications of photovoltaic energy (PV) and solar heating (SH)

Farm applications	Energy type*	Fields	Livestock	Other uses
Water pumping	PV	Wells, ponds, streams, irrigation	Wells, ponds, streams	Domestic uses
Building needs	PV		Security and task lighting, ventilation, feed, product handling equipment, refrigeration	Battery charging, task lighting, ventilation fans, AC needs, refrigeration
	SH		Air cooling, air/space heating, water heating	Domestic uses of solar heat
Farm and ranch	PV	Feeder/sprayer, irrigation sprinkler controls, security and task lighting	Electric fences, feeder/sprayer	Electric fences, invisible fences, battery charging, compressor for fish farming, fans for crop drying, greenhouse heating
	SH			Crop drying, greenhouse heating

*PV - Photovoltaic energy, SH – Solar heating

Source: Expanded from NREL (1997)

Appropriate applications for PV systems on farms, ranches and orchards include:

- » Solar water pumping.
- » Electric fencing to contain livestock.
- » Lighting
- » Small motors
- » Solar powered vehicles
- » Solar milking machine

a. **Water pumping:** Photovoltaic (PV) water pumping systems may be the most cost-effective water pumping option in locations where there is no existing power line. When properly sized and installed, PV water pumps are very reliable and require little maintenance. The size and cost of a PV water pumping system depends on the local solar resource, the pumping depth, water demand, and system purchase and installation costs. Although today's prices for PV panels make most crop irrigation systems too expensive, PV systems are very cost-effective for remote livestock water supply, pond aeration, and small irrigation systems.

If the water is required at times when the sun is not shining, there are two means of storing the energy:

- » Chemical storage (battery) where the PV module keeps the battery bank charged so it can power the pump on demand. Battery storage is a good choice when space is an issue or if the system needs constant pressure and batteries can last for three to eight years.
- » Mechanical storage (elevated reservoir) where the PV module powers the pump during sunlight hours to ensure that a storage reservoir is full for later demand. Systems that use this type of storage are called solar direct systems, and an elevated reservoir keeps the system simple and eliminates the need for battery maintenance.

There are two commonly used configurations for PV water pumping systems which are shown schematically using Fig 16.5. These systems require two non-interchangeable solar controllers. In the case of battery-connected solar water pumping systems (with batteries), the solar controller is used to regulate the flow of electricity from the PV module to either the pump or to the batteries. In battery-powered systems, the controller regulates the amount of current flowing into the battery from the PV module to maximize its available power. It also prevents the batteries from damage by over-charging, as reported by Chel and Tiwari (2010). In solar direct water pumping systems (no batteries), the controller optimizes the power from the module to the pump to produce as much water flow as possible. Pressure switches, expansion tanks, float valves, foot valves and inlet filters are important components of the PV water pumping system to protect the system to ensure trouble-free operation.

Today, many water pumps on the market are specifically designed to be powered by a solar photovoltaic (PV) panel, but any pump with a direct current motor can operate with PV. Submersible, piston, rotary vane, centrifugal and diaphragm are some of the different types of pumps available, which can overcome significant vertical and horizontal distances that can reach hundreds of meters. Often, direct current pumps can move the same quantity of water as a traditional pump, but operate with half the power, which means the cost to power a solar pump can be half of the cost to power an alternating current unit. The improved motor efficiency of the direct current pump has the most to do with this.

Fig 16.5. PV integrated water pumping application

An advantage of using direct solar radiation as a power source for irrigation is that it is available at the site of application without the employment of a distribution system (Kenna *et al.*, 1985; Whiffen *et al.*, 1992). Plant water demand and the quantity of water pumped by a photovoltaic-powered water pumping system are both directly correlated to daily solar insolation.

Photovoltaic power for irrigation is cost-competitive with traditional energy sources for small, remote applications, if the total system design and utilization timing is carefully considered and organized to use the solar energy as efficiently as possible. In the future, when the prices of fossil fuels rise and the economic advantages of mass production reduce the peak, watt cost of the photovoltaic cell, photovoltaic power will become more cost-competitive and more common (Helikson *et al.*, 1991).

b. Solar electric fence: Solar powered electric fences are highly effective and dependable for large fields and cattle farms (Fig. 16.6). These fences typically consist of a SPV unit as a source of power, an energizer that produces high voltage impulses (8 kv) emitted at intervals of 0.9 to 1.2 seconds, along with a 12 V battery. The impulse carries 10 mA of current and delivers a shock lasting for a fraction of a second. The batteries can be recharged using readymade solar fence chargers. Battery operated solar fences may cost from Rs. 45,000-50,000 per acre. Cheaper versions costing as low as Rs. 10,000-25,000 per acre have been developed using locally made materials in some places in India.

Fig. 16.6. Solar powered electric fence

c. **Lighting:** Even when utility power is available nearby, using PV to charge batteries for lighting may be the cheapest option for outbuildings. The cost of a transformer and running wires to where the light is needed can add up. A simple PV system can operate low- or high-pressure sodium lights, as well as fluorescent and incandescent bulbs.

d. **Small motors:** Electric motors with small power needs can be very handy in remote areas or in places where running an electrical line is a problem. PV-powered automatic gate openers use a 35 cm by 32.5 cm PV panel to charge the battery. PV is also used to run aeration fans in grain storage bins and to power automatic supplement feeders.

e. **Solar powered vehicles:** Tractors are primarily diesel-powered but look for technologies toward electric assist using solar to greatly expand in future years. The technology is there and is in fact in place with some powered tools in agriculture of all sizes, but look for it to greatly expand in the industry in coming years.

Some research has shown that agricultural electric vehicles could avert 23.3 tons of greenhouse gases each year.

Solar electric mowers (Fig. 16.7) are available with cordless and rechargeable battery options. These lawn mowers do not emit toxic fumes and do not require frequent refueling to run. Only a few hours of recharging from a solar-powered battery charger are required. It is also possible to convert an existing fuel or electric lawn mower into a solar mower.

Fig. 16.7. Solar powered small mower

Similarly, tractors and planting machines are available with solar panels on top to power them (Fig. 16.8). Solar-powered tractors can easily handle non-energy intensive operations like planting and harvesting. Their operating costs would be a fraction of those of conventional tractors. However, the technology is relatively new in India and used in a few places along with conventional tractors. May be in a decade, we will have fully solar-powered agricultural machineries.

Fig. 16.8. Solar powered small tractor

Electronic sensors used to determine soil moisture, precipitation, and location-specific weather data can also be made to work with solar energy. They can also be programed to be remotely operated.

f.   **Solar milking machine:** Milking machine for cows operated on solar power instead of diesel or electric power is another innovation (Fig. 16.9). A SPV module connected to batteries powers the machine. A mobile milking machine along with solar panels and battery backup is available for Rs. 70,000. Some states like Karnataka provide subsidies of up to 50% for these machines. There are manual milking machines that can be either hand operated or connected to solar power as well.

Fig. 16.9. Solar powered milking machine

Other appropriate applications for PV systems on farms, ranches and orchards include:

»   Power for feed or product grinding.

»   Electric-powered egg collection and handling equipment.

»   Product refrigeration.

»   Livestock feeder and sprayer motors and controls.

»   Compressors and pumps for fish farming.

»   Battery charging.

16.4.2. Solar thermal applications

The second most widely used application of solar energy is to produce heat, which has applications for various agricultural processes as follows:

» Drying crops and grains
» Water heating
» Greenhouse heating

16.4.2.1. Drying crops and grains: Using the sun to dry crops and grain is one of the oldest and most widely used applications of solar energy. The simplest and least expensive technique is to allow crops to dry naturally in the field, or to spread grain and fruit out in the sun after harvesting. The disadvantage of these methods is that the crops and grain are subject to damage by birds, rodents, wind, and rain, and contamination by wind-blown dust and dirt. More sophisticated solar dryers protect grain and fruit, reduce losses, dry faster and more uniformly, and produce a better-quality product than open-air methods.

The basic components of a solar dryer are an enclosure or shed, screened drying trays or racks, and a solar collector (Fig. 16.10). In hot, arid climates the collector may not even be necessary. The southern side of the enclosure itself can be glazed to allow sunlight to dry the material. The collector can be as simple as a glazed box with a dark colored interior to absorb the solar energy that heats air. The air heated in the solar collector moves, either by natural convection or forced by a fan, up through the material being dried. The size of the collector and rate of airflow depends on the amount of material being dried, the moisture content of the material, the humidity in the air, and the average amount of solar radiation available during the drying season.

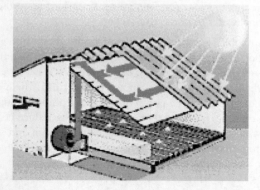

Fig. 16.10. Grains dried using solar powered dryer.

While the cost of a solar collector can be high, using the collector at other times of the year, such as for heating farm buildings, may make a solar dryer more cost-effective. It is possible to make small, very low-cost dryers out of simple materials. These systems can be useful for drying vegetables and fruits for home use.

Solar applications protect grain and fruit, reduce losses, dry faster and more uniformly, and produce a better-quality product than open-air methods.

16.4.2.2. Water heating: Solar space-heating systems can be used in livestock, dairy and other agriculture operations that have significant space and water heating requirements. Livestock and dairy operations often have substantial air and water heating requirements. Modern pig and poultry farms raise animals in enclosed buildings, where it is necessary to carefully control temperature and air quality to maximize the health and growth of the animals. These facilities need to replace the indoor air regularly to remove moisture, toxic gases, odors and dust. Heating this air, when necessary, requires large amounts of energy. With proper planning and design for harsh winter conditions, solar air/space heaters such as the Trombe wall (Chel *et al.*, 2008) can be incorporated into farm buildings to pre-heat incoming fresh air. These systems can also induce or increase natural ventilation levels during summer months.

Solar water heating systems can provide low to medium temperature hot water for pen cleaning. Commercial dairy farms use large amounts of energy to heat water to clean equipment, as well as to warm and stimulate cows' udders. Heating water and cooling milk can account for up to 40% of the energy used on a dairy farm. Solar water heating systems may be used to supply all or part of these hot water requirements. Water heating can account for as much as 25% of a typical family's energy costs and up to 40% of the energy used in a typical dairy operation. A properly sized solar water heating system could cut those costs in half. Hot water is also needed for pen and equipment cleaning and a host of other agricultural uses.

Solar water heating systems use the sun to heat either water or a heat-transfer fluid, such as a water-glycol antifreeze mixture, in collectors most commonly mounted on a roof. The heated water is then stored in a tank similar to a conventional gas or electric water tank. When water is drawn from the water heater, it is replaced with the solar-heated water from that tank. Some systems use an electric pump to circulate the fluid through the collectors. Solar water heaters can operate in any climate. Performance varies depending, in part, on how much solar energy is available at the site, but also on the temperature of incoming water. The colder the water, the more efficiently the system operates. In almost all climates, there is need for a conventional backup system. In fact, many building codes require a conventional water heater as the backup.

In its simplest form, a solar water heater consists of an absorber, a storage tank, insulation, piping and a transparent cover. Solar energy heats the absorber surface and a heat transfer fluid (indirect) or water (direct) flowing through tubes attached to the absorber. If a heat-transfer fluid is used, there is a heat exchanger that then heats the water as shown in Fig. 16.11. The heated water is transferred to the insulated storage tank either with a pump or without a pump through natural convection. A transparent cover (glass or plastic) is placed above the absorber to reduce heat losses due to radiation and also on account of wind flowing over the absorber. The bottom and sides of the absorber are covered with insulation to reduce both types of heat losses. The absorber, cover and insulation are placed within a plastic or metal container.

Fig. 16.11. Solar water heating system

16.4.2.3. Greenhouse heating: Another agricultural application of solar energy is greenhouse heating. Commercial greenhouses typically rely on the sun to supply their lighting needs, but are not designed to use the sun for heating. They rely on gas or oil heaters to maintain the temperatures necessary to grow plants in the colder months. Solar greenhouses, however, are designed to utilize solar energy for both heating and lighting. A solar greenhouse has thermal mass to collect and store solar heat energy, and insulation to retain this heat for use during the night and on cloudy days. In the northern hemisphere, a solar greenhouse is oriented to maximize southern glazing exposure. Its northern side has little or no glazing, and is well insulated. To reduce heat loss, the glazing itself is also more efficient than single pane glass, and various products are available ranging from double pane to "cellular" glazing. A solar greenhouse reduces the need for fossil fuels for heating. A gas or oil heater may serve as a back-up heater, or to increase carbon dioxide levels to induce higher plant growth. A solar greenhouse may be an underground pit, a shed-type structure, or a Quonset hut (Fig. 16.12).

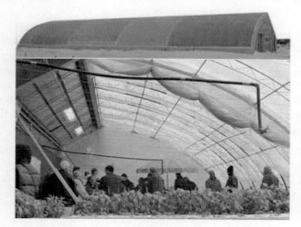

Fig. 16.12. Solar powered greenhouse heating.

Solar greenhouses differ from conventional greenhouses in the following five ways:

» Have glazing oriented to receive maximum solar heat during the winter.
» Use heat-storing materials to retain solar heat.
» Have large amounts of insulation where there is little or no direct sunlight.
» Use glazing material and glazing installation methods that minimize heat loss.
» Rely primarily on natural ventilation for summer cooling.

A greenhouse for crop production is a highly sophisticated structure, which aims at providing ideal conditions for satisfactory plant growth and production throughout the year. For satisfactory plant growth, the growth factors, namely light intensity, temperature, humidity and air composition are maintained at optimum levels inside the greenhouse for higher productivity. The maintained chamber inside the greenhouse is also referred to as a controlled environment greenhouse.

The climatic requirements for plant growth can therefore be summarized as follows; Verlodt (1990):

» Plants grown under protected cultivation are mainly adapted to average temperatures ranging from 17 to 27 °C. Considering the warming-up effect of solar radiation in the greenhouse, the above temperature range can be possible without any heating arrangement in it when the outside ambient temperature prevails in the range from 12 to 22 °C.

» If the mean daily outside temperature is below 12 °C, the greenhouse is to be heated, particularly at night. When the mean daily temperature is above 22° C, especially during summer, artificial cooling is necessary or cultivation in the greenhouse is to be stopped. Natural ventilation is sufficient when ambient mean temperatures range from 12 to 22 °C.

» The absolute maximum temperature for plants should not be higher than 35–40 °C.

» The minimum threshold for soil temperature is 15 °C.

» Verlodt (1990) suggests a threshold of the average night temperature as 15–18.5 °C for heat-requiring plants such as tomato, pepper, cucumber, melon and beans.

» The safe ranges of relative humidity are from 70–80%.

16.4.3. Benefits

Renewable energies represent a corner stone to steer our energy system in the direction of sustainability and supply security. Generating electricity, heat or biofuels from renewable energy sources has become a high priority in the energy policy strategies at a national level as well as on a global scale. Challenging goals for these "new" supply options to meet our energy demands have been set, e.g. at the European level by the commitment of meeting 20% of the overall energy demand from renewable energy sources by 2020. Solar energy is one of the renewable energy resources widely used in the agriculture sector for various applications (Gustav *et al.*, 2008).

So solar farming is not only environmentally friendly, but also reliable and cost effective. Maintenance cost is low, since there will be fewer moving parts and lesser impact of oil corrosion. The Indian government is also helping farmers by providing subsidies and loans to acquire solar powered agricultural tools. Agricultural experts are being encouraged to guide farmers in operating them. The time has come to switch to solar power on Indian farms!

Solar technology offers farmers an opportunity to stabilize their energy costs. Some solar strategies involve an initial investment, but once those costs are recovered, fuel is free. This allows farmers to budget more effectively and to save money, as they can avoid the high (and unpredictable) expenses of municipal energy sources.

PV systems also require much less maintenance than other traditional farm energy sources - and they are quite efficient. Greenhouse nurseries, for example, enable year-round crop growth. Photovoltaic panels can also dry crops more quickly and evenly than other methods. Additionally, solar collectors can lower costs when they are used in dairy and livestock operations, as the sun's energy can power water heaters and other necessary equipment.

For many agriculture needs, the alternative is solar energy. Modern, well-designed, simple-to-maintain solar systems can provide the energy that is needed at the given location and for the given time period. These are systems that have been tested and proven around the world to be cost-effective and reliable, and they are already raising levels of agricultural productivity worldwide (Gustav *et al.*, 2008).

The advantages of solar energy applications are as follows:

» No fuel, low running costs.

» Modular nature.

» Long life.

» Reliability.

» Low maintenance.

» Clean energy, avoids greenhouse gas emissions.

16.5. WIND ENERGY SYSTEMS

Other renewable energy sources such as wind energy can also be used in agriculture. Small wind systems can provide power that can be used directly or stored in batteries. These systems are very reliable in areas that get enough consistent wind. The systems can be very cost-effective and reliable for many power needs on farms and ranches.

Wind energy is basically harnessing of wind power to produce electricity. The kinetic energy of the wind is converted to electrical energy. When solar radiation enters the earth's atmosphere, different regions of the atmosphere are heated to different degrees because of earth curvature. This heating is higher at the equator and lowest at the poles. Since air tends to flow from warmer to cooler regions, this causes what we call winds, and it is these airflows that are harnessed in windmills and wind turbines to produce power.

Wind energy is used to turn mechanical machinery to do physical work, such as crushing grain or pumping water by windmills (a much older technology). Most modern wind power is generated in the form of electricity by converting the rotation of turbine blades into electrical current by means of an electrical generator.

Farmers are in a unique position to benefit from the growth in the wind industry. To tap this market, farmers can lease land to wind developers, use the wind to generate power for their farms, or become wind power producers themselves. Farmers can generate their own power from the wind. Small wind generators (Fig. 16.13), ranging from 400 watts to 40 kilowatts or more, can meet the needs of an entire farm or can be targeted to specific applications. In Texas and the West, for example, many ranchers use wind generators to pump water for cattle. Electric wind generators are much more efficient and reliable than the old water-pumping fan-bladed windmills. They may also be cheaper than extending power lines and are more convenient and cheaper than diesel generators.

Fig. 16.13. Wind energy generators

"Net metering" enables farmers to get the most out of their wind turbines. When a turbine produces more power than the farm needs at that moment, the extra power flows back into the electricity system for others to use, turning the electric meter backwards. When the turbine produces less than the farm is using, the meter spins forward, as it normally does. At the end of the month or year, the farmer pays for the net consumption or the electric company pays for the net production. Net metering rules and laws are in place in most states.

Appropriate applications for wind energy on farms, ranches and orchards are as follows:

16.5.1. Water pumping

Wind turbine electricity generation can be used to raise the living standard of rural dwellers by improving agricultural productivity (Fig. 16.14). Wind turbine has significant benefit in the areas where there is a shorter rainy season and hence demand for pumped water. After installing wind turbine water pumps in a farm, one can raise higher value crops throughout the year and also supply water to the livestock. There is the requirement of appropriate training for the local farmers to use wind turbine-based water pump irrigation. At present, mostly fossil fuel-powered water pumps are used in the farms. However, very few wind-powered water pumps are installed in the world, e.g. a wind turbine water pump in Nigeria in Goronyo in Katsina State, Kedada in Bauchi State and in Sokoto State. Presently, a 5-kW pilot wind turbine/generator is installed in Sayya Gidan-Gada village in Sokoto State, Nigeria. Other applications of wind power using water pumps are: domestic water supply, water supply for livestock, drainage, salt ponds, fish farms, *etc.*

 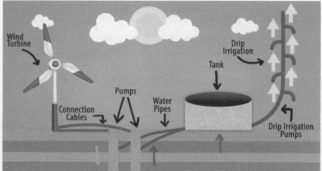

Fig. 16.14. Water pumping windmill

16.5.2. Electricity generation

The demand for electricity is growing with the increase in population, especially in rural areas which are not connected to the electrical grid. Therefore, provision of electricity to the remote rural communities can be made cheaply at the start from a wind power system as compared with other options, e.g. extension of grid power lines or other types of fossil fuel which are hard to transport due to poor road networks to the rural communities. Wind energy needs to be promoted first to the farms which need electricity to contribute significant improvement in the crop yield from agriculture (Fig. 16.15). Small, highly efficient wind turbines can be installed in rural farms. The cost of installing one wind turbine is close to that of putting up electrical poles, overhead power lines and other equipment necessary to connect to the electrical grid. The advantage is that the farm owner owns the generating equipment and is freed from paying monthly electrical bills.

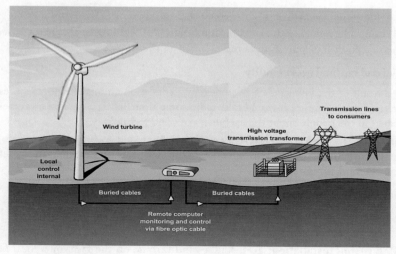

Fig. 16.15. Wind farms as a source of electrical energy for the farmers

16.5.3. Grinding grains and legumes

Wind energy can be used for grinding grains and legumes. Long before the invention of electricity, early wind turbines did very useful work for grinding. Windmills were used in many places in Europe over the last several centuries to turn heavy granite disks called millstones. The millstones were used to crush dry grains such as wheat, barley and corn to make flour or meal. This technique can be applied on farms for production of flour.

16.5.4. Environmental impact

Wind energy farming is an environmentally-friendly option, with the following features:

  » It is pollution-free: reduces air and noise pollution.
  » It does not require fuel for operation.
  » It does not produce toxic or radioactive waste.
  » It does not create greenhouse gases (each mega-watt hour generated by wind energy helps to reduce 0.8 to 0.9 tons/year of greenhouse gas emissions that are produced by coal or fossil fuel generation each year).
  » Reduces concentrations of CO_2, SO_2, NO_X, thereby reducing acid rain.
  » Concerns about noise pollution and "visual pollution" of the landscape.
  » When large arrays of wind turbines are installed on farmland, only about 2% of the land area is required for the wind turbines. The rest is available for farming, livestock, and other uses.
  » Birds could be killed when they run into the turbine.

16.5.5. Economic feasibility

  » Landowners, farmers and ranchers often receive payments for the use of their land, which enhances their income and increases the value of the land.
  » Operational and maintenance cost is low.

» Zero input fuel cost.

» It is domestic, reducing the need for importation of fossil fuels. This helps in reducing gas emission from transportation of fuels.

16.5.6. Social and cultural acceptability

» As a result of the low level of awareness and education among the populace, especially in the rural communities; it would take quite some time before a general awareness could be created in the people about the usefulness of wind technology.

» There could be tension over land between land owners and government or private companies over the installation of wind turbines due to the lack of awareness about the benefits of wind farms in local communities.

16.5.7. Shortcomings

The shortcomings of wind energy use in agriculture are numerous; however, a few have been highlighted below:

16.5.7.1. High cost of installation: To install a wind turbine on the farm by a local farmer is an expensive venture. Many are involved in subsistence agriculture and so cannot afford to install one. For example, a 10-kW turbine, which is the most common size for a home, will typically cost US$ 22,000–29,000. This amount is very large for the average farmer. However, the installation can be done by the local government or private companies who lease the land from the owner and in turn pay rent to the farmer.

16.5.7.2. Lack of spare parts: Presently, wind turbines and components are not manufactured in all parts of the world and so getting spare parts to replace any damaged or old parts can be a problem. To get these components would result in importation. This would result in the emission of poisonous gases during transportation of these parts.

16.5.7.3. Lack of skilled technical experts to repair the turbine in case of damage: Due to lack of many wind turbines in many parts of the world, there are not many trained and skilled technical experts that can repair these turbines when there is damage. This means that there would be a need to bring in foreign experts to repair and maintain wind turbines. This is not a sustainable approach as the maintenance of these turbines would depend largely on the importation of expatriates who would likely be paid more than what a local engineer could charge.

16.5.7.4. Low level of public awareness: The public awareness of renewable energy sources and technologies in many parts of the world and benefits, both economic and environmental, is generally low. Consequently, the public is not well-equipped to influence the government to begin to take more decisive initiatives in enhancing the development, application, dissemination and diffusion of renewable energy resources and technologies in the national energy market.

16.5.7.5. General absence of comprehensive national energy policy: Many parts of the world lag in formulation of a comprehensive energy policy; only sub-sectoral policies have been formulated. Since such a policy is pivotal to using energy-efficient renewable energy technologies, this has, to a large extent, contributed to the lack of attention paid to the renewable energy technologies.

16.6. BIOMASS ENERGY SYSTEMS

Biomass is a renewable energy resource derived from the carbonaceous waste of various human and natural activities. It is derived from numerous sources, including the by-products from the wood industry, agricultural crops, raw material from the forest, household wastes *etc.*

Biomass does not add carbon dioxide to the atmosphere as it absorbs the same amount of carbon in growing as it releases when consumed as a fuel. Its advantage is that it can be used to generate electricity with the same equipment that is now being used for burning fossil fuels. Biomass is an important source of energy and the most important fuel worldwide after coal, oil and natural gas. Bio-energy, in the form of biogas, which is derived from biomass, is expected to become one of the key energy resources for global sustainable development. Biomass offers higher energy efficiency through form of Biogas than by direct burning.

Biomass energy has the potential to supply a significant portion of energy needs, while revitalizing rural economies, increasing energy independence, and reducing pollution. Farmers would gain a valuable new outlet for their products. Rural communities could become entirely self-sufficient when it comes to energy, using locally grown crops and residues to fuel farm operations and tractors and to heat and power farm homes and buildings.

Opportunities for biomass energy are growing. Incentives are available to develop advanced technologies and crops to produce energy, chemicals, and other products from biomass. A number of states also provide incentives for biomass energy.

16.6.1. Resources on the farm

16.6.1.1. Biomass residues: Agricultural activities generate large amounts of biomass residues. While most crop residues are left in the field to reduce erosion and recycle nutrients back into the soil, some could be used to produce energy without harming the soil. Other wastes such as whey from cheese production and manure from livestock operations can also be profitably used to produce bioenergy while reducing disposal costs and pollution.

16.6.1.2. Energy crops: Crops grown for energy could be produced in large quantities, just as food crops are. While corn is currently the most widely used energy crop, native trees and grasses are likely to become the most popular in the future. These perennial crops require less maintenance and fewer inputs than do annual row crops, so they are cheaper and more sustainable to produce.

16.6.1.3. Grasses: Switch grass appears to be the most promising herbaceous energy crop. It produces high yields and can be harvested annually for several years before replanting. Other native varieties that grow quickly, such as big bluestem, reed canary grass, and wheat grass, could also be profitable.

16.6.1.4. Trees: Some fast-growing trees make excellent energy crops, since they grow back repeatedly after being cut off close to the ground. These short-rotation woody crops can grow to 40 feet in less than eight years and can be harvested for 10 to 20 years before replanting. In cool, wet regions, the best choices are poplar and willow. In warmer areas, sycamore, sweet gum, and cottonwood are best.

16.6.1.5. Oil plants: Oil from plants such as soybeans and sunflowers can be used to make fuel. Like corn, however, these plants require more intensive management than other energy crops.

16.6.2. Protecting the land

With thoughtful practice and management, perennial energy crops can improve the soil quality of land that has been overused for annual row crops. The deep roots of energy crops enhance the structure of the soil and increase its organic content. Since tilling occurs infrequently, the soil suffers little physical damage from machinery. One study estimates that converting a corn farm of average size to switch grass could save 66 truckloads of soil from erosion each year.

Perennial energy crops need considerably less fertilizer, pesticide, herbicide, and fungicide than annual row crops. Reduced chemical use helps protect ground and surface water from poisons and excessive aquatic plant growth. Furthermore, deep-rooted energy crops can serve as filters to protect waterways from chemical runoff from other fields and prevent sedimentation caused by erosion.

Finally, perennial energy crops can create more diverse habitats than annual row crops, attracting a wider variety of species such as birds, pollinators, and other beneficial insects, and supporting larger populations. Furthermore, the long harvest window for energy crops enables farmers to avoid nesting or breeding seasons.

16.6.3. Converting biomass to energy

Most biomass is converted to energy the same way it always has been—by burning it. The heat can be used directly for heating farm buildings, crop drying, dairy operations, and agro-industrial processes. It can also be used to produce steam and generate electricity. For example, many electric generators and businesses; burn biomass by itself or with other fuels in conventional power plants.

Biomass can also be converted into liquids or gases to produce electricity or transportation fuels. Ethanol is typically produced through fermentation and distillation, in a process much like that used to make beer. Soybean and canola oils can be chemically converted into a liquid fuel called biodiesel. These fuels can be used in conventional engines with little, if any, modification.

Biomass can be converted into a gas by heating it under pressure and without oxygen in a "gasifier." Manure too can be converted using a digester. The gas can then be burned to produce heat, steam, or electricity.

Other biogas applications are still in development, but show great potential. One promising technology is direct combustion in an advanced gas turbine to run a generator and produce electricity. This process is twice as efficient as simply burning raw biomass to produce electricity from steam. Researchers are also developing small, high-speed generators to run on biogas. These "micro-turbines" have no more than three moving parts and generate as little as 30 kilowatts, which could power a medium-sized farm. Several companies are also considering converting gasifier biomass into ethanol as a less expensive alternative to fermentation.

Alternatively, biogas can be processed into hydrogen or methanol, which can then be chemically converted to electricity in a highly efficient fuel cell. Fuel cells can be large enough to power an entire farm or small enough to power a car or tractor.

An innovative experiment in Missouri provides one example of the possibilities. Corn is used to produce ethanol, and the waste from the process is fed to cows for dairy production. Cow manure fertilizes the corn and is also run through a digester to produce biogas. A fuel cell efficiently converts the biogas into electricity to run the operation. The end products are ethanol, electricity, and milk. All the waste products are used within the project to lower costs.

16.6.4. Biomass briquetting

The process of densifying loose agro-waste into a solidified biomass of high density, which can be conveniently used as a fuel, is called Biomass Briquetting. Briquette is also termed as "Bio-coal". It is pollution free and eco-friendly. Some of the agricultural and forestry residues can be briquetted after suitable pre-treatment.

A list of commonly used biomass materials that can be briquetted are given below: corn cob, jutes tick, saw dust, pine needle, bagasse, coffee spent, tamarind, coffee husk, almond shell, groundnut shells, coir pith, bagasse pith, barley straw, tobacco dust, rice husk, deoiled bran.

16.6.5. Applications

16.6.5.1. Water pumping and electricity generation: Using biomass gas, it is possible to operate a diesel engine on dual fuel mode-part diesel and part biomass gas. Diesel substitution of the order of 75 to 80% can be obtained at nominal loads. The mechanical energy thus derived can be used either for energizing a water pump set for irrigational purpose or for coupling with an alternator for electrical power generation - 3.5 KW - 10 MW.

16.6.5.2. Heat generation: A few of the devices, to which gasifier could be retrofitted, are dryers for drying tea, flower, spices, kilns for baking tiles or potteries, furnaces for melting non-ferrous metals, boilers for process steam, *etc*. Direct combustion of biomass has been recognized as an important route for generation of power by utilization of vast amounts of agricultural residues, agro-industrial residues and forest wastes. Gasifiers can be used for power generation and available up to a capacity 500 kW. The Government of India through MNES and IREDA is implementing power-generating system based on biomass combustion as well as biomass gasification

16.6.6. High efficiency wood burning stoves

These stoves save more than 50% fuel wood consumption. They reduce drudgery of women saving time in cooking and fuel collection and consequent health hazards. They also help in saving firewood leading to conservation of forests. They also create employment opportunities for people in the rural areas.

16.6.6.1. Bio-fuels: Unlike other renewable energy sources, biomass can be converted directly into liquid fuels- biofuels- for our transportation needs (cars, trucks, buses, airplanes, and trains). The two most common types of biofuels are ethanol and biodiesel.

Ethanol is an alcohol, similar to that used in beer and wine. It is made by fermenting any biomass high in carbohydrates (starches, sugars, or celluloses) through a process similar to brewing beer. Ethanol is mostly used as a fuel additive to cutdown a vehicle's carbon monoxide and other smog-causing emissions. Flexible-fuel vehicles, which run on mixtures of gasoline and up to 85% ethanol, are now available.

Biodiesel, produced by plants such as rapeseed (canola), sunflowers and soybeans; can be extracted and refined into fuel, which can be burned in diesel engines and buses. Biodiesel can also be made by combining alcohol with vegetable oil, or recycled cooking greases. It can be used as an additive to reduce vehicle emissions (typically 20%) or in its pure form as a renewable alternative fuel for diesel engines.

16.6.6.2. Biopower: Biopower, or biomass power, is the use of biomass to generate electricity. There are six major types of biopower systems: direct-fired, cofiring, gasification, anaerobic digestion, pyrolysis, and small - modular.

Most of the biopower plants in the world use direct-fired systems. They burn bioenergy feedstocks directly in boiler to produce steam. This steam drives the turbo-generator. In some industries, the steam is also used in manufacturing processes or to heat buildings. These are known as combined heat and power facilities. For example, wood waste is often used to produce both electricity and steam at paper mills.

Gasification systems use high temperatures and an oxygen-starved environment to convert biomass into a gas (a mixture of hydrogen, carbon monoxide, and methane). The gas fuels a gas turbine, which runs an electric generator for producing power.

Methane can be produced from biomass through a process called anaerobic digestion. Anaerobic digestion involves using bacteria to decompose organic matter in the absence of oxygen. In landfills -scientific waste disposal site - wells can be drilled to release the methane from the decaying organic matter. The pipes from each well carry the gas to a central point where it is filtered and cleaned before burning. Methane can be used as an energy source in many ways. Most facilities burn it in a boiler to produce steam for electricity generation or for industrial processes.

Pyrolysis occurs when biomass is heated in the absence of oxygen. The biomass then turns into liquid called pyrolysis oil, which can be burned like petroleum to generate electricity. A biopower system that uses pyrolysis oil is being commercialized.

A small, modular system generates electricity at a capacity of 5 megawatts or less. This system is designed for use at the small-town level or even at the consumer level. For example, some farmers use the waste from their livestock to provide their farms with electricity. Not only do these systems provide renewable energy, they also help farmers meet environmental regulations.

16.6.7. Potential

An Oak Ridge National Laboratory (ORNL) study found that farmers could grow 188 million dry tons of switch grass on 42 million acres of cropland in the United States at a price of less than $50 per dry ton delivered. This level of production would increase total U.S. net farm income by nearly $6 billion. ORNL also estimates that about 150 million dry tons of corn stover and wheat straw are available annually in the United States at the same price, which could increase farm income by another $2 billion. This assumes about 40 per cent of the total residue is collected and the rest is left to maintain soil quality.

The energy crop production is limited to areas where these crops can be produced without irrigation and where sufficient research has been done to provide reliable information on yields and management requirements.

16.7. HYDRO ENERGY

The potential energy of falling water, captured and converted to mechanical energy by water-wheels, powered the start of the industrial revolution. Wherever sufficient head or change in elevation could be found, rivers and streams were dammed and mills were built. Water under pressure flows through a turbine causing it to spin. The turbine is connected to a generator, which produces electricity.

16.7.1. Small hydro power

Small hydro power is a reliable, mature and proven technology. It is non-polluting, and does not involve setting up of large dams or problems of deforestation, submergence and rehabilitation. India has an estimated potential of 10,000 MW.

16.7.2. Micro hydel

Hilly regions of India, particularly the Himalayan belts, are endowed with rich hydel resources with tremendous potential. The MNES has launched a promotional scheme for portable micro hydel sets for these areas. These sets are small, compact and light weight. They have almost zero maintenance cost and can provide electricity/power to small cluster of villages. They are ideal substitutes for diesel sets run in those areas at high generation cost.

Micro (up to 100 kW) mini hydro (101-1000 kW) schemes can provide power for farms, hotels, schools and rural communities, and help create local industry.

16.8. TIDAL AND OCEAN ENERGY

16.8.1. Tidal energy

Tidal electricity generation involves the construction of a barrage across an estuary to block the incoming and outgoing tide. The head of water is then used to drive turbines to generate electricity from the elevated water in the basin as in hydro-electric dams.

Barrages can be designed to generate electricity on the ebb side, or flood side, or both. Tidal range may vary over a wide range (4.5-12.4 m) from site to site. A tidal range of at least 7 m is required for economical operation and for sufficient head of water for the turbines.

16.8.2. Ocean energy

Oceans cover more than 70% of Earth's surface, making them the world's largest solar collectors. Ocean energy draws on the energy of ocean waves, tides, or on the thermal energy (heat) stored in the ocean. The sun warms the surface water a lot more than the deep ocean water, and this temperature difference stores thermal energy.

The ocean contains two types of energy: thermal energy from the sun's heat, and mechanical energy from the tides and waves.

Ocean thermal energy is used for many applications, including electricity generation. There are three types of electricity conversion systems: closed-cycle, open cycle, and hybrid. Closed cycle systems use the ocean's warm surface water to vaporize a working fluid, which has a low boiling point, such as ammonia. The vapor expands and turns a turbine. The turbine then activates a generator to produce electricity. Open-cycle systems actually boil the sea water by operating at low pressures. This produces steam that passes through a turbine / generator. The hybrid systems combine both closed-cycle and open-cycle systems.

Ocean mechanical energy is quite different from ocean thermal energy. Even though the sun affects all ocean activity, tides are driven primarily by the gravitational pull of the moon, and waves are driven primarily by the winds. A barrage (dam) is typically used to convert tidal energy into electricity by forcing the water through turbines, activating a generator.

16.9. GEOTHERMAL ENERGY

It is a type of energy that comes from the earth. It is trapped in the Earth's crust at a depth of 10 km in the form of hot springs, geysers, and other natural phenomena. About 250 hot springs with temperatures ranging from 90 to 130 degrees Celsius have been discovered in areas like Puga Valley in Ladakh, Manikaran in Himachal Pradesh, and Tattapani in Chhattisgarh, indicating that India's geothermal potential is largely found along the Himalayas. The National Aerospace Laboratory in Bangalore has established a pilot project near Manikaran for research and development as well as data collection in order to construct larger geothermal power plants.

16.10. RECOMMENDATIONS AND POLICY DEVELOPMENT

There is high potential for solar, wind and biomass energy in various parts of the world, especially in India. Hence, the following recommendations are hereby made to further help formulate strategies and measures in implementing all renewable energy projects.

» To embark on a massive education program to enlighten the citizens, particularly in the rural communities, about the benefits of renewable energy sources.

» To develop, promote and harness the renewable energy resources of the country and incorporate all viable ones into the national energy mix.

» Provide various opportunities to the communities and allow them to choose the best renewable source of energy suitable for them.

» To promote decentralized energy supply, especially in rural areas, based on renewable energy resources.

» To de-emphasize and discourage the indiscriminate use of fossil fuels and wood.

» To promote efficient methods in the use of wind energy resources.

» To keep abreast of international developments in renewable energy technologies and applications.

16.11. CONCLUSION

The only realistic solution to the problem of non-renewable energy is to find sources of renewable energy to replace today's dwindling supplies of affordable and usable fossil energy. Solar energy is the only source of truly renewable energy – renewable at least for the next few billion years. Windmills, falling water, solar collectors, and photovoltaic cells are all sources of renewable solar energy. The most common solar energy collectors are green plants. After all, plants were the original collectors of today's fossil energy. So, it is only logical to look to agriculture as a renewable source of alternative energy for the future. However, we need to be realistic about the extent to which energy from agriculture can replace our current use of fossil energy. While the energy experts may not agree on specific quantities or percentages, the overall limits on energy from agriculture are fairly basic and straight forward.

The use of renewable energy is rapidly growing. Various technologies exist that are applicable to agricultural use of renewable energy. Technology is constantly improving and costs are reducing. Funding models and incentives are also available. Long-term benefits are likely to accrue. Job creation and local manufacturing potential exists.

Use of renewable energy (RE) in rural remote areas of many developing countries could help farmers to increase agricultural productivity as well as to earn more money by value addition to their produce (e.g. controlled drying of fruits and vegetables, cheese production from milk, off-seasonal production of fruits and vegetables with irrigation, etc.). The potential for using REs in the agricultural value chain is plentiful and often has many advantages compared to conventional technologies like diesel generators. A high level of integration of RE into an agricultural process can lead to high efficiencies, low environmental impact and low production costs. There is always some sort of RE resource available in any location, it is just important to choose the adequate source or a good combination of sources.

Chapter - 17

The Way Ahead

17.1. INTRODUCTION

Agriculture in the 21st century has multiple challenges. It has to produce more food and fiber to feed a growing population with a smaller rural labor force, more feed stocks for a potentially huge bioenergy market, contribute to overall development in the many agriculture- dependent developing sustainable production methods and adapt to climate change (FAO, 2012). Agriculture continues to be a source of livelihood for majority of Indian population and contributed about 14 per cent to the gross domestic product (GDP) of the country in 2014-15. About 60% population of India is dependent on agriculture for their livelihood. India experienced tremendous increase in agricultural productivity and food supplies that propelled the country into food self-sufficiency from a situation of acute food shortages in the 1960s and 1970s. Between 1966-67 and 2016-17 production of food grains increased by three-fold (from 95 to 275 million tons), of fruits and vegetables by seven-fold (from 40 to 268 million tons) and of milk by eight-fold (from 20 to 163 million tons). On the whole, agricultural sector during this period grew at an annual rate of around 3 per cent that helped millions of rural people escape poverty (Datt and Ravallion, 1998; Datt *et al.*, 2016).

Despite such a revolutionary progress in agriculture the economic condition of farmers remains deplorable on account of several factors, such as the declining scale of production, deceleration in technological gains, increasing frequency of extreme climatic events (e.g., droughts, floods and heat-waves), rising input prices, increasing volatility in food prices and lack of income opportunities in non-agricultural sectors. The per capita income of farmers is just one-fifth of the national per capita income, and a majority of them, especially those at the bottom of land distribution, are stuck in a low-income trap.

Anticipating that the consequences of continued agrarian distress could be disastrous for the nation's food security and economic growth, the Union government in its annual budget of 2016-17 set a target to double farmers' incomes by 2022-23. In the absence of strategic investments and innovations in farm sector and its linkages with non-farm sector it would be difficult to achieve the target in such a short period of time. If the target has to be achieved by 2022-23, farmers' income has to grow at least at 10 per cent a year.

Towards this end, both the Union and state governments have taken several measures, such as promotion of micro-irrigation systems, crop insurance, horticulture and animal husbandry; restoration of soil health; linking farmers to markets and incentives for development of value chains; raising support prices of important crops; increasing flow of institutional credit, and creation of employment opportunities outside agriculture. The underlying assumption is that all the farmers, irrespective of their economic status and geographical location, would benefit from these interventions.

The pathway for enhancing farmers' income requires consideration of different dimensions related to enhancement in agricultural production along with providing efficient markets and improved marketing facilities. As area expansion is limited, production enhancement can be done through bridging yield gaps in crops through adoption of efficient and effective cultivation practices, crop diversification with focus on high value crops, further improvements in the total factor productivity, proper irrigation management along with other factors leading to productivity improvements (GOI, 2007; Evenson *et al.*, 1998; Chand *et al.*, 2012; Birthal *et al.*, 2007). The losses in India's agricultural produce is estimated to be Rs. 926,510 million, approximately US$ 13 billion (MoFPI, 2016), indicating that there is need for post-harvest management, better infrastructure and proper management to prevent these losses. The efficient marketing network would be the key factor for monetization of the output and realization of better gains.

17.2. RAINBOW REVOLUTION

The agricultural policy of 2000 envisaged holistic development of Indian agriculture and aimed to achieve through rainbow revolution. Economic survey 2015-16 observed, "Indian agriculture is in a way, a victim of its own past success, especially the Green revolution". It suggested an integrated development program to make the agricultural sustainability and rainbow revolution as a concept was developed eventually. Many revolutionary events occurred in the field of agriculture which further led to growth and development across the globe. However, it is also necessary to understand the drawbacks of various revolutions while learning from the strengths out of it.

Rainbow revolution is a proper solution, because it integrates all agriculture and allied sectors in a sustainable manner. This demands proper coordination, implementation and monitoring of the support policies in addition to the allocation of resources. The farmers had low to medium nature of favorable attitude towards IPM and green manuring i.e. organic farming and the level of sensitivity to minimize global warming problem was medium among the agricultural research scholars. Motivating factors like education, economic status, cosmopoliteness, extension media exposure, risk orientation and level of aspiration had positive and highly significant association with farmers' perception towards using agrochemicals. Appropriate and effective institutional vehicles with PPP are essential to develop and integrate all on farm and non-farm sector policy and interventions in a sustainable manner. Agricultural extension and rural advisory services are necessary to bring positive attitude, knowledge, technologies, and services towards sustainability among human resources involved in agricultural activities.

17.3. KEY CHALLENGES

17.3.1. Growing competition for land

The competition for land has been intensifying due to its growing demand in non-agricultural activities like housing, manufacturing and infrastructure. From 6.2 per cent in 1975-76 the share of non-agricultural land in the total land area has increased to 8.7 per cent in 2014-15. On the other hand, the net sown area has almost stagnated around 141 million hectares.

17.3.2. Smallholder farmers

Between 1981 and 2011 the number of landholdings increased from 89 million to 140 million and their average size declined from 1.84 hectares to 1.15 hectares. The number of landholdings measuring less than or equal to one hectare almost doubled raising their share to 67 per cent in the total land holdings. If this trend continues unabated, for majority of the households the agriculture-based livelihoods are likely to be untenable.

17.3.3. Growing stress on water resources

According to an estimate by the World Resources Institute about 54 per cent of India's land is water stressed (WRI, 2015). Of the potential availability of water of 1869 billion cubic meters (bcm) about 1123 bcm (690 surface and 433 groundwater) is utilizable. Agriculture, with a share of 85 per cent in the total water demand (813 bcm), is the largest consumer of water. Irrigation alone accounts for 90 per cent of groundwater withdrawal and is responsible for its over-exploitation. According to a study by the World Bank (2005), if the rate of groundwater extraction remains unabated about 60 per cent aquifers in the country would go dry by 2025.

17.3.4. Climate change

Climate change is emerging as a big threat to sustainable development of agriculture and agriculture-based livelihoods. By 2035, agricultural productivity under changed climate is likely to be 6 per cent less (Birthal *et al.*, 2014a), ranging from 1 per cent to 10 per cent across crops (Birthal *et al.*, 2014b). The consequences of extreme climatic events (droughts, floods, cyclones and heat waves), frequency of which is predicted to increase (World Bank, 2013; Birthal *et al.*, 2015b) will be more severe. Not only that changes in climate would induce greater risks of insect pests and diseases, in terms of their increased resurgence and resistance to pesticides, and also emergence of new pest strains.

17.3.5. Energy shortage

Agriculture uses considerable amount of energy, directly and indirectly, and it is becoming energy-intensive. Between 1980-81 and 2006-07, commercial energy consumption in Indian agriculture increased six-fold, from 3.04 thousand MJ (mega joules) to 18.48 thousand MJ per hectare, the maximum increase being for electricity and diesel (Jha *et al.*, 2012). Prices of electricity, petrol and diesel have been increasing, and their growing demand will push up energy prices further. While the rising prices of energy inputs will add to the cost of production, their lesser use will adversely affect agricultural productivity.

17.3.6. Limited R and D system

India's agricultural R and D system has focused on raising crop yields, but less on stress-tolerance and natural resource management. Further, linkages between research and its dissemination systems are weak. Sixty per cent farm households do not have access to information on modern agricultural technologies and practices (Birthal *et al.*, 2015c). The outreach of government extension system is limited only to 7 per cent of the households. There is a need for re-orientation of agricultural research agenda encompassing multiple disciplines and a greater focus on basic and strategic research, and improving outreach and efficiency of agricultural services.

17.3.7. Financing smallholders

Credit requirements of smallholders are trivial, the financial institutions because of the higher transaction costs and lending risks associated with small amounts hesitate to provide loans to them. There is a bias in the allocation of institutional credit; the high-value, high-growth poor-poor segments of agriculture such as animal husbandry (including fisheries) and horticulture together receive about 5 per cent of the total institutional credit or 15 per cent of the total investment credit as against their share of more than 50 per cent in the total value of agricultural output.

17.3.8. Improper markets

Farmers are vulnerable to the exploitation by traders in informal as well as formal markets (Negi *et al.*, 2018; Meenakshi and Banerji, 2005). Further, the increasing volatility in agricultural prices and consumer concerns for safe and quality food are posing a significant challenge to farmers' participation in market-oriented agri-food systems. There is an apprehension that in the absence of appropriate institutional arrangements and social safety nets, the small-scale producers, entrepreneurs and processors will be more affected by globalisation.

17.4. STRATEGIES FOR RAINBOW REVOLUTION

The comprehensive strategies for Rainbow Revolution are presented below:

» Production technologies
- Reducing yield gaps
- Improved crop varieties and hybrids
- Enhancing crop density
- Efficient water management (microirrigation)
- Integrated nutrient management
- Fertigation
- Plant bioregulators
- Resource use efficiency
- GM crops
- Protected cultivation
- Precision agriculture
- Diversification towards high value crops and livestock
- Integrated crop-livestock farming systems
- Renewable energy
- Integrated pest management
- Increase in total factor productivity

» Post-harvest management and value addition
- Reducing post-harvest losses
- Encouraging value addition and processing

» Transfer of technology

» Marketing and price realization
- Improving price realization to the farmers
- Reducing price volatility

» Enhancing non-farm income

- Creating non-farm opportunities
- Reducing dependence on agriculture
» Enabling policies
  - Natural resource management
  - Infrastructure support
  - Enable credit environment
  - R and D and extension services
  - Risk management-integrated approaches
  - Marketing strategies
  - Scaling up and aggregation
  - Off farm/non-farm/wage income
  - Strengthening/widening partnership
  - Crop insurance
  - Linkages among the organizations and stakeholders
  - Prioritization of areas for investment
  - Centre-state linkages
  - Making the farmers' party to the mission

17.4.1. Increasing productivity

» Use of high yielding, nutritious, pest and disease resistant varieties including use of GM crop cultivars. Only 30% of hybrid/HYV seeds used in cultivation is marketed one. Hence, the space available for use of HYV/hybrid seeds in the country is self-evident.

» Development of hi-tech nurseries to raise quality planting material and selling to the fellow farmers at reasonable price.

» Use of precision agriculture - application of precise amount of inputs such as seeds, water, fertilizers, pesticides, at correct time to the crops for economizing on inputs, increasing productivity and maximizing yields.

» Efficient use of irrigation – Only 50% of the requirement for irrigation can be met by the year 2030. The reason is rate of depletion of water resources is three times more than the rate of creation of new irrigation capacities. That is why significance is being given to water conservation works in scheme like GOI *Pradhan Mantri Krishi Sinchai Yojana*. Focus on proper management of water assets may act as a tipping point. Many studies have shown that practice of participatory irrigation management (PIM) i.e. community management of water assets led to substantial increase of area under irrigation and reduction in water use. Drip irrigation including fertigation (reduce fertilizer requirement) and chemigation (reduce pesticide requirement) should be followed.

» Optimum use of fertilizers including bio-fertilizers based on soil testing. Integrated use of both organic manures and chemical fertilizers. Issue of soil health card for more than 90% of farm holdings in India no doubt is going to be a game changer with regard to judicious use of inputs.

» Growing more than one crop per year wherever irrigation facilities are available.

» Diversification towards high value crops including horticultural crops and livestock.

» Using integrated pest management (IPM) technologies for management of crop pests. Reduced use of chemicals and increased use of eco-friendly technologies like cultural and biological methods and host resistance. Across the world, IPM is getting popularized among farmers as it is 50% cheaper than chemical pesticides, ensures higher productivity by 10-15 % and leaves very less harmful residues. Moreover, shifting to IPM can get at least 10-15% higher selling price for farmers if agricultural products grown through IPM is certified as such.

» Inter-linking of rivers, construction of check dams and farm ponds to enhance irrigation facilities.

» Peri-urban agriculture to supply fresh vegetables and fruits to nearby cities.

» Farm mechanization – use of small tractors, laser land levellers, precision seeders/ planters, fertilizer and pesticide applicators. In a standard farming practices, almost 50% of the cost of production is the labor component. Of late, rural labor market is showing an irreversible trend of rise in wage rate and shortage of labor. On the other hand, use of diverse agricultural machineries, perhaps on the line of agricultural machinery cooperatives in China, can reduce cost of production and increase farm productivity each by 20-30%.

» Prevention of post-harvest losses using warehouses (godowns) for storage of grains, cold chains for perishables, and food processing for value addition. Cooperatives can play a significant role in providing high tech storage and packaging facilities *etc.*

» Integrated farming systems – simultaneous rising of crops along with cattle, poultry, mushroom cultivation, bee-keeping, fisheries, sericulture, *etc*. In specific areas of horticulture, pisciculture, poultry farming *etc*., the reward is very high to consummate with risk involved. Compared with traditional farming, pisciculture is three times more rewarding. Similarly, there is lot of market potential in the cultivation of aromatic and herbal plants.

17.5. FUTURE PROSPECTS

17.5.1. Targeting growth strategies

Rainbow revolution in agriculture is quite challenging but it is needed and is attainable. Three-pronged strategy focused on (i) development initiatives, (ii) technology, and (iii) policy reforms in agriculture is needed to achieve the rainbow revolution.

» The rates of increase in sources underlying growth in output need to be accelerated to meet the goal.

» The country need to increase use of quality seed, fertilizer and power supply to agriculture.

» Area under irrigation has to be expanded and area under double cropping should be increased.

» Besides, area under fruits and vegetables is required to increase.

» In the case of livestock, improvement in herd quality, better feed, increase in artificial insemination, reduction in calving interval and lowering age at first calving are the potential sources of growth.

» Three-fourths of the marginal farmers who stay at the bottom of income distribution, should be at the forefront of any developmental strategy.

» The efforts and investments should concentrate more on eastern region that has lagged behind in agricultural and economic development and is home to about 60 per cent of the total low-income marginal farmers.

» There are prospects of raising farmers' income within agriculture by improving resource use efficiency and diversifying production towards high-value crops and animal husbandry.

» For majority of households, farming alone is not sufficient to escape low-income trap, and recourse has to be with the development of rural non-farm sector.

With these strategies in mind, some important interventions for sustainable improvements in farming and farmers' income include:

The agricultural policy should emphasise exploiting sustainable intensive agriculture. Currently, only about 40 per cent of the country's net cropped area is cultivated more than once. This needs to be raised by improving farmers' access to reliable irrigation sources and seeds of short-duration high-yielding crops/varieties befitting the existing cropping systems.

Irrigation is important for raising farm productivity; and given the acute scarcity of water, the recourse has to be with improving water-use efficiency, which is 30-40 per cent for surface irrigation and 60-70 per cent for groundwater irrigation. Modern methods of irrigation, viz. sprinkler and drip systems can reduce water and energy consumption by 12-84 per cent and 29-45 per cent, respectively and without any yield penalty. Currently, only about 9 million hectares of area is irrigated through micro-irrigation systems, much less than the potential of 42 million hectares (Palanisami *et al.*, 2011).

The conservation agriculture technologies, such as zero-tillage and laser-levelling and water scheduling devices (e.g., tensiometer) improve water and nutrient-use efficiency (Jat *et al.*, 2009; Vatta *et al.*, 2018), reduce cost of production and improve crops' resilience to extreme climatic events, such as droughts and heat waves.

There is a need for improving non-renewable energy-use efficiency and exploiting renewable energy resources, for example, solar and wind power. Vatta *et al.* (2018) have shown that scheduling of irrigation through tensiometer can save electric power by 13 per cent. Also there is a considerable scope to convert agricultural waste into energy. Agriculture generates 686 million tons of crop residues, of which one-third is available for electricity generation, and if this amount is utilized for electricity generation, it can meet 17 per cent of the total primary energy demand in the country (Hiloidhari *et al.*, 2014). India has also huge livestock population producing 2,600 million tons of dung a year that can be used to produce 263,702 million cubic meter of biogas or 476 terawatt hours of electricity (Kaur *et al.*, 2017). Solar farming is a potential source of energy as well as of income for farmers.

In order to enhance farmers income, the primacy of cereals has to give way to diversification towards high-value high-growth pro-poor sectors, such as horticulture and animal husbandry (Birthal *et al.*, 2015d; Birthal and Negi, 2012). The sustained rise in per capita income and expanding urbanization are triggering rapid growth in demand for high-value food commodities including animal products (Joshi and Kumar, 2016).

To make markets competitive and remunerative for farmers, the Union government has taken several initiatives to develop value chains through institutional arrangements (e.g., farmer producer organisations (FPOs) and contract farming). These institutions need to be promoted aggressively so as to improve transparency in price discovery, to reduce trade cost associated with small surpluses, and to reduce farmers' dependence on informal traders and input dealers for credit.

Of late, the Union government has decided to fix minimum support prices (MSP) of important crops 1.5 times of the cost of production This is unlikely to benefit much the farmers as the government procures only rice and wheat and small quantities of pulses and coarse cereals. A number of commodities, including fruits, vegetables and milk that make sizable proportion of farmers' income remain outside the purview of price policy.

In view of the limited outreach of the government extension system, the mobile and internet can serve as an important vehicle for dissemination of information. Farmers need a bundle of information, and, therefore, the need is to develop backend infrastructure for compilation, processing and dissemination of information.

There is a crucial link between infrastructure and farmers' income, but lack of complementarities among different infrastructures restrict farmers capturing benefits of the existing infrastructures. This implies a need for holistic development of infrastructure taking into consideration the complementarities among different types.

To overcome the problem of surplus cattle due to farm mechanization, the option is to promote of sex semen technology that provides farmers a choice of desired sex of the offspring. At present, the technology is imported, and is costlier, and there is an urgent need to invest in research on sex semen technology.

In the long-run, boost to farmers' income must come from technological breakthroughs that push yield frontiers, enhance resource use efficiency, reduce cost of production and improve resilience of agriculture to climate change. This means more allocation of resources for agricultural research, improvements in efficiency of research and reorientation of research agenda to address the emerging challenges. Currently, India spends only about 0.6 per cent of its agricultural GDP on agricultural research and development (Beintema *et al.*, 2012). This needs to be raised at least to 1 per cent of the agricultural GDP.

For sustained rise in farmers' incomes, the constraint due to ubiquitous smallholdings can be mitigated through the strategies for broad-based growth of rural non-farm sector would be required. There is considerable scope for rural industrialization as agriculture generates considerable surpluses for manufacturing of value-added products. The expanding rural non-farm sector will create opportunities for investment in ancillary industries related to inputs, equipment, machines and support services, and generate incomes for investment in farm production. Investment in human capital and value chains will be a key to rural industrialization.

17.5.2. Options for future

The future additional or promising yields would be possible by allocating more area to the crops, which have comparative yield advantage, and improving their productivity through advanced technologies. The need for more food has to be met through higher yields per unit of land, water, energy and time. Further, explore the possibilities to increase yields by using modern inputs to boost yields levels and growth of output. India is fortunate to have a large arable land, which many other nations do not have. Sow seed of success is a reformist blueprint for second green revolution. Investment in infrastructure instead of direct subsidies; special program for irrigation; diversification and intensification of cropping pattern would definitely boost yields.

Fortunately, as we approach the new millennium we are experiencing three major revolutions in science and technology in a fundamental manner.

» The gene revolution: It provides a molecular understanding of the genetic basis of living organisms, as well as the ability to use this understanding to develop new processes and products for agriculture, the environment and for human and animal health.

» The information and communication revolution: It allows rapid growth in the systematic assimilation and dissemination of relevant and timely information, as well as an improved access ability to the universe of knowledge and communication through low cost electronic networks.

» The Eco-technology revolution – It promotes the blending of the best in traditional knowledge and technology with frontier technologies, renewable energy and new materials.

In principle, these three types of advances when coupled with improvements in management science and governance greatly increase the power of a scientific approach to genetic and agronomic improvement and the integrated management of local and regional developmental policies.

17.6. CONCLUSION

There is a need to attain rainbow revolution covering the entire spectrum of activities in agriculture, which will make India a developed nation free of poverty, hunger, malnutrition, and make the country environmentally safe. Major approaches include re-orienting focus from intensification to diversification, from sustenance to commercialization, and turning the agricultural units to enterprises. Different central and state level programs have been floated to execute and monitor the outreach of technologies, soil health, farm credit and market to the farmers. Price supports are triggered for many of the crops and entrepreneurship is inculcated to the farming community.

» To secure future of agriculture and to improve livelihood of 60% of India's population dependant on agriculture, adequate attention needs to be given to improve the welfare of farmers and raise agricultural income. Respectable income in farm sector will also attract youth towards farming profession and ease the pressure on non-farm jobs, which are not growing as per the expectations.

» Research Institutes should come up with technological breakthroughs for shifting production frontiers and raising efficiency in use of inputs. Evidence is growing about scope of agronomic practices like precision farming to raise production and income of farmers substantially. Similarly, modern machinery such as laser land leveller, precision seeder and planter, and practices like System of Rice Intensification (SRI), direct seeded rice, zero tillage, raised bed plantation and ridge plantation allow technically highly efficient farming. They require strong extension for the adoption by farmers.

» ICAR and SAUs should develop models of farming system for different types of socio-economic and bio-physical settings combining all their technologies in a package with focus on farm income. This would involve combining technology and best practices covering production, protection and post-harvest value addition for each sub-system with other sub-systems like crop sequences, crop mix, livestock, horticulture, and forestry. Such shift requires interdisciplinary approach to develop on knowledge of all disciplines.

» About one third of the increase in farmers' income is easily attainable through better price realization, efficient post-harvest management, competitive value chains and adoption of allied activities. This requires comprehensive reforms in market, land lease and rising of trees on private land. Agriculture has suffered due to absence of modern capital and modern knowledge. There is a need to liberalize agriculture to attract responsible private investments in production and market. Similarly, FPOs and FPCs can play big role in promoting small farm business. Ensuring MSP alone for farm produce through competitive market or government intervention will result in sizeable increase in farmers' income in many states.

Most of the development initiatives and policies for agriculture are implemented by the States. States invest much more than the outlay by the Centre on many development activities like irrigation. Progress of various reforms related to market and land lease are also State subjects. Therefore, it is essential to mobilize States and UTs to own and achieve the goal of Rainbow Revolution. If concerted and well-coordinated efforts are made by the Centre and all the States and UTs, the country can achieve the goal of Rainbow Revolution.

> » As most of the farmers in the country lie in the marginal and small category with very small holding size which makes the diffusion of advanced technologies difficult. The holdings are tiny and scattered particularly in the hilly areas. Thus, land consolidation coupled with other suitable reforms need to be effectively implemented. Further, the climatic risks are resulting in decline in productivity and creating distorting impact on prices. Thus, risk management is an essential component to be studied in detail.

> » Despite of all strengths, an inherent drawback in these approaches has been that many of them operate in isolates, lacking a unified framework that numerically explains various factors that contribute to Rainbow Revolution. Of course, strategies could only be region-specific and could operate only under given agro-climatic forces. The sources would ultimately vary at disaggregated geographies, demanding numerous approaches at different environments. Still, the macro forces that explain the short-future, and possible contributions of factor productivity, labor transformation, terms of trade and market volatility demands high place in achieving the goal. There is a need to devise strategies to Rainbow Revolution, balancing both macro- and meso- environments. The study covers possible contributions of different sources at both national level and at state level.

> » As with any crisis of our time, Rainbow Revolution will require the efforts of all stakeholders. Through increased collaboration and partnerships, we can leverage the resources, expertise, and tools of the collective whole. The "Green Revolution" demonstrated the potential for science to bring countries from famine to a surplus of food. The world must again embrace collective innovation to achieve global food and nutrition security through "Rainbow Revolution". We will need to support the full array of innovative solutions that are available to farmers, including agricultural biotechnology, for Rainbow Revolution.

> » India has already achieved resilience in agriculture (including the horticultural sector) through effective agricultural technology generation, and suggests that the country is now on the threshold of a "rainbow revolution" that will ensure both household nutrition security and prosperity for its people.

> » To achieve this, the country need to reformulate the agricultural policies, which must focus on area allocation policy and agricultural productivity policies. Agriculture is being in the state list, the center is needed to take consensus of state governments. The area allocation under different crop sectors and principal crops must be based on spatial and temporal comparative yield advantage with same existing allocation of land. Further, there is a need to make comprehensive integrated policy of cropping and farming systems with incorporation of crops, fisheries and livestock farming. Perhaps, these may require cooperation and coordination from farmers. Farmer may not be inclined to change the adopted structured land use-cropping pattern but through price support mechanisms and other incentives this could be achieved. Further, the country needs to formulate national productivity policies suited to the crops, which have comparative yield advantage. Technologies have to be developed to increase the productivity of the major crops and a strong extension need to be put in place to reach these technologies to the farmers.

> » Self-reliance in food grains is perhaps our most significant achievement. If the current production is an indication, meeting the projected doubling of food production to feed 950 billion people by 2050 AD will not be difficult. Integration of the philosophies of different revolutions in a sustainable manner would lead the country to attain Rainbow Revolution.

References

Abate, T. (ed.). 2012. *Four Seasons of Learning and Engaging Small Framers: Progress of Phase 1.* International Crops Research Institute for the Semi-Arid-Tropics, PO Box 39063, Nairobi, Kenya, 258 pp.

Abd El Moneem, K.M.H., Fawaz, S.B.M., Saeed, F.A. and El Shehaby, A.I. 2005. Effect of clove size and certain micronutrients on Fusarium & basal rot of garlic. *Assiut J. Agril. Sciences* 36: 163-175.

Abdallah, 1998. Improving vegetable transplants using soil solarization. II. Onion (*Allium cepa*). *Ann. of Agril Sci., Special Issue* 3: 831-843.

Abdel A.S.M and Haroun, M.S. 1990. Efficiency of some herbicides on weed control and yield of onion (*Allium cepa* L.). *Egyptian J. Agron.* 15: 35-44.

Acamovic, T, Sinurat, A, Natarajan, A. *et al.* 2005. Poultry. In E. Owen, A. Kitalyi, N. Jayasuriya and T. Smith (eds.) *Livestock and Wealth Creation: Improving The Husbandry of Animals Kept by Resource-Poor People in Developing Countries.* Nottingham University Press, Nottingham, UK.

Adesogan, A.T., Havelaar, A.H., McKune, S.L. *et al.* 2020. Animal source foods: Sustainability problem or malnutrition and sustainability solution? Perspective matters. *Global Food Security* 25:100325.

Aghora, T.S. and Pathak, C.S. 1991. Heterosis and combining ability in a line x tester cross of onion (*Allium cepa* L.). *Veg. Sci.* 18: 53-58.

Agricultural Statistics at a Glance. 2015. Government of India Ministry of Agriculture & Farmers Welfare Department of Agriculture, Cooperation & Farmers Welfare, Directorate of Economics and Statistics, New Delhi.

Agriwatch. 2018. *Cotton News: USDA Estimates 2018-19 India's Cotton Output Down by 2%.* Accessed February 21, 2019. http://www.agriwatch.com/newsdetails.php? USDA -estimates-2018-19-Indias-Cotton-output-down-by-2%&st=NEWS&commodityid=24 &sid=471851.

Ahir, R.P. and Maharishi R.P. 2008. Effect of pre-harvest application of fungicides and biocontrol agent on black mold (*Aspergillus niger*) of onion in storage. *Ind. Phytopath.* 61: 130–131.

Ahmad, S. and Karimullah. 1998. Relevance of management practices in downy mildew in onion. *Sarhad J. Agri.* 14: 161-162.

Ameta, O.P., Sumeriya, H.K., Mahla, M. and Vyas, A.K. 2001. Yellow revolution: past trends and future potential. *Intensive Agric* 39(5–6): 3–6.

Annapurna, K., Bojappa, K.M. and Bhargava, B.S. 1988. Leaf sampling guide for sapota (*Manilkara achras* M. Foseberg) cv. Cricket Ball. *Crop Res.* 1: 69-75.

Anon. 1994. "Notices." *Federal Register 59, No. 99 (1994).* http://www.aphis.usda.gov/brs/aphisdocs2/93_25801p_com.pdf.

Anon. 2011. "How'd We 'Make' a Non-browning Apple?" *Arctic Apples* (blog), December 7, 2011. Accessed August 11, 2015, http://www.arcticapples.com/blog/julia/how-did-we-make-nonbrowning-apple.

Anon. 2012. *Del Monte Inquiry Letter to APHIS BRS*, July 30, 2012. Accessed August 11, 2015, http:// www.aphis.usda.gov/biotechnology/downloads/reg_loi/del_monte_inquiry_letter.pdf;

Anon. 2013. "Simplot Says It Has Made a Better Potato." *Idaho Statesman*, May 9, 2013. Accessed July 31, 2015, http://www.idahostatesman.com/2013/05/09/2569097/a-better-potato.html.

Anon. 2014a. *"Anti-GMO Activists in Bangladesh Tell Lies to Farmers and the Media,"* April 8, 2014. Accessed July 31, 2015, http://btbrinjal.tumblr.com/post/82090416816/anti-gmo-activists-in-bangladesh-tell-lies-to;

Anon. 2014b. "Super-rice Defies Triple Whammy of Stresses." *New Scientist*, February 28, 2014. Accessed July 31, 2015, https://www.newscientist.com/article/dn25147-super-rice-defies-triple-whammy-of-stresses/.

Anon. 2015a. *"Roundup Ready Soybean Patent Expiration."* Monsanto. Accessed July 31, 2015, http:// www.monsanto.com/newsviews/pages/roundup-ready-patent-expiration. aspx.

Anon. 2015b. "Press Release: Non-browning Arctic® Apples to Be Granted Approval." *Arctic Apples* (blog), February 13, 2015. Accessed August 11, 2015, http:// www.arcticapples.com/ blog/joel/press-release-nonbrowning-arctic%C2%AE-apples-be-granted-approval.

Anon. 2015c. "Press Release: Arctic® Apples Receive Canadian Approval." *Arctic Apples* (blog), March 20, 2015. Accessed August 11, 2015, http://www.arcticapples. com/ blog/joel/press-release-arctic%C2%AE-apples-receive-canadian-approval.

Anon. 2015d. Haro von Mogel, K. 2015. *"Q&A with Haven Baker on Simplot's Innate™ Potatoes."* Biology Fortified, May 8, 2013. Accessed July 31, 2015, http://www. biofortified.org/2013/05/ qa-with-haven-baker-innate-potatoes/.

Anon. 2015e. *"Patent Application Title: Pineapple Plant Named Rose (EF2-114)."* PatentDocs. Accessed August 11, 2015, http://www.faqs.org/patents/app/ 20130326768.

Anon. 2015f. *"Myths & Facts,"* *AquaBounty Technologies*. Accessed July 31, 2015, http:// aquabounty. com/press-room/myths-facts/.

Anon. 2015g. *"Who Invented Golden Rice and How Did the Project Start?"* Golden Rice Project. Accessed August 11, 2015, http://www.goldenrice.org/Content3-Why/why3_ FAQ. php#Inventors.

Anon. 2015h. "Cotton Crop Scraps Become Healthy Snacks." *Sciencemuseum*, November 24, 2006. Accessed August 11, 2015, http://www.sciencemuseum.org.uk/antenna/non-toxiccotton/.

Anon. 2018. *Horticultural Statistics at a Glance 2018*. http://agricoop.nic.In/sites/ default/ files/.

Arya, P.S. and Bakshi, B.R. 1999. Onion based cropping systems studies under mid-hill conditions of Himachal Pradesh. *Adv. Horti. Forestry* 6: 79-85.

Aulakh, M.S., Sidu, B.S., Arora, B.R. and Singh, B. 1985. Content and uptake of nutrients by pulses and oilseed crops. *Indian J Ecol* 12(2): 238–242.

Ayala-Zavala, J.F., Vega-Vega, V., Rosas-Domínguez, C. *et al.* 2011. Agro-industrial potential of exotic fruit by-products as a source of food additives. *Food Research International* 44(7): 1866- 1874.

Ayyappan, S. 2012. Indian fisheries: Issues and the way forward. *National Academy of Science Letters* 35: 1-6.

Ayyappan, S., Jena, J.K., Gopalakrishnan, A. and Pandey, A.K. 2013. *Handbook of Fisheries and Aquaculture*. Directorate of Knowledge Management in Agriculture, Indian Council of Agricultural Research, New Delhi, India, 1116 pp.

BAHS (Basic Animal Husbandry Statistics). 2019. Department of Animal Husbandry, Dairying and Fisheries. Ministry of Agriculture, Government of India, Available from, http://dadf.gov.in/sites/default/filess/BAHS%20%28Basic%20Animal%20 Husbandry% 20Statistics-2019%29.pdf.

Bajaj, K.L., Kaur, G. and Singh, T. 1979. Lachrymatory factor and other chemical constituents of some varieties of onion. *Journal of Plant Food* 3: 119-203.

Balasubramayam, V.R.A., Dhake, V. and Moitra, P. 2000. Micro Irrigation and fertigation of V-12 Onion (Abstr.). *International Conference on Micro and Sprinkler Irrigation Systems*, Jalgaon, p.77.

Barnwal M.K. and Prasad, S.M. 2005. Influence of date of sowing on Stemphylium blight disease. *J. Res., Birsa Agriculture University* 17: 63-67.

Barnwal, M.K., Prasad, S.M. and Kumar, S. 2006. Cost effective fungicidal management of Stemphylium blight of onion. *J. Res., Birsa Agricultural University* 18: 153-155.

BBC News. 2013. *"Uganda's Genetically Modified Golden Bananas."* BBC News, March 27, 2013. Accessed August 11, 2015, http://www.bbc.com/news/world-africa-21945311.

Beintema, N., Stads, G.J., Fuglie, K. and Heisey, P. 2012. *ASTI Global Assessment of Agricultural R&D Spending*. International Food Policy Research Institute, Washington D.C., U.S.A.

Belton, B., Padiyar, A., Ravibabu, G. and Rao, K.G. 2017. Boom and bust in Andhra Pradesh: Development and transformation in India's domestic aquaculture value chain. *Aquaculture* 470: 196-206.

Bender, A. 1992. *Meat and Meat Products in Human Nutrition in Developing Countries*. FAO Food and Nutrition Paper #53. *Food Policy and Nutrition Division of FAO* 2: 1–88.

Bender, D.A. 1993. Onions. In: W.F. Bennett (ed.) *Nutrient Deficiencies and Toxicities in Crop Plants*, pp. 131–135. APS Press, St. Paul, Minnesota.

Bhambal, S.B. 1987. *Effect of Foliar Application of Micronutrients on Growth, Yield, Fruit Quality and Leaf Nutrient Status of Pomegranate* (Punica granatum *L.*) cv. Ganesh. M. Sc. (Ag.) thesis, Mahatma Phule Agri. Univ., Rahuri, Maharashtra.

Bhargava, B.S. and Chadha, K.L. 1993. Leaf nutrient guide for fruit crops. In K.L. Chadha and O.P. Pareek (eds.) *Advances in Horticulture*, Vol. 2, pp. 973-1029. Malhotra Publishing House, New Delhi.

Bhargava, B.S., Raturi, G.B. and Hiwale, S.S. 1990. Leaf sampling technique in ber (*Zizyphus mauritiana* Limk) for nutritional diagnosis. *Singapore J. Prim. Ind.* 18: 85.

Bhonde S.R., Sharam, S.B. and Chougale, A.B.1997. Effect of biofertilzer in combination with nitrogen through organic sources on yield and quality of onion. *NHRDF News Lett.* 17(2): 1-3.

Biji, K.B., Ravishankar, C.N., Mohan, C.O. and Srinivasa Gopal, T.K. 2015. Smart packaging systems for food applications: A review. *Food Sci. Technol.* 52(10): 6125-6135.

Birthal, P.S., Joshi, P.K., Roy, D. and Thorat, A. 2007. *Diversification in Indian Agriculture towards High-Value Crops: The Role of smallholders*. IFPRI Discussion Paper 00727, International Food Policy Research Institute, Washington, D.C.

Birthal, P.S., Khan, M.T., Negi, D.S. and Agarwal, S. 2014b. Impact of climate change on yields of major food crops in India: Implications for Food Security. *Agricultural Economics Research Review* 27(2): 145-155.

Birthal, P.S., Kumar, S., Negi, D.S. and Roy, D. 2015c. The impact of information on returns from farming: evidence from a nationally representative farm survey in India. *Agricultural Economics* 46:1-13.

Birthal, P.S. and Negi, D.S. 2012. Livestock for higher, sustainable and inclusive agricultural growth. *Economic and Political Weekly* 47(26-27): 89-99.

Birthal, P.S., Negi, D.S., Khan, M.D. and Agarwal, S. 2015b. Is Indian agriculture becoming resilient to droughts?. *Food Policy* 56: 1-12.

Birthal, P.S., Negi, D.S., Kumar, S. *et al.* 2014a. How sensitive is Indian agriculture to climate change?. *Indian Journal of Agricultural Economics* 69(4): 474-487.

Birthal, P.S., Roy, D. and Negi, D.S. 2015d. Assessing the impact of crop diversification on farm poverty in India. *World Development* 72: 70-92.

Blaise, C.D., Ravindran, C.D. and Singh, J.V. 2006. Trends and stability analyses to interpret results of long-term effects of application of fertilizers and manure to rainfed cotton. *Journal of Agronomy and Crop Science* 192: 319-330.

Blaise, D. 2006. Balanced fertilization for high yield and quality cotton In: D.K. Benbi, M.S. Brar, and S.K. Bansal (eds.) *Balanced Fertilization for Sustaining Crop Productivity*, pp. 255-271. Proc. Intl. Symp. PAU, Ludhiana, IPI.

Blaise, D. 2011. Tillage and green manure effects on Bt transgenic cotton (*Gossypium hirsutum* L.) hybrid grown on rainfed vertisols of central India. *Soil Tillage Res.* 114: 86-96.

Blaise, D. and Prasad, R. 2005. Integrated plant nutrient supply: An approach to sustained cotton production. *Indian J. Fert.* 1: 37-46.

Blaise, D. and Ravindran, C.D. 2003. Influence of tillage and residue management on growth and yield of cotton grown on a Vertisol over 5 years in a semi-arid region of India. *Soil and Tillage Research* 70: 163-173.

Borlaug, N.E. 1972. "Nobel Lecture, December 11, 1970." In *Nobel Lectures, Peace 1951–1970*, edited by Frederick W. Haberman. Elsevier Publishing Company, Amsterdam, 1972. http://www.nobelprize.org/nobel_prizes/peace/laureates/1970/borlaug-lecture. html.

Borlaug, N.E. 2000. Ending world hunger: The promise of biotechnology and the threat of antiscience zealotry. *Plant Physiology* 124(2): 487-490.

Bowonder, B. 1979. Impact analysis of the green revolution in India. *Technol Forecast Soc Chang.* 15: 297–313.

Brar, K.S., Sidhu, A.S. and Chadha, M.L. 1993. Screening onion varieties for resistance to *Thrips tabaci* Lind. and *Helicoverpa armigera* (Hubner). *J. Insect Sci.* 6: 123-124.

Brown, L.R., Christopher F., Hilary, F. *et al.* (eds.). 2001. *State of the World 2001: A World-Watch Institute Report on Progress Toward a Sustainable Society.* Norton, New York.

Carpenter, J. 2014. *"How Many Pounds of GM Foods Are Produced Each Year in the U.S.A."* GMO Answers, May 8, 2014. Accessed July 31, 2015, https://gmoanswers.com/ ask/how-many-pounds-gm-foods-are-produced-each-year-usa.

Castle, L.A., Wu, G. and McElroy, D. 2006. Agricultural input traits: Past, present and future. *Curr. Opin. Biotechnol.* 17: 105-112.

Chand, R. 2016. "*Doubling Farmers' Income: Strategy and Prospects*". Presidential Address delivered at 76th Annual Conference of Indian Society of Agricultural Economics, Assam Agricultural University, Jorhat, Assam.

Chand, R., Kumar, P. and Kumar, S. 2012. Total factor productivity and contribution of research investment to agricultural growth in India. *Agricultural Economics Research Review* 25(2): 181-194.

Charles, D. 2015. "*A Top Weed Killer Could Cause Cancer. Should We Be Scared?*," March 24, 2015. NPR-*The Salt* (blog). http://www.npr.org/sections/thesalt/2015/03/24/394912399/a-top-weedkiiller-probably-causes-cancer-should-we-be-scared.

Chaturvedi, S.K. 2009. Pulses research and development in achieving millennium development goals. National symposium on "*Achieving Millennium Development Goals: Problems and Prospects.*" Bundelkhand University, Jhansi, India, pp. 1-5.

Chel, A., Nayak, J.K. and Kaushik, G. 2008. Energy conservation in honey storage building using Trombe wall. *Energy Build.* 40: 1643–1650.

Chel, A. and Tiwari, G.N. 2010. Stand-alone photovoltaic (PV) integrated with earth to air heat exchanger (EAHE) for space heating/cooling of adobe house in New Delhi (India). *Energy Convers. Manage.* 51: 393–409.

Chopade, S.O, Bansode, P.N and Hiwase, S.S. 1998. Studies on fertilizer and water management to onion. *PKV Res. J.* 22: 44-46.

Chopra, S. 2010. Horticultural interventions for food security challenges. In: *Souvenir of the Fourth Indian Horticulture Congress*, New Delhi.

Christopher, B., Barange, M., Subasinghe, R. *et al.* 2015. Feeding a billion by 2050 - Putting fish back on the menu. *Food Security* 7: 261-274.

CMFRI. 2020. *Marine Fish Landings in India 2019*. Technical Report, CMFRI Booklet Series No. 24/2020. ICAR-Central Marine Fisheries Research Institute, Kochi, India, 15 pp.

Cranshaw, W.S. 2015. "*Bacillus thuringiensis.*" Colorado State University. Accessed July 31, 2015, http://www.ext.colostate.edu/pubs/insect/05556.html.

DAC&FW. 2017b. *Horticultural Statistics at a Glance 2017*. Ministry of Agriculture & Farmers Welfare, Government of India, New Delhi.

Danforth Center. 2015. "*Virus Resistant Cassava for Africa.*" Danforth Center. Accessed August 12, 2015, http://www.danforthcenter.org/docs/default-source/newsmedia/ infographics/infographic-horozontal.pdf?sfvrsn=0.

Darshan Singh, D., Sidhu, A.S., Thakur, J.C. and Singh, D. 1986. Relative resistance of onion and garlic cultivars to *Thrips tabaci* Lind. *J. Res. Punjab Agril. Uni.* 23: 424-427.

Dastagiri M.B. 2004. *Demand and Supply Projections for Livestock Products in India*. Policy Paper 21.

Datt, G. and Ravallion, M. 1998. Farm productivity and rural poverty in India. *Journal of Development Studies* 34(4): 62-85.

Datt, G., Ravallion, M. and Murgai, R. 2016. *Growth, Urbanization and Poverty Reduction in India*, Policy Research Working Paper 7568, The World Bank, Washington D.C., U.S.A.

Datta, A. 2013. Genetic engineering for improving quality and productivity of crops. *Agriculture & Food Security* 2: 15.

Dawe, D. 1998. Re-energizing the green revolution in rice. *American Journal of Agricultural Economics* 80: 948–953.

De Janvry, A. and Subbarao, K. 1986. *Agricultural Price Policy and Income Distribution in India.* Studies in Economic Planning, Oxford University Press, New Delhi.

Debajit S., Akhtar, M.S., Pandey, N.N. *et al.* 2011. *Nutrient Profile and Health Benefits of Cold-Water Fishes,* Bulletin No. 17. ICAR Directorate of Coldwater Fisheries Research, Bhimtal, Uttarakhand, India, 40 pp.

Delgado, C. 2003. Rising consumption of meat and milk in developing countries has created a new food revolution. *Journal of Nutrition* 133 (11, sup 2): 3907S–3910S.

Delgado, C., Rosegrant, M. and Meijer, S. 2001. Livestock to 2020: the revolution continues. *International Agricultural Trade Research Consortium. http://www.iatrcweb.org/ oldiatrc/Papers/ Delgado.pdf* (Accessed October 22, 2002).

Department of Animal Husbandry, Dairying and Fisheries. 2017. *Basic Animal Husbandry and Fisheries Statistics.* Government of India, New Delhi. Accessed September 20, 2018. http://www.dahd. nic.in/sites/default/filess/Tables%20of%20BAH%26amp% 3BFS% 202017%20% 281% 29.pdf.

Department of Animal Husbandry, Dairying and Fisheries. 2017. *National Action Plan for Egg and Poultry—For Doubling Farmers' Income by 2022.* Ministry of Agriculture and Farmers Welfare, Government of India, New Delhi.

DES. 2013. Agricultural statistics at a glance 2013. Directorate of Economics and Statistics. Ministry of Agriculture, Government of India. http://www.nhrdf.com/contentPage. asp?sub_ section_ code=104.

Desh, B.B. 2002. *Status of S in Groundnut Growing Red and Laterite Soils of Orissa and its Integrated Management in Groundnut-Sesame and Groundnut-Finger Millet System.* Ph.D. Dissertation, Orissa University of Agriculture and Technology, Bhubaneswar, India.

Devulkar, N.G., Bhanderi , D.R., More, S.J. and Jethava, B.A. 2015. Optimization of yield and growth in onion through spacing and time of planting. *Green Farming* 2: 305-307.

Directorate of Cotton Development. 2017. *Status Paper of Indian Cotton.* Ministry of Agriculture and Farmers Welfare, Government of India, Nagpur.

Discovery. 2013. "Bananas Get Pepper Power." *Discovery*, February 11, 2013. Accessed July 31, 2015, http://news.discovery.com/earth/bananas-peppers-genes.htm.

Djurfeldt, G. and Jirström, M. 2005. The puzzle of the policy shift—The early green revolution in India, Indonesia, and the Philippines. In Göran Djurfeldt, Hans Holmén, Magnus Jirström, and Ron Larsson (eds.) *The African Food Crisis: Lessons from The Asian Green Revolution.* CABI, Wallingford.

Dodamani, B.M., Hosmani, M.M. and Hunshal, C.S. 1993. Management of chilli + cotton + onion intercropping systems for higher returns. *Farming Systems* 9: 52-55.

DOGR. 2011. *Annual report 2010-11.* Directorate of Onion and Garlic Research, Rajgurunagar, Pune, India. 80p.

DOGR. 2012. *Annual report 2011-12.* Directorate of Onion and Garlic Research, Rajgurunagar, Pune, India. 72p.

DOGR. 2013. *Annual report 2012-13*. Directorate of Onion and Garlic Research, Rajgurunagar, Pune, India. 92 pp.

Dow AgroSciences. 2014. *"USDA Allows Commercialization of Dow Agrosciences' Enlist™ Corn, Soybean Traits: Farmer Voice Supports USDA's Action on Enlist."* Dow AgroSciences, September 17, 2014. Accessed July 31, 2015, http://newsroom.dowagro. com/press-release/usda-allows-commercialization-dow-agrosciences-enlist-corn-soybean -traits.

Dumas, A., Dijkstra, J. and France, J. 2008. Mathematical modelling in animal nutrition: A centenary review. *J. Agric. Sci.* 146: 123–142.

Economist. 2013. "Genetically Modified Trees: Into the Wildwood." *Economist*, May 4, 2013. Accessed August 11, 2015, http://www.economist.com/news/science-and-technology/21577033-gm-species-may-soon-be-liberated-deliberately-wildwood.

Edwards, P., Zang, W., Belton, B. and Little, D.C. 2019. Misunderstandings, myths and mantra in aquaculture: Its contribution to world food supplies has been systematically over reported. *Mar. Policy* 106: 103547.

Elangovan, M., Suthanthirapandian, I.R and Sayed, S. 1996. Intercropping of onion in chilli. *Annals of Agricultural Science* 34: 839-857.

Enting, H., Kooij, D., Dijkhiuzen, A.A. *et al.* 1997. Economic losses due to clinical lameness in dairy cattle. *Livestock Products Science* 49: 259-267.

Evenson, R.E., Carl, P. and Mark, W.R. 1998. *Agricultural Research and Productivity Growth in India*, Vol.109. Intl. Food Policy Res. Inst., Washington, DC, USA.

Evenson, R.E., Pray, C.E. and Rosegrant, M.W. 1999. *Agricultural Research and Productivity Growth in India*. Research Report 109, International Food Policy Research Institute, Washington, D.C.

FAO. 2007. Poultry. *Proceedings of The International Conference Poultry in The Twenty-First Century: Avian Influenza And Beyond*, Bangkok, Thailand.

FAO. 2012. *World Agriculture Towards 2030/2050: The 2012 Revision*. ESA Working paper No. 12-03.

FAO. 2014. *Innovation in Family Farming*. Food and Agriculture Organization of United Nations, Rome.

FAO. 2015a. *"Why Is Provitamin A Important for Health?"* Golden Rice Project. Accessed August 11, 2015, http://www.goldenrice.org/Content3-Why/why3_FAQ.php# Inventors.

FAO. 2015b. *Myanmar Floods Deal Major Blow to Country's Agriculture*. Food and Agriculture Organization (FAO), Rome.

FAO. 2018. *Transforming the Livestock Sector through the SDGs*. FAO, Rome. http:// www.fao.org/ 3/ CA1177EN/ca1177en.pdf.

FAO. 2020. *The State of World Fisheries and Aquaculture*. Food and Agriculture Organisation of the United Nations, Rome, Italy, 244 pp.

Farghali, M.A. and Zeid, M.I.A. 1995. Phosphorus fertilization and plant population effects on onion grown in different soils. *Assiut J. Agric. Sci.* 26(4): 187-203.

Ferguson, E.L., Gibson, R.S., Opare-Obisau, C. *et al.* 1993. The zinc nutrition of preschool children living in two African countries. *J Nutr* 123: 1487–1496.

Fournier, F., Guy B., and Robin S. 1995. Effect of *Thrips tabaci* (Thysanoptera: Thripidae) on yellow onion yields and economic thresholds for its management. *Entom. Soc. Amer.* 88: 1401-1407.

GAIN Report. 2008. *India Livestock and Products Annual 2008* (IN8098). ThermoFisher Scientific.

Ghosh, B., Bose, T.K. and Mitra, S.K. 1986. Chemical induction of flowering and control of fruit drop in litchi (*Litchi chinensis* Sonn.). *Proc. 22ⁿᵈ Int. Hort. Cong.*, California, Abstr. 1189.

Gibson, R.S.1994. Content and bioavailability of trace elements in vegetarian diets. *Am J Clin Nutr* 59: 1223S–1232S.

Giger, E., Prem, R. and Leen, M. 2009. Increase of agricultural production based on genetically modified food to meet population growth demands. *School of Doctoral Studies (European Union) Journal* 1: 98-124.

Gillis, J. 2009. "Norman Borlaug, Plant Scientist Who Fought Famine, Dies at 95."*New York Times*, September 13, 2009. http://www.nytimes.com/2009/09/14/business/energy-environment/14borlaug.html?pagewanted=all.

GOI. 2007. *Agriculture Strategy for Eleventh Plan: Some Critical Issues*. Planning commission, Govt. of India, New Delhi.

GOI. 2007. *Report of the Working Group on Horticulture, Plantation Crops and Organic Farming for the XI Five Year Plan (2007-12)*. Planning Commission, Govt. of India, New Delhi.

GOI. 2009. *Report of the Task Force on Irrigation*. Planning Commission of India, Government of India, New Delhi.

GOI. 2014. *Mission for Integrated Development of Horticulture: Operational Guidelines*. Horticulture Mission, Govt. of India, New Delhi.

GoI. 2018. *Handbook on Fisheries Statistics* 2018. Department of Animal husbandry, Dairying and Fisheries, Ministry of Agriculture, Government of India, Krishi Bhavan, New Delhi.

GoI. 2020b. *PIB Release on New Schemes for Fisheries*. Ministry of Fisheries, Animal Husbandry and Dairying, Govt. of India, New Delhi, India.

Gollin, D., M. Morris, M.W. and Byerlee, D. 2005. Technology adoption in intensive post-green revolution systems. *American Journal of Agricultural Economics* 87(5): 1310–1316.

Gonsalves, D. 2015. "Transgenic Papaya in Hawaii and Beyond." *AgBioForum* 7(1–2) (2004). Accessed July 31, 2015, http://www.agbioforum.org/v7n12/v7n12a07-gonsalves.htm;

Gonsalves, D. *et al.* 2015. "*Papaya Ringspot Virus.*" American Phytopathological Society. Accessed July 31, 2015, http://www.apsnet.org/edcenter/intropp/lessons/viruses/Pages/ PapayaRingspotvirus.aspx.

Gopal, J. 2014. Pre-and post-harvest losses in onion. *Proceedings National Conference on Pre-/Post -Harvest Losses & Value Addition in Vegetables*, IIVR, Varanasi, pp. 25-29.

Gulati, A. and Juneja, R. 2018. *From Plate to Plough: Timidity and Technology*. The Indian Express. Accessed December 6, 2018. https://indianexpress.com/article/opinion/ columns/indian-cotton-growers-farmers-cotton-crop-cotton-farming-narendra-modi-govt-5292899/.

Gulati, A. and Verma, S. 2016. *From Plate to Plough: A Clear Trend towards Non-Vegetarianism in India. The Indian Express*. Accessed September 15, 2018. https:// indianexpress.com/article/opinion/ columns/india-diet-indian-palate-non-vegetarian-vegetarianism-3099363/.

Gupta, R.B.L. and Pathak, V.K. 1987. Management of purple blotch *Alternaria porri* (Ellis:) Clif. of onion by summer ploughing and alteration of date of sowing. *Z. Microbiol.* 142: 163-166.

Gupta, R.P., Srivastava, P.K. and Pandey, U.B. 1986. Control of purple blotch of onion seed crop. *Indian Phytopath*. 39: 303-304.

Gupta, R.P., Srivastava, P.K. and Sharma, R.C. 1996a. Chemical control of purple blotch and Stemphylium blight diseases of onion. *NHRDF News Letter* 16: 14-16.

Gupta, R.P., Srivastava, P.K. and Sharma, R.C. 1996b. Effect of foliar spray of different fungicides on the control of Stemphylium blight disease and yield of onion bulb. *NHRDF News Letter* 16: 13-14.

Gupta, S.C. and Gangwar, S. 2012. Effect of molybdenum, iron and microbial inoculants on symbiotic traits, nutrient uptake and yield of chickpea. *Journal of Food Legumes* 25(1): 45-49.

Gustav R., Anne H., Thomas F. *et al*. 2008. Potentials and prospects for renewable energies at global scale. *Energy Policy* 36: 4048–4056.

Hallikeri, S.S., Halemani, H.L., Patil, B.C. and Nanadagavi, R.A. 2011. Influence of nitrogen management on expression of cry protein in Bt-cotton (*Gossypium hirsutum*). *Indian Journal of Agronomy* 56: 62-67.

Hamilton, B.K., Yoo, K. and MPike, L. 1998. Changes in pungency of onions by soil type, sulphur nutrition and bulb maturity. *Scientia Horticulturae* 74(4): 249-256.

Hanumashetti, S.I., Rao, M.M. and Bankapur, V.M. 1981. Preliminary studies on the effect of growth regulators and chemicals on the improvement of colouration in Gulabi grapes. *Curr. Res.* 10: 45-46.

Hayes, B.J., Bowman, P.J., Chamberlain, A.J. and Goddard, M. E. 2009. Genomic selection in dairy cattle: Progress and challenges. *J. Dairy Sci*. 92: 433–443.

Hazell, P.B.R. 2003. Green revolution, curse or blessing? In Joel Mokyr (ed.) *The Oxford Encyclopedia of Economic History*. Oxford University Press, Oxford, UK.

Hegde, D.M. and Sudhakara Babu, S.N. 2009. Declining factor productivity and improving nutrient use efficiency in oilseeds. *Indian J Agron* 54(1): 1–8.

Helikson, H.J., Haman, D.Z. and Baird, C.D. 1991. *Pumping Water for Irrigation Using Solar Energy*. University of Florida, Florida Cooperation Extension Services, Institute of Food and Agriculture Sciences, Fact sheet EES-63, USA.

Hiloidhari, M., Das, D. and Baruah, D.C. 2014. Bioenergy potential from crop residue biomass in India. *Renewable and Sustainable Energy Reviews* 32: 504-512.

Hisham El-Osta, S. and Mitchell, J.M. 2000. Technology adoption and its impact on production performance of dairy operations. *Review of Agricultural Economics* 22(2): 477-498.

Hokkanen, H.M.T. 1998. Ecological impact of transgenic, insect resistant crops. In *Génie Génétique: perspectives, inconnues et risques. Les applications en agriculture et dans l'alimentation*. Colloque international, Groupe des Verts au Parlement Européen Bruxelles, p. 8.

IARI. 2010. *Annual Report 2011-12*. Indian Agricultural Research Institute, New Delhi, India. 200 pp.

Ibrahim, S.T., Khalil, H.E and Kamel, A.S. 2005. Growth and productivity of sugar beet, onion and garlic grown alone and associations under different inter and intraspacing. *Annals of Agricultural Science* 43 :497-516.

IIPR. 2011. *Vision 2030*. Indian Institute of Pulses Research, Kanpur.

Indira, A., Bhagavan, M.R. and Virgin, I. 2005. *Agricultural Biotechnology and Biosafety in India: Expectations, Outcomes and Lessons*. Stockholms Environment Institute, Stockholm.

IPCC (Intergovernmental Panel on Climate Change). 2007. *Climate Change 2007: Impacts, Adaptation and Vulnerability. Summary for Policy Makers*. See http://www.ipcc. ch/publications_and_data/ar4/wg2/en/spm.html.

IRRI. 2015. "Why Is Golden Rice Needed in the Philippines Since Vitamin A deficiency Is Already Decreasing?" IRRI. Accessed August 11, 2015, http://irri.org/index.php? option =com_k2&view=item&id=12352&lang=en.

IRRI. 2015a. "*About Golden Rice.*" IRRI. Accessed August 11, 2015, http://irri.org/index. php?option=com_k2&view=item&id=10202&lang=en.

IRRI. 2015b. "*Does Golden Rice Contain Daffodil Genes?*" IRRI. Accessed August 11, 2015, http://irri. org/golden-rice/faqs/does-golden-rice-contain-daffodil-genes.

ISAAA. 2008. *Global Status of Commercialized Biotech/GM Crops: 2008 The First Thirteen Years, 1996 to 2008*. ISAAA Brief 39-2008: Executive Summary. Ithaca, NY.

ISAAA. 2015. "*Pocket K No. 10: Herbicide Tolerance Technology: Glyphosate and Glufosinate.*" ISAAA. Accessed July 31, 2015, https://isaaa.org/resources/ publications/pocketk/10/default.asp.

ISAAA. 2016. *Global Status of Commercialized Biotech/GM Crops: 2016*. ISAAA *Brief* No. 52, ISAAA, Ithaca, NY.

Islam, M.K., Alam, M.F. and Islam, A.K.M.R. 2007. Growth and yield response of onion (*Allium cepa* L.) genotypes to different levels of fertilizers. *Bangladesh Journal of Botany* 36(1): 33-38.

Jabeda, A., Wagner, S., McCracken, J. *et al.* 2012. Targeted microRNA expression in dairy cattle directs production of β-lactoglobulin-free, high-casein milk. *PNAS* 109(42): 16811-16816.

Jalota, S.K., Buttar, G.S., Sood, A. *et al.* 2008. Effects of sowing date, tillage and residue management on productivity of cotton (*Gossypium hirsutum* L.)–wheat (*Triticum aestivum* L.) system in northwest India. *Soil and Tillage Research* 99: 76-83.

Jat, M.L., Gathala, M.K., Ladha, J.K. *et al.* 2009. Evaluation of precision land levelling and double zero-till systems in the rice–wheat rotation: water use, productivity, profitability and soil physical properties. *Soil and Tillage Research* 105(1): 112-121.

Jayasankar, P. 2018. Present status of freshwater aquaculture in India-A review. *Indian J. Fish.* 65(4): 157-165.

Jena, D., Sahoo, R., Sarang, D.R. and Singh, M.V. 2006. Effect of different sources and levels of S on yield and nutrient uptake by groundnut-rice cropping system in an Inceptisol of Orissa. *J Indian Soc Soil Sci* 54(1): 126–129.

Jha, G.K., Pal, S. and Singh, A. 2012. Changing energy-use pattern and the demand projections for Indian agriculture. *Agricultural Economics Research Review* 25(1): 61-68.

Jha, S.N., Vishwakarma, R.K., Ahmad, T. *et al.* 2015. *Assessment of Quantitative Harvest and Post-Harvest Losses of Major Crops and Commodity in India*. ICAR—All-India Co-ordinated Research Project on Post-Harvest Technology, ICAR-CIPHET.

John Innes Centre. 2014. "*Bumper Harvest for GM Purple Tomatoes.*" John Innes Centre, January 25, 2014. https://www.jic.ac.uk/ news/2014/01/gm-purple-tomatoes/.

Johnson, M., Hazell, P.B.R. and Gulati, A. 2003. The role of intermediate factor markets in Asia's green revolution: Lessons for Africa? *American Journal of Agricultural Economics* 85(5): 1211–1216.

Joshi, P.K. and Kumar, P. 2016. Food demand and supply projections for India. In F. Brower and P.K. Joshi (eds.) *International Trade and Food Security: The Future of Indian Agriculture*. CAB International, Wallingford, U.K.

Kaladharan, P., Johnson, B., Nazar, A.K. *et al.* 2019. Perspective plan of ICAR-CMFRI for promoting seaweed mariculture in India. *Mar. Fish. Inf. Serv. T&E Ser.* 240: 17-22.

Kalra, C.L., Beerh, J.K., Manan, J.K. *et al.* 1986. Studies on influence of cultivars on the quality of dehydrated onion (*Allium cepa* L.). *Indian Food Packer* 40: 20-27.

Kalt, W. 2002. Health functional phytochemicals of fruits. *Horticulture Review* 27: 269-315.

Karanja, F., Gilmour, D. and Fraser, I. 2012. Dairy productivity growth, efficiency change and technological progress in Victoria, Paper presented at *Annual Conference of Australian Agricultural and Resource Economics Society*, Fremantle, Western Australia.

Katiha, P.K., Jena, J.K., Pillai, N.G.K. *et al.* 2005. Inland aquaculture in India: Past trend, present status and future prospects. *Aquac. Econ. Manage.* 9: 237-264.

Kaur, G., Brar, Y.S. and Kothari, D.P. 2017. Potential of livestock generated biomass: Untapped energy source in India. *Energies* 10: 3-15.

Kenna, J.P., Gillett, W.B., Power, I.T. and Halcrow, S.W. 1985. *Handbook on Solar Water Pumping*. World Bank, Washington, DC and IT Publications, London.

Khan, A.A., Jilani, G., Akhtar, M.S. *et al.* 2009. Phosphorous solubilizing bacteria: Occurrence, mechanisms and their role in crop production. *Journal of Agriculture and Biological Sciences* 1: 48-58.

Khan, M.I., Shah, M.H., Raja, W. and Teeli, N.A. 2006. Effect of intercropping on the soil fertility and economics of sunflower and companion legumes under temperate conditions of Kashmir. *Environment Ecology* 245(1): 171–173.

Khanduja, S.D. and Garg, Y.K. 1984. Macro-nutrient element composition of leaves from jujube (*Zizyphus mauritiana* Lamk.) tree in north India. *Indian J. Hort.* 41: 22-24.

Khura, T.K., Indra-Mani and Srivastava, A.P. 2011. Design and development of tractor drawn onion (*Allium cepa*) harvester. *Indian Journal of Agricultural Science* 81: 528-532.

Khurana, S.C and Bhatia, A.K. 1991. Intercropping of onion and fennel with potato. Indian *Journal of Weed Science* 23: 64-66.

King, D.A., Peckham, C., Waage, J.K. *et al.* 2006. Infectious diseases: Preparing for the future. *Science* 313: 1392–1393.

King, S.L., Kratochvil, J.A. and Boyson, W.E. 2001. *Stabilization and Performance Characteristics of Commercial Amorphous-Silicon PV Modules*. Sandia National Laboratories. http://photovoltaics. sandia.gov/docs/PDF/kingkrat.pdf.

Knapp, S. 2008. Potatoes and poverty. *Nature* 455: 170–171

Kothari, S.K., Singh U.B., Sushil Kumar and Kumar S. 2000. Inter-cropping of onion in menthol mint for higher profit under subtropical conditions of north Indian plains. *Journal of Medicinal Aromatic Plant Science* 22: 213-218.

Kranthi, K.R., Naidu, S., Dhawad, C.S. *et al.* 2005. Temporal and intra-plant variability of cry1Ac expression in Bt cotton and its influence on the survival of the cotton bollworm, *Helicoverpa armigera* (Noctuidae: Lepidoptera). *Current Science* 89: 291-297.

Krishna, V., Erenstein, O., Sadashivappa P. and Vivek, B.S. 2014. Potential economic impact of biofortified maize in the Indian poultry sector. *International Food and Agribusiness Management Review* 17(4): 111-140.

Kumar, V., Patil, R.G. and Patel, J.G. 2011. Efficient water management technology for sustainable cotton production in central India. In: K.R. Kranthi, M.V. Venugopalan, R.H. Balasubrahmanya, S. Kranthi, S.B. Singh and D. Blaise (eds.) *World Cotton Research Conference -5 Book of Papers*, pp. 376-385. Excel Publishers, New Delhi.

Kupferschmidt, K. 2013. "Activists Destroy 'Golden Rice' Field Trial." *Science Magazine*, August 9, 2013. Accessed August 11, 2015. *http://news.sciencemag.org/asiapacific/ 2013/08/ activists-destroy-golden-rice-field-trial.*

Lal, K. 2000. Foot and mouth disease: Present status and future strategies for control in India. *Indian Farming* 50: 28-31.

Lal, K.K. and Jena, J.K. 2019. Fish genetic resources - India. In: R.K. Tyagi, D.H.N. Munasinghe, K.H.M.A. Deepananda, F. Niranjan and R.K. Khetarpal (eds.) *Regional Workshop on Underutilized Fish and Marine Genetic Resources and their Amelioration - Proceedings and Recommendations.* Asia-Pacific Association of Agricultural Research Institutions (APAARI), Bangkok, Thailand, 55 pp.

Lambert, A.D., Smith, J.P. and Dodds, K.L. 1991. Shelf life extension and microbiological safety of fresh meat - A review. *Food Microbiology* 8: 267-297.

Landes, M.R. 2010. *Growth and Equity Effects of Agricultural Marketing Efficiency Gains in India.* DIANE Publishing.

Layrisse, M., Martinez-Torres, C., Mendez-Costellaro, H. *et al.* 1990. Relationship between iron bioavailability from diets and the prevalence of iron deficiency. *Food and Nutr. Bull.* 12: 301–309.

Leaf, M.J. 1984. *Song of Hope: The Green Revolution in a Punjab Village.* Rutgers University Press, New Brunswick, NJ.

Leakey, R. and Kranjac-Berisavljevic, G. 2009. Impacts of AKST (Agricultural Knowledge Science and Technology) on development and sustainability goals. In B.D. McIntyre, H.R. Herren, J. Wakhungu and R.T. Watson (eds.) *Agriculture at A Crossroads*, pp. 145–253. Island Press, Washington, DC.

Lim, X.Z. 2014. "*Is Glyphosate, Used with Some GM Crops, Dangerously Toxic to Humans?*," April 30, 2014. Accessed August 8, 2015, http://www. Geneticliteracy project.org/2014/04/30/is-glyphosate-used-with-some-gm-crops-dangerously-toxic-to-humans/.

Lipton, M.L. 1985. Research and design of a policy frame in agriculture. In T. Rose (ed.) *Crisis and Recovery in Sub-Saharan Africa.* Development Center, OECD, Paris.

Lopez-Pereira, M.A. 1993. Economics of quality protein maize as a feedstuff. *Agribusiness* 9 (6): 557-568.

Lu, F.M. 1990. Colour preference and using silver mulches to control onion thrips, *Thrips tabaci* Lindeman. *Chinese J. Entom.* 10: 337-342.

Mahajan V., Lawande, K.E., Krishnaprasad, V.S.R. and Srinivas, P.S. 2011. Bhima Shubra and Bhima Shweta - new white onion varieties for different seasons. In *Souvenir & Abstract: National Symposium on Alliums: Current Scenario and Emerging Trends*, p. 162.

Maini, S.B., Diwan, B. and Anand, J.C. 1984. Storage behaviour and drying characteristics of commercial cultivars of onion. *Journal of Food Science and Technology* 21: 417-419.

Mani, V.P., Chauhan, V.S., Joshi, H.C. and Tandon, J.P. 1999. Exploiting gene effects for improving bulb yields in onion. *Ind. J. Genet. Pl. Breed.* 59: 511-514.

Manjula, K. and Saravanan, G. 2015. Poultry industry in India under globalised environment—opportunities and challenges. *International Journal of Scientific Research* 4(8): 391-393.

Marikhur, R.K., Dhar, A.K. and Kaw, M.R. 1977. Downy mildew of *Allium cepa* and its control with fungicides in Kashmir Valley. *Ind. Phytopath.* 30: 576-577.

Mathur, K., Sharma, S.N. and Sain, R.S. 2006. Onion variety RO-59 has higher yield and resistance to purple blotch and Stemphylicum blight. *J. Mycol. Pl. Pathol.* 36: 49-51.

Maxham, A. 2014. "Masked Eggplant Thugs Plant a Field of Lies." *Voices for Reason* (blog), April 17, 2014. https://ari.aynrand.org/blog/2014/04/17/masked-eggplant-thugs-plant-a-field-of-lies.

McIntire, J., Bourzat, D. and Pingali, P. 1992. *Crop-Livestock Interactions in sub-Saharan Africa.* World Bank, Washington, DC.

Meenakshi, J.V. and Banerji, A. 2005. The unsupportable support price: An analysis of collusion and government intervention in paddy auction markets in north India. *Journal of Development Economics* 76(2): 377-403.

Mehta, R. 2003. *The WTO and the Indian Poultry Sector.* Asia Pacific School of Economics and Government, The Australian Natl. Univ. http://aspem.anu.edu.au.

Mehta, R. and Nambiar, R.G. 2007. *The Poultry Industry in India.* The Food and Agriculture Organization (FAO), Bangkok.

Mitra, J., Shrivastava, S.L. and Rao, P.S. 2012. Onion dehydration: A review. *Journal of Food Science and Technology* 49: 267-277.

Mohammad, M.J. and Zuraiqi S. 2003. Enhancement of yield and nitrogen and water use efficiencies by nitrogen drip fertigation of garlic. *J. Pl. Nutr.* 26: 1749- 1766.

Mohanty, B., Vivekanandan, E., Mohanty, S. *et al.* 2017. The impact of climate change on marine and Inland fisheries and aquaculture in India. In: B.F. Philips and M. Perez-Ramirez (eds.) *Climate Change Impacts on Fisheries and Aquaculture: A Global Analysis*, Vol. 2, pp. 569-602. Wiley-Blackwell, New Jersey, USA.

Mollah., M.R.A., Rahman., S.M.L., Khalequzzaman., K.M. *et al.* 2007. Performance of intercropping groundnut with garlic and onion. *International Journal of Sustainable Crop Production* 2: 31-33.

Moloney, C. 2016. India's major agricultural produce losses estimated at Rs 92,000 crore. *Business News*, August 11, 2016.

Monsanto. 2015. "*YieldGard Biotech Maize: Increasing Yields Protection Against the Maize Stalk Borer Through Biotechnology.*" Monsanto. Accessed July 31, 2015, http://www. monsantoafrica.com/products/farmersguides/yieldgard.asp.

Mottet, A., Haan, C., Falcucci, A. *et al.* 2017. Livestock: On our plates or eating at our table? A new analysis of the feed/food debate. *Global Food Security* 14: 1-8.

Munilkumar, S. and Nandeesha, M.C. 2007. Aquaculture practices in Northeast India: Current status and future directions. *Fish Physiol. Biochem.* 33(4): 399-412.

Murkute, A.A. and Gopal, J. 2013. Taming the glut. *Agriculture Today* 16:28-30.

Murphy, S.P., Beaton, G.H., Calloway, D.H. 1992. Estimated mineral intakes of toddlers: Predicted prevalence of inadequacy in village populations in Egypt, Kenya, and Mexico. *Am J Clin Nutr* 56: 565–572.

Nagalaxmi, K., Annap, P., Venkatshwarlu, G. *et al.* 2015. Mislabelling in Indian seafood: An investigation using DNA barcoding. *Food Control* 59: 196-200. DOI:10.1016/j. foodcont. 2015.05.018.

Naik, R,, Annamali, SJ.K. and Ambrose, D.C.P. 2007. Development of batch type multiplier onion peeler. *Proceedings of the International Agricultural Engineering Conference on Cutting-edge Technologies and Innovations on Sustainable Resources for World Food Sufficiency*. Bangkok, Thailand.

Nair, L. 2014. Emerging trends in Indian aquaculture. *J. Aquat. Biol. Fish*. 2(1): 1-5.

Nalayini, P., Raj, S.P. and Sankaranarayana, K. 2011. Growth and yield performance of cotton (*Gossypium hirsutum*) expressing *Bacillus thuringiensis* var: *kurstaki* as influenced by polyethylene mulching and planting techniques. *Indian Journal of Agricultural Sciences* 81: 55-59.

Namukwaya, B., Tripathi, L., Tripathi, J.N. *et al.* 2012. Transgenic banana expressing *Pflp* gene confers enhanced resistance to *Xanthomonas* wilt disease. *Transgenic Research* 21(4): 855-862.

Narayanamoorthy, A. 2008. Drip irrigation and rainfed crop cultivation nexus: The case of cotton crop. *Indian Journal of Agricultural Economics* 63: 487-501.

NARL. 2015. "*About the National Agricultural Research Laboratories.*" NARL. Accessed August 11, 2015, http://www.narl.go.ug/.

NDDB. 2015. *Handbook of Good Animal Husbandry Practices [Internet]*. Available from: www. dairyknowledge. in/article/handbook-good-dairyhusbandry-practices [Accessed: May 15, 2021].

Negi, D.S., Birthal, P.S., Roy, D. and Khan, M.T. 2018. Farmers' choice of market channels and producer prices in India: Role of transportation and communication networks. *Food Policy* 81(C): 106-121.

New Scientist. 2006. "Edible Cotton Breakthrough May Help Feed the World." *New Scientist*, November 20, 2006. Accessed August 11, 2015, https://www.newscientist.com/ article/dn10612-edible-cotton-breakthrough-may-help-feed-the-world/.

New, M.B., Valenti, W.C., Jidwal, J.H. *et al.* 2010. *Freshwater Prawns: Biology and Farming*. Wiley-Blackwell, New Jersey, USA, 531 pp.

NFDB. 2019. *Aquaculture Technologies Implemented by NFDB*. National Fisheries Development Board, Department of Fisheries Ministry of Fisheries, Animal Husbandry and Dairying, Govt. of India, Hyderabad, India, 58 pp.

Nielsen, R.L. 2015. "*A Compendium of Biotech Corn Traits.*" Purdue University, May 2010. Accessed July 31, 2015, http://www.kingcorn.org/news/timeless/BiotechTraits.html.

Nimbalkar, V., Verma, H., Singh, J. and Kansal, S. 2020. Awareness and adoption level of subclinical mastitis diagnosis among dairy farmers of Punjab, India. *Turkish Journal of Veterinary and Animal Sciences* 44: 845-852.

NRCOG. 2004. *Annual Report 2003-04*. National Research Centre on Onion and Garlic, Rajgurunagar, Pune. 52pp.

NREL. 1997. *Twenty Years of Clean Energy*. National Renewable Energy Laboratory, Washington D.C.

OECD/FAO. 2020. Dairy and dairy products. In *OECD-FAO Agricultural Outlook 2020-2029*, OECD Publishing, Paris.

Osborn, D.E. 2003. *Overview of Amorphous Silicon (a–Si) Photovoltaic Installations at SMUD*. ASES 2003: America's Secure Energy, Austin, TX, 8 pp.

Palanisami, K., Mohan, K., Kakumanu, K.R. and Raman, S. 2011. Spread and economics of microirrigation in India: Evidence from nine states. *Economic and Political Weekly* 46(26-27): 81-86.

Palanisami, K. and Raman, S. 2012. *Potential and Challenges in Up-scaling Micro-irrigation in India Experiences from Nine States.* Water Policy Research HIGHLIGHT IWMI-TATA Water Policy Programme.

Palti, J. 1989. Epidemiology, production and control of downy mildew of onion caused by *Perasospora destructor*. *Phytoparasitica* 17: 1.

Pandotra, V.R. 1965. Purple blotch disease of onion in Punjab II: Studies on the life history, viability and infectivity of the causal organism *Alternaria porri*. *Proc. Ind. Acad. Sci. Sec. B* 61: 326-330.

Panwar, A.S., Singh, N.P., Munda, G.C. and Patel, D.P. 2001. Groundnut – production technology for hill region. *Intensive Agric* 39(5–6): 7–9.

Pasricha, N.S. and Tandon, H.L.S. 1993. Fertilizer management in oilseeds. In: H.L.S. Tandon (ed.) *Fertilizer Management in Commercial Crops*, pp. 65–66. Fertilizer Development and Consultation Organisation, New Delhi, India.

Pathak, C.S., Singh, D.P., Deshpande, A.A. and Sreedhar, T.S. 1986. Sources of resistance to pruple blotch in onion. *Veg. Sci.* 13: 300-303.

Patton, L. 2015. "McDonald's Pursuit of the Perfect French Fry." *Bloomberg Business*, April 19, 2012. Accessed July 31, 2015, http://www.bloomberg.com/bw/articles/2012-04-19/mcdonalds-pursuit-of-the-perfect-french-fry.

Pawar, D.D., Dingre, S.K., Bhakre, B.D. and Surve, U.S. 2013. Nutrient and water use by Bt cotton (*Gossypium hirsutum*) under drip fertigation. *Indian Journal of Agronomy* 58: 237-242.

Perkins, J.H. 1997. *Geopolitics and the Green Revolution: Wheat, Genes, and the Cold War.* Oxford University Press, New York.

Perkowski, M. 2013. *"Del Monte Gets OK to Import Biotech Pineapple"* Capital Press, April 25, 2013 (updated May 23, 2013). Accessed August 11, 2015, http://www. capitalpress. com/content/mp-transgenic-pineapple-041613.

Perry, B. and Sones, K. 2009. *Global Livestock Disease Dynamics Over the Last Quarter Century: Drivers, Impacts and Implications*. FAO, Rome, Italy. (Background paper for the SOFA 2009).

Peter, K. 1999. *Informatics on Turmeric and Ginger.* Indian Inst. of Spices Research, Calicut, Kerala, India.

Peter, K.V. 1999. Spices research and development – An updated review. *National Seminar on Sustainable Horticultural Production in Tribal Regions*, Central Hort. Expt. Stn., Ranchi, pp. 48-54.

Pingali, P.L. and Raney, T. 2005. *From the Green Revolution to the Gene Revolution: How will the Poor Fare?* ESA Working Paper No. 05-09.

Potrykus, I. 2015. "The 'Golden Rice' Tale." *AgBioWorld*. Accessed August 11, 2015, http://www.agbioworld.org/biotech-info/topics/goldenrice/tale.html.

Prasanna, B.M., Vasal, S.K., Kassahun, B. and Singh, N.N. 2001. Quality protein maize. *Current Science* 81(10): 1308-1319.

Press Information Bureau (PIB). 2017. *India Becomes Second Largest Fish Producing Country in the World*. Ministry of Agriculture and Farmers Welfare, Government of India, New Delhi. Accessed May 13, 2018. http://www.pib.nic.in/newsite/mbErel. aspx?relid= 173699.

Raheja, P.C. 1973. *Mixed cropping*. Indian Council of Agricultural Research, New Delhi, India, Technical Bulletin (Agric.) No. 42, pp 24–26.

Rahman, M.A., Chiranjeevi, C.H. and Reddy, I.P. 2000. Management of leaf blight disease of onion. *Proc. National Symposium on Onion Garlic Production and Post-Harvest Management. Challenges and Strategies*, Nasik (India), pp. 147-149.

Rajkumar, U., Rama Rao, S.V. and Sharma, R.P. 2010. Backyard poultry farming-changing the face of rural and tribal livelihoods. *Indian Farming* 59: 20-24.

Ramaswamy, N. 1971. *Studies on the Effect of Nitrogen on the Growth and Development of 'Robusta'* (Musa cavendishi *L.*). M. Sc. (Ag.) thesis, Annamalai Univ., Annamalinagar.

Rana, M.K. 2010. Fruits and Vegetables: A potential source of non-nutrients bioactive substances (Functional foods). *Processed Food Industries* 27: 26-33.

Rani, V. and Srivastava, A.P. 2012. Design and development of onion detopper. *AMA Agriculture Mechanization in Asia, Africa and Latin America* 43: 69-73.

Ranjan, R., Megaranjan, S., Xavier, B. *et al.* 2018. Broodstock development, induced breeding and larval rearing of Indian pompano, *Trichinotus mookalee* (Civier 1832) - A new candidate species for aquaculture. *Aquaculture* 495: 265-272.

Rao, M.M. 1997. Studies on the improvement of finger size in Munavalli (Musa AAB) banana. In M.M. Rao and G.S. Sulikeri (eds.) *Research and Development in Fruit Crops in North Karnataka*, pp. 59-61. Univ. of Agri. Sci., Dharwad, Karnataka.

Rao, M.M., Narasimhan, P., Nagaraja, N. and Anandaswamy, B. 1968. Effect of naphthalene acetic acid and p-chlorophenoxy acetic acid on control of berry drop in Anab-e-Shahi grape. *J. Food Sci. Tech.* 5: 127-128.

Rathod, P. and Chander, M. 2016. Adoption status and factors influencing adoption of livestock vaccination in India: An application of multinomial logit model. *Indian Journal of Animal Sciences* 86(9): 1061-1067.

Raveloson, C. 1990. Situation et contraintes de l'aviculture villageoise à Madagascar In: *CTA Seminar Proceedings, Smallholder Rural Poultry Production, Thessaloniki, Greece* 2: 135-138.

Reddy, K.C. and Reddy, K.M. 2005. Differential levels of vermicompost and nitrogen on growth and yield in onion (*Allium cepa* L.) radish (*Raphanus sativus* L.) cropping system. *J. Res., ANGRAU* 33:11-17.

Remiro, D. and Kirmati, H. 1975. Control of seven curls disease of onion with benomyl. *Sum. Phytopathology*, pp. 51-54.

Reuther, W., Batchelor, H.J. and Webber, H.J. 1962. *Citrus Industry, Vol. II.* Univ. of California Press, Berkeley, California.

Rice, X. 2011. "Ugandan Scientists Grow GM Banana as Disease Threatens Country's Staple Food." *Guardian*, March 8, 2011. Accessed July 31, 2015, http://www.theguardian.com/world/2011/mar/09/gm-banana-crop-disease-uganda.

Rosen, R.J. 2013. "Genetically Engineering an Icon: Can Biotech Bring the Chestnut Back to America's Forests?" *Atlantic*, May 21, 2013. Accessed August 11, 2015, http://www. theatlantic.com/ technology/archive/2013/05/genetically-engineering-an-icon-can-biotech-bring-the-chestnut-back-to-americas-forests/276356/.

Ruiz, R.S. and Escaff, M. 1992. Nutrición y fertilización de la cebolla. Serie La Platina-Instituto de Investigaciones Agropecuarias. *Estación Experimental La Platina (Chile)* 37: 69-73.

Rumpel, K.S and Dysko, J. 2003. Effect of drip irrigation and fertilization timing and rate on yield of onion. *J. Veg. Crop Prod.* 9: 65-73.

Ryan, C. 2015 . "The Dose Makes the Poison." *Cami Ryan* (blog), March 5, 2014. Accessed July 31, 2015, https://doccamiryan.wordpress.com/2014/03/05/the-dose-makes-the-poison/.

Saimbhi, M.S. and Bal, S.S. 1996. Evaluation of different varieties of onion for dehydration. *Punjab Vegetable Grower* 31: 45-46.

Salakinkop, S.R. 2011. Enhancing the productivity of irrigated *Bt* cotton (*Gossypium hirsutum*) by transplanting technique and planting geometry. *Indian Journal of Agricultural Sciences* 81: 150–153.

Sample, I. 2012. "GM Cow Designed to Produce Milk Without an Allergy-causing Protein." *Guardian*, October 1, 2012. http://www.theguardian.com/science/2012/oct/ 01/ gm-cow-milk-alllergy-protein.

Samra, J.S., Thakur, R.S. and Chadha, K.L. 1978. Comparison of some mango cultivars in terms of their macronutrient status in fruiting and non-fruiting terminals. *Indian J. Hort.* 35: 144-187.

Sankar, V., Lawande, K.E. and Tripathi, P.C. 2008. Effect of micro-irrigation practices on growth and yield of garlic. *J. Spices Arom. Crops* 17: 232-234.

Sankar, V, Qureshi, A.A, Tripathi, P.C. and Lawande, K.E., 2005. Production potential and economics of onion based cropping systems under western Maharashtra region (Abstr.). Paper presented on *National Symposium on Current Trends in Onion, Garlic, Chillies and Seed Spices– Production, Marketing and Utilization, (SYMSAC-II)*, Rajgurunagar, Pune, p 79.

Sankar, V., Thangasamy A., and J. Gopal. 2014. *Improved Cultivation Practices for Onion*. Tech Bulletin No. 21, Directorate of Onion and Garlic Research, Rajgurunagar, 23 pp.

Sankar, V., Tripathi, P.C., Qureshi, A. and Lawande, K.E. 2005. Fertigation studies in onion and garlic (Abstr.). *National Symposium on Current Trends in Onion, Garlic, Chillies and Seed Spices– Production, Marketing and Utilization, (SYMSAC-II)*, Rajgurunagar, Pune, pp. 62 & 80.

Sankar, V., Veeraragavathatham, D. and Kannan, M. 2009. Organic farming practices in white onion (*Allium cepa* L.) for the production of export quality bulbs. *Journal of Eco-friendly Agriculture* 4: 17-21.

Sarsavadia, P.N., Sawhney, R.L., Pangavhane, D.R. and Singh S.P. 1999. Drying behaviour of brined onion slices. *Journal o f Food Engineering* 40: 219-226.

Scholten, B. and Basu, P. 2009. White counter-revolution? India's dairy cooperatives in a neoliberal era. *Human Geography* 2 (1): 17–28.

Scott, N.R. 2006. *Impact of Nanoscale Technologies in Animal Management*. Wageningen Academic Publishers, The Netherlands.

Scott, S.J., McLeod, P.J., Montgomery, F.W. and Hander, C.A. 1989. Influence of reflective mulch on incidence of thrips (Thysanoptera: Thripidae: Phlaeothripidae) in stacked tomatoes. *J. Entom. Sci.* 24: 422-427.

Sen, A.K. 1981. *Poverty and Famines: An Essay on Entitlement and Deprivation.* Clarendon, Oxford, UK.

Sentenac, H. 2014. "GMO Salmon May Soon Hit Food Stores, but Will Anyone Buy It?" *FoxNews*, March 11, 2014. Accessed July 31, 2051. http://www.foxnews.com/leisure/ 2015,2014/ 03/11/ gmo-salmon-may-soon-hit-food-stores-but-will-anyone-buy-it/.

Shaji, C., Sajal, K.K. and Vishal, T. 2014. Storm surge studies in the North Indian Ocean: A review. *Indian J. Geo-Mar. Sci.* 43(2): 125-147.

Shantharam, S. 2010. Setback to Bt brinjal will have long-term effect on Indian science and technology. *Curr. Sci.* 98(8): 996-997.

Sharangi, A.B, Pariari, A, Datta, S and Chatterjee, R .2003. Effect of boron and zinc on growth and yield of garlic in New Alluvial Zone of West Bengal. *Crop Research* 25: 83-85.

Sharma, A., Sharma, P., Brar, M.S. and Dhillon, N.S. 2009. Comparative response to sulphur application in raya (*Brassica juncea*) and wheat (*Triticum aestivum*) grown on light textured alluvial soils. *J Indian Soc Soil Sci* 57(1): 62–65.

Sharma, I.M. 1997. Screening of onion varieties/lines against purple blotch caused by *Alternaria porri* under field conditions. *Pl. Dis. Res.* 12: 60-61.

Sharma, O.P., Bantewad, S.D., Patange, N.R. *et al.* 2015. Implementation of integrated pest management in pigeon pea and chickpea pests in major pulse-growing areas of Maharashtra. *Journal of Integrated Pest Management* 15(1): 12.

Sharma, P.K., Kumar, S., Yadav, G.L. *et al.* 2007. Effect of last irrigation and field curing on yield and post-harvest losses of rabi onion (*Allium cepa*). *Annals of Biology* 23: 145-148.

Sharma, R.P. and Chatterjee, R.N. 2009. Backyard poultry farming and rural food security. *Indian Farming* 59: 36-37, 48.

Shende, D.G. 1977. *Effect of N, P and K on Growth, Yield and Quality of Pomegranate* (Punica granatum L.). M. Sc. (Ag.) thesis, Mahatma Phule Agri. Univ., Rahuri, Maharashtra.

Shinoj, P., Gopalakrishnan, A. and Jena, J.K. 2020. *Demographic Change in Marine Fishing Communities in India.* FAO, Bangkok.

Shiva, V. 1991. *The Violence of the Green Revolution: Third World Agriculture, Ecology, and Politics.* Third World Network, Penang, Malaysia.

Shiva, V. 1993. *Monocultures of the Mind: Perspectives on Biodiversity and Biotechnology.* Zed Press, London.

Shrivastava, G.K., Khanna, P., Tomar, H.S. and Tripathi, R.S. 2000. Sorghum cultivation and its production technology for eastern Madhya Pradesh. *Intensive Agric* 38(5–6): 1–5.

Shukla, P.K. and Nayak, S. 2015. Challenges in export of poultry and poultry products. In: Souvenir, *32nd Annual conference of IPSA and National symposium,* College of Avian Sciences and Management, Tiruvazhamkunnu, Palakkad, Kerala, pp. 95-108.

Siddiqui, M.W., Ayala-Zavala, J.F. and Dhua, R.S. 2015. Genotypic variation in tomatoes affecting processing and antioxidant attributes. *Critical Review in Food Science and Nutrition* 55(13): 1819-1835.

Siddiqui, M.W. and Dhua, R.S. 2010. Eating artificially ripened fruits is harmful. *Current Science* 99(12): 1664-1668.

Siddiqui, M.W., Momin, C.M., Acharya, P. *et al*. 2013. Dynamics of changes in bioactive molecules and antioxidant potential of *Capsicum chinense* Jacq. cv Habanero at nine maturity stages. *Acta Physiologea Plantarum* 35 (4): 1141-1148.

Simm, G. 1998. *Genetic Improvement of Cattle and Sheep*. CABI Publishing, Wallingford, UK.

Simm, G., Bu¨nger, L., Villanueva, B. and Hill, W.G. 2004. Limits to yield of farm species: Genetic improvement of livestock. In R. Sylvester-Bradley and J. Wiseman (eds.) *Yields of Farmed Species: Constraints and Opportunities in the 21st Century*, pp. 123–141. Nottingham University Press, Nottingham, UK.

Singh, D. 1996. Comparative study of autumn v/s spring sugarcane crop in different cropping systems. *Indian Sugar* 46 (9): 727-729.

Singh, R. 2015. "*Papaya Ringspot*." The American Phytopathological Society. Accessed July 31, 2015, http://www.apsnet.org/publications/imageresources/Pages/ fi00157.aspx.

Singh, R., Nandal, T.R., Singh, R. 2002. Studies on weed management in garlic (*Allium sativum* L.). *Ind. J. Weed Sci*. 34: 80-81.

Singh, R.A. 1999. A case study: Farming system in Farrukhabad and Kannauj districts (UP). *Agric Ext Rev* 11(2): 22–28.

Singh, R.B. 2009. Serving farmers to render India prosperous. *Agric Today* 12(2): 24–26.

Singh, R.K., Ghosh, P.K., Bandyopadhyay, K.K. *et al*. 2006. Integrated plant nutrient supply for sustainable production in soybean-based cropping system. *Indian J Fertilizers* 1(11): 25–32.

Sinha, V.R.P., Gupta, M.V., Banerjee, M.K. and Kumar, D. 1973. Composite fish culture in Kalyani. *J. Inland Fish. Soc. India* 5: 201-208.

Smith, P., Martino, D., Cai, Z. *et al*. 2007. Agriculture. In B. Metz, O.R. Davidson, P.R. Bosch, R. Dave and L.A. Meyer (eds.) *Climate Change 2007: Mitigation*, pp. 497-540. Contribution of Working Group III to the Fourth Assessment Report of the Intergovernmental Panel on Climate Change, Cambridge University Press, Cambridge, UK.

Smith, P., Martino, D., Cai, Z. *et al*. 2008. Greenhouse gas mitigation in agriculture. *Phil. Trans. R. Soc. B* 363: 789 –813.

Smith. C.J.S., Watson, C.F., Ray, J. *et al*. 1988. Antisense RNA inhibition of polygalacturonase gene expression in transgenic tomatoes. *Nature* 334: 724-726.

Srihari, D. and Rao, M.M. 1996. Induction of flowering in "off" phase mango trees by soil application of paclobutrazol. *Vth Int. Mango Symp.*, Tel Aviv, Israel.

Srinivas, P.S. and Lawande K.E. 2006. Maize barrier as a cultural method for management of thrips in onion (*Allium cepa* L.). *Ind. J. Agril. Sci*. 76: 167-171.

Srinivas, P.S. and Lawande, K.E. 2007. Seedling root dip method for protecting onion plants from thrips. *Ind. J. Pl. Prot.* 35: 206-209.

Srivastava, P.K. and Pandey, V.B. 1995. *Compendium of Onion Diseases*. Tech. Bull. No.7, NHRDF, Nasik, 26 pp.

Srivastava, P.K., Sharma, R.C. and Gupta, R.P. 1995. Effect of different fungicides on the control of purple blotch and Stemphylium blight diseases in onion and seed crop. *NHRDF Newsletter* 15: 6-9.

Srivastava, P.K., Tiwari, B.K., Srivastava, K.J. and Gupta, R.P. 1996. Chemical control of purple blotch and basal rot diseases in onion bulb crop during Kharif. *NHRDF Newslett.* 16: 7-9.

Srivastava, R., Agarwal, A., Tiwari, R.S. and Kumar, S. 2005. Effect of micronutrients, zinc and boron on yield, quality and storability of garlic (*Allium sativum*). *Ind. J. Agri. Sci.* 75: 157-159.

Staal, S.J., Pratt, A.N. and Jabbar, M. 2008. *Dairy Development for the Resource Poor—Part 1: A Comparison of Dairy Policies and Development in South Asia and East Africa.* PPLPI Working Paper No. 44-1. International Livestock Research Institute, Addis Ababa, Ethiopia.

Steinfeld, H., Gerber, P., Wassenaar, T. *et al.* 2006. *Livestock's Long Shadow: Environmental Issues and Options.* FAO, Rome, Italy.

Stuertz, M. 2002. "Green Giant." *Dallas Observer*, December 5, 2002. http://www. dallasobserver. com/news/green-giant-6389547.

Sugha, S.K., Develash, R.K. and Tyagi, R.D. 1992. Performance of onion genotypes against purple blotch pathogen. *South Ind. Hort.* 40: 297.

Swaminathan, M.S. 2000. For an evergreen revolution. In *The Hindu Survey of Indian Agriculture 2000*, pp. 9-15.

Syda Rao, G., Imelda Joseph, Philopose, K.K. and Suresh Kumar, M. 2014. Cage culture in India. *Aquac. Int.* 22: 961-962.

Tamil Selvan, C., Valliappan, K. and Sundararajan, R. 1990. Studies on the weed control efficiency and residues of oxyfluorfen in onion (*Allium cepa* L.). *Intl. J. Trop. Agri.* 8: 123-128.

Tandon, H.L.S. and Sekhon, G.S. 1988. *Potassium Research and Agricultural Production in India.* FDCO, New Delhi, India, viii + 144 pp.

Taryn, G., Frank, A., James, A. *et al.* 2020. A global blue revolution: Aquaculture growth across regions, species and countries. *Rev. Fish. Sci. Aquac.* 28: 107-116.

Thilakavathy, S. and N. Ramaswamy. 1998. Effect of inorganic and biofertilizer treatments on yield and quality parameters of multiplier onion. *NHRDF Newslett.* 18: 18-22.

Thind, H.S., Buttar, G.S., Singh, A.M. *et al.* 2012. Yield and water use efficiency of hybrid Bt cotton as affected by methods of sowing and rates of nitrogen under surface drip irrigation. *Archives of Agronomy and Soil Science* 58: 199-211.

Thornton, P.K. and Gerber, P. 2010. Climate change and the growth of the livestock sector in developing countries. *Mitigation Adapt. Strateg. Glob. Change* 15: 169 –184.

Thornton, P.K., van de Steeg, J., Notenbaert, A. and Herrero, M. 2009. The impacts of climate change on livestock and livestock systems in developing countries: A review of what we know and what we need to know. *Agric. Syst.* 101: 113 –127.

Toor, S.S., Singh, S., Garcha, A.I.S. *et al.* 2000. Gobhi sarson as an intercrop in "Autumn Sugarcane" for higher returns. *Intensive Agric* 38(5–6): 29–30.

Trindade-Santos, I., Moyes, F. and Magurran, A.E. 2020. Global change in the functional diversity of marine fisheries exploitation over the past 65 years. *Proc. R. Soc. Biol. Sci. Ser. B.* 287: 2020-2089.

Tripathi, L., Mwaka, H., Tripathi, J.N. and Tushemereirwe, W.K. 2010. Expression of sweet pepper *Hrap* gene in banana enhances resistance to *Xanthomonas campestris* pv. *musacearum. Molecular plant pathology* 11(6): 721–731.

Tripathi, L., Mwangi, M., Abele, S. and Aritua, V. 2009. *Xanthomonas* wilt: A threat to banana production in East and Central Africa. *Plant Disease* 93(5): 450-451.

Tripathi, P.C. and Lawande, K.E. 2009. A new gadget for onion grading. *AG Journal* 90: 1-4.

Tripathi, P.C., Sankar, V. and Lawande, K.E. 2010. Influence of micro-irrigation methods on growth, yield and storage of rabi onion. *Indian Journal of Horticulture* 67: 61-65.

Tripathi, P.C., Sankar, V., Mahajan, V. and Lawande, K.E. 2011. Response of gamma irradiation on post-harvest losses in some onion varieties. *Indian Journal of Horticulture* 68: 556-560.

University of Illinois. 2011. *"News Release: Stalk Borers and Corn Borers."* University of Illinois Extension, June 20, 2011. Accessed July 31, 2015, http://web.extension.illinois. edu/state/newsdetail. cfm?NewsID=21044.

Unnikrishnan, A.S., Manimurali, M. and Ramesh Kumar, M.R. 2006. *Sea-Level Changes along the Indian Coast.* National Institute of Oceanography, Goa, India.

Upton, M, 2007. Scale and structures of the poultry sector and factors inducing change: Inter country differences and expected trends. In *Proc. Poultry in the 21st Century: Avian Influenza and Beyond.* International Poultry Conference, Bangkok, pp. 49- 79.

USDA. 2015 . *"Adoption of Genetically Engineered Crops in the U.S."* USDA. Accessed July 31, 2015, http://www.ers.usda.gov/data-products/adoption-of-genetically-engineered-crops-in-the-us/recent-trends-in-ge-adoption.aspx.

USDA, Foreign Agricultural Service. 2017. *Production, Supply and Distribution Online Data.* Accessed July 9, 2018. https://apps. fas.usda.gov/psdonline/app/index.html#/ app/downloads.

Valenti, W.C., Kimpara, J.M., Preto, B.D.L. and Moraes Valenti, P. 2018. Indicators of sustainability to assess aquaculture systems. *Ecol. Indic.* 88: 402-413.

Vatta, K., Sidhu, R.S., Lall, U. *et al.* 2018. Assessing the economic impact of a low-cost water-saving irrigation technology in Indian Punjab: the Tensiometer. *Water International* 43(2): 305–321.

Vekaria, R.S., Pandya, R.D. and Thumar, D.N. 2000. Knowledge and adoption behaviour of rainfed groundnut growers. *Agric Ext Rev* 12(1): 23–27.

Venugopalan, M.V., Blaise, D., Yadav, M.S. and Deshmukh, R. 2011. Fertilizer response and nutrient management strategies for cotton. *Indian J. Fert.* 7: 82-94.

Verlodt, H. 1990. *Greenhouses in Cyprus, Protected Cultivation in the Mediterranean Climate.* FAO, Rome.

Verma, L.R., Pandey, U.B., Bhonde, S.R. and Srivastava, K.J. 1999. Quality evaluation of different onion varieties for dehydration. *News Letter National Horticultural Research and Development Foundation* 19 (2/3): 1-6.

VIB News. 2015 . *"MON810 Scientific Background Report."* VIB (report). Accessed July 31, 2015, http://www.vib.be/en/news/Documents/VIB_Dossier_MON810_ENG.pdf.

Vick, B.D., Neal, B., Clark, R.N. and Holman, A. 2003. *Water Pumping with AC Motors and Thin-film Solar Panels.* ASES Solar 2003: America's Secure Energy, Austin, TX, 6 pp.

Vijayan, K.K. 2019. Domestication and genetic improvement of Indian white shrimp, *Penaeus indicus*: A complimentary native option to exotic pacific white shrimp, *Penaeus vannamei*. *J. Coast. Res.* 86(spl.): 270-276.

Vinay Singh, J., Bisen R.K., Agrawal H.P. and Singh, V. 1997. A note on weed management in onion. *Veg. Sci.* 24: 157-158.

Voosen, P. 2011. "*Crop Savior Blazes Biotech Trail, but Few Scientists or Companies are Willing to Follow.*" *New York Times*, September 21, 2011. Accessed July 31, 2015, http://www.nytimes. com/gwire/2011/09/21/21greenwire-crop-savior-blazes-biotech-trail-but-few-scien-88379.html.

Wagner, H. 2003. "*Researchers Get to the Root of Cassava's Cyanide-Producing Abilities.*" Ohio State University. http://researchnews.osu.edu/archive/cassava.htm;

Walton, D. 2015. "*GMO Myth: Farmers 'Drown' Crops in 'Dangerous' Glyphosate. Fact: They Use Eye Droppers.*" Genetic Literacy Project, January 22, 2015. Accessed July 31, 2015, http://www. geneticliteracyproject.org/2015/01/22/gmo-myth-farmers-drown-crops-in-dangerous-glyphosate-fact-they-use-eye-droppers/.

Warade, S.D, Shinde, S.V. and Gaikwad, S.K.1996. Studies on periodical storage losses in onion. *Allium News Letter* 10: 37-41.

Wattiaux, M.A. 2011. *Mastitis: The Disease and its Transmission. Dairy Essentials, Babcock.* Available from: https://www.yumpu.com/en/document/read/29801928/23-15 (Accessed: May 15, 2021).

Whiffen, H.J.H., Haman, D.Z. and Baird, C.D. 1992. Photovoltaic-powered water pumping for small irrigation systems. *Appl. Eng. Agric.* 8: 625–629.

WHO. 2015. "*Micronutrient Deficiencies.*" World Health Organization. Accessed July 31, 2015, http:// www.who.int/nutrition/topics/vad/en/.

World Bank. 2005. *India's Water Economy: Bracing for a Turbulent Future.* World Bank, Washington, D.C., U.S.A.

World Bank. 2007. *World Development Report 2008: Agriculture for Development.* World Bank, Washington, D.C.

World Bank. 2013. *Turn Down the Heat: Climate Extremes, Regional Impacts, and the Case for Resilience.* The World Bank, Washington, D.C., U.S.A.

WRI (Water Resources Institute). 2015. *India's Growing Water Risks.* Available at: http://www.wri. org/blog/2015/02/3-maps-explain-india%E2%80%99s-growing-waterrisk.

Wu, E.W.K. and Lau, I.P.L. 2008. *The Potential Application of Amorphous Silicon Photovoltaic Technology in Hong Kong.* Hong Kong Electrical and Mechanical Service Department. http://www.emsd. gov.hk/emsd/e_download/wnew/conf_papers/ emsd paper _ inal.pdf (Accessed June 12, 2010).

Yadav, I.C., Devi, N.L., Syed, J.H. *et al.* 2015. Current status of persistent organic pesticides residues in air, water, and soil, and their possible effect on neighboring countries: A comprehensive review of India. *Sci. Total Environ.* 51(1): 123–137.

YouTube. 2015. "*Hawaii Snapshot—a Legendary Scientist and His Papaya Dreams.*" YouTube. Accessed August 10, 2015, https://www.youtube.com/watch?t=95&v=fn_ 0KdbTlR8.

Subject Index